INTRODUCTION TO ENGINEERING DESIGN

BOOK 3

POSTAL SCALES

James W. Dally
University of Maryland, College Park

David N. Rocheleau
University of South Carolina, Columbia

In consultation with:

Thomas M. Regan
University of Maryland, College Park

College House Enterprises, LLC
Knoxville, Tennessee

THANKS FOR PERMISSIONS

The authors, consultant and publisher thank the copyright and trademark owners for providing permission to use copyrighted materials and trademarks. We apologize for any errors or omissions in obtaining permissions. Where appropriate, we referenced similar treatments and gave credit for prior work. Errors or omissions in obtaining permissions or in giving proper references are not intentional. We will correct them at the earliest opportunity after the error or omission is brought to our attention. Please contact the publisher at the address given below.

The manuscript was prepared by the author in Word 97 using 12 point Times New Roman font. The book was printed from camera ready copy by Publishing and Printing Inc., Knoxville, TN.

College House Enterprises, LLC.
5713 Glen Cove Drive
Knoxville, TN 37919, U. S. A.
Phone   (423) 558 6111
FAX   (423) 584 1766
email   jimdally@aol.com

ISBN 0-9655911-2-3

To my wife and children
Anne Tziritas Dally
Lisa Dally Wilson
William James Dally
and
Michelle Dally Johnston

and
To my wife Beth Ann and my daughter Emma Jane
who bring happiness, completeness, and balance to my life.

James W. Dally obtained a Bachelor of Science and a Master of Science degree both in Mechanical Engineering from the Carnegie Institute of Technology. He obtained a Doctoral degree in Mechanics from the Illinois Institute of Technology. He has taught at Cornell University, Illinois Institute of Technology, the U. S. Air Force Academy, and served as Dean of Engineering at the University of Rhode Island. He is currently a Glenn L. Martin Professor of Engineering at the University of Maryland.

Professor Dally has also worked at the Mesta Machine Co., IIT Research Institute, NIST, NRL, and IBM's Federal Systems Division. He is a fellow of the American Society for Mechanical Engineers, Society for Experimental Mechanics, and the American Academy of Mechanics. He was appointed as an honorary member of the Society for Experimental Mechanics in 1983 and elected to the National Academy of Engineering in 1984. In 1992 he was selected by his peers to receive the Senior Faculty Outstanding Teaching Award. He was a recipient of the Distinguished Scholar Teacher Award at the University of Maryland in 1994. He was also a member of the team receiving the 1996 Outstanding Educator Award sponsored by The Boeing Co.

Professor Dally has authored or co-authored several other books including: *Experimental Stress Analysis, Photoelastic Coatings, Instrumentation for Engineering Measurements, Packaging of Electronic Systems, Production Engineering and Manufacturing,* and *Introduction to Engineering Design, Books 1 and 2*. He has written about 200 scientific papers and holds five patents.

David N. Rocheleau received all of his degrees in Mechanical Engineering—a Bachelor of Science from the University of Vermont, a Master of Science from the University of Illinois at Urbana, and a Doctoral degree from the University of Florida. He has taught as a visiting instructor at the University of Illinois and is currently teaching in the Department of Mechanical Engineering at the University of South Carolina as an Assistant Professor.

Upon completing his Bachelor's degree, Professor Rocheleau served as a Field Engineer with Schlumberger Well Services in Midland, Texas, performing measurements to aid in the production and completion of oil and gas reservoirs. Following his Master's degree, he developed computer-aided engineering software for mechanical engineers at Computervision, Inc., Bedford, Massachusetts.

Professor Rocheleau received the Pi Tau Sigma 1997 Mechanical Engineering Professor of the Year Award. His research work is in the area of applied mechanisms design, computer-aided-engineering, and robotics. His major sponsored research funding is in the mechanical systems development of autonomous robots for the Department of Energy. He has published more than 25 technical papers and serves as the coordinator for the Freshman Year Program in the College of Engineering at the University of South Carolina.

CODE:

INSTRUCTOR AND STUDENT EXPECTATIONS

Learning and teaching require trust and mutual understanding between the student and the instructor. Trust and understanding lead to an enhanced learning environment. If we all recall a listing of expectations, a better relation can be developed between the instructor and his or her class that significantly promotes the learning process. While many of these expectations may be self evident, we believe the list will remind everyone of his or her obligations for learning, increased understanding, communication, and respect. The list, shown below, was developed by the students and instructors in the Introduction to Engineering Design course over the past eight years.

Students expect the instructor to:

- Respect all students.
- Be fair in grading.
- Provide leadership.
- Be committed to teaching and advising.
- Provide encouragement rather than discouragement.
- Clearly define course requirements and the grading algorithms.
- Balance course workload with credit hours.
- Schedule office hours and be available to help.
- Provided candid and timely feedback on assignments.
- Arrive before the scheduled class time and prepare the classroom.

Instructors expect the student to:

- Show respect to everyone involved in the program.
- Be responsible for your own progress and learning.
- Be dedicated to understanding and learning.
- Stay current with materials and issues covered in class.
- Be a positive and creative team member.
- Be inquisitive and compete within the framework of a team.
- Be interested in engineering and product design.
- Attend class or notify the instructor in advance if you intend to be absent.
- Arrive on time for class with a positive attitude.

PREFACE

This book is the third in a series dealing with Introduction to Engineering Design. College House Enterprises, LLC is preparing a new text each academic year for the first year engineering students at the University of Maryland @ College Park. We are encouraged that the book is also being adopted by other Colleges and Universities.

The procedure that is followed in offering a design experience to first year students is to:

- Teach the class in relatively small sections.
- Divide the class into product development teams with five or six members per team.
- Assign a major project entailing the development of a prototype that will require the entire semester.
- The teams all develop the same product.
- In the product realization process, the students:
  - Design
  - Manufacture or procure parts for the prototype.
  - Assemble the prototype.
  - Test and evaluate the prototype.
- In developing the prototype, we have the opportunity to emphasize:
  - Communication skills.
  - Team building skills.
  - Graphics.
  - Software applications including CAD.
  - Design methods and procedures.

The textbook is used to support the students during the semester long project. Some of the material may be covered in class or in a computer laboratory. Other material is covered with reading assignments. In other instances, the student uses the text as a reference document in independent study. Exercises are provided at the end of each chapter that may be used for assignments when the demands of the project on the students' time are not excessive.

The course, Introduction to Engineering Design, has evolved over a period of eight years and reflects the ideas, opinions, and experiences of many faculty members who have participated in teaching the course at the University of Maryland @ College Park. The book is also based on practices that have been successful in teaching design to first year students. Through presentations at meetings sponsored by the American Society for Engineering Education and informal discussions with colleagues from other colleges of engineering, the course was focused beyond our personal intuition. Also, we were fortunate to be part of the ECSEL coalition because it gave us excellent opportunities to experiment with a curriculum that had not changed significantly for more than twenty five years. The Engineering Education Coalitions funded by The National Science

Foundation gave us credibility as we tried to make significant changes to the content in the first course in engineering. It was and still is, in many colleges of engineering, essential to develop a course which provides students a much more satisfying educational experience in their first encounter with the engineering curriculum.

The book is organized into six parts to present the various topics that the students need as they proceed through a significant portion of product realization process. We introduce the product realization process by considering the design of a postal scale with either an analog or a digital readout. By assigning a demanding overarching project, we employ a holistic approach that avoids compartmentalization of knowledge. Design of a product is treated by the instructor as an opportunity to integrate a spectrum of knowledge about many topics. We have found that developing a product motivates the students. They learn much more on their own initiative than we could ever teach them in a course. The motivational benefit of hands-on participation in the design of their own product and building and testing the prototype significantly speeds the learning process.

Part I of the textbook covers material on postal scales. We briefly introduce the product development process and the students role in a development team in producing a scale with analog and digital readouts. The product is described in Chapter 2 together with theory on mass and weight. Some history about measuring weight is included prior to introducing the product specification. A few concepts for the design of scales are introduced briefly. Much more detail on the design of weighing machines is given in Chapters 3 and 4. We cover the mechanical scales in Chapter 3 by giving some of the theory for scales based on equilibrium, springs, hydraulics, and electronics. Much more detail on the electronic scales is given in Chapter 4. We recognize that some of the analysis presented in Chapters 3 and 4 may be beyond the scope of the ordinary first year student; however, we believe that some may understand and lead others on their team to learn and understand this material.

Part II presents three chapters on engineering graphics. We begin in Chapter 5 with a relatively complete treatment of three-view drawings. Pictorial drawings including isometric, oblique, and perspective are covered in Chapter 6. A treatment of Tables and Graphs is given in Chapter 7. The emphasis in Part II is on manual preparation of the graphics. CAD preparation of drawings and spreadsheet generation of graphs is covered in Part III.

Part III describes three different software programs useful to the first year engineering student. A feature based solid modeling program, Pro/ENGINEER, is described in considerable detail with several example drawings. A spreadsheet program, Microsoft Excel, is described in Chapter 9. The coverage is intended to provide the students with entry level skills. With spreadsheets we expect the student to be able to perform calculations and to plot curves and prepare several types of charts. In Chapter 10, we provide a brief description of a computer graphics presentation program, namely Microsoft PowerPoint. Our experience is that the student can learn to master this program in one to two hours if they make use of the chart wizard incorporated in the program. The coverage of the capabilities of PowerPoint is related to preparing design briefings on a postal scale.

Part IV is related to the product development process. We first describe development teams in considerable detail. We find that the idea of working on teams is new to most students who have been educated in the secondary school system in the U. S. They have been trained in high school to act as individuals and to avoid cooperation in learning. In Chapter 11, we introduce a number of useful topics such as team member traits, positive and negative team behavior, and effective team meeting. Experience with several thousand students has shown that many of them initially have trouble adapting to the team concept. (Males appear to have more trouble than females.) However, over the semester they slowly learn how to work as effective team members and they appreciate the opportunity to do so. The social bonding that takes place on the team with the first semester Freshman students is interesting to observe. Chapter 12 deals with the product development process. We have tried to incorporate a very wide range of material in this chapter. For this reason the coverage often is brief and we refer the student to more thorough higher level books on the topic. However, we cover that part of the product development process which is important to the assigned project. We start with the product specification and cover material on design concept generation and then concept selection. Methods are introduced that the student can effectively employ to generate design concepts and then to select the best concept though the use of a systematic design trade-off analysis.

Part V treats the very important topic of communications. A chapter on technical reports describes many aspects of technical writing. The most important lesson here is that a technical report is different than a paper for the History or English Departments. An effective professional report is written for a predefined audience with specific objectives. We describe the technical writing process and give many suggestions to facilitate composing, revision, editing and proofreading. The final chapter in the book covers design briefings. We draw a distinction between speeches, presentations and group discussions. Then we focus on the technical presentation and indicate the importance of preparing excellent visual aids. We make extensive use of PowerPoint in illustrating the types of visual aids to be employed in a design briefing. Finally, we include a discussion of the delivery of the presentation and the need for extensive rehearsing.

Part VI contains three chapters dealing with engineering and society. Chapter 15, provides a historical perspective on the role engineering played in developing civilization and on improving the lives of the masses. In this chapter, we move from the past into the present and indicate the current relationship between business, consumers and society. Safety, risk and performance are covered in Chapter 16. Here we discuss failure, and it implications on the safety and well being of those using our products. Theoretical methods are introduced to determine both component and system reliability. Finally, in Chapter 17, we discuss ethics, character and engineering. A large number of topics are covered so the instructor can select from among them. The code of ethics recommended by the Accreditation Board for Engineering and Technology (ABET) is included in Appendix B. A description of the Challenger accident is also covered because it is an excellent case history covering safety related conflicts between management and engineers.

In the next few years, it is imperative that we think very seriously about ABET's new criteria for program accreditation. The new Criteria 2000 are very different from the current criteria for accreditation. We currently accredit our engineering programs based on input, and many pages of instructions are provided by ABET to guide in the assessment of that input. However, Criteria 2000 are based primarily on educational outcomes. Very little information given in Criteria 2000 pertains to program input. The guide provided by ABET for assessment of educational outcomes is very terse. It appears that it will be a program responsibility in each College of Engineering to specify the intended educational outcomes for their program, to define the metrics used to measure these outcomes, and to plan and place into effect methods of measurement of these metrics over extended periods of time.

We have considered Criteria 2000 in writing this book, and believe a first course in engineering design should have the expected educational outcomes listed below:

Communication Skills

1. Engineering Graphics

   - Understand the role of graphics in engineering design.
   - Understand orthographic projection in producing multi-view drawings.
   - Understand three-dimensional representation with pictorial drawings.
   - Understand dimensioning and section views.
   - Demonstrate capability of preparing drawings using both manual and computer methods.
   - Demonstrate understanding by incorporating appropriate high-quality drawings in the design documentation.

2. Design Briefings:

   - Within a team format, present a design review for the class using appropriate visual aids.
   - Participation of each team member in the briefing.

3. Design Reports:

   - The team's design is documented in a professional style report incorporating time schedules, costs, parts list, drawings and an analysis.

Team Experience

   - Develop an awareness of the challenges occurring in teamwork.
   - Demonstrate teamwork in the product realization process through a systematic design concept selection process involving participation of all the team members.
   - Demonstrate planning from conceptualization to the evaluation of the prototype.
   - Understand and demonstrate share responsibility among team members.
   - Demonstrate teamwork in preparing design reports and presenting design briefings.

Software Applications

   - Demonstrate entry level skills in using spreadsheets for calculations and data analysis.
   - Show a capability to prepare graphs and charts with a spreadsheet.
   - Show a capability to prepare professional quality visual aids.
   - Understand entry level skills in employing a CAD program.
   - Demonstrate these computer skills in preparing appropriate materials for design briefings and design reports.

Design Project:

- The design project is the overarching theme of the course.
- Utilize all the skills listed above to assist in the product development process.
- Demonstrate competence in defining design objectives.
- Generate design concepts that meet the design objectives.
- Understand the basis for design for manufacturing, assembly and maintenance.
- Manage the team and the project effectively.

Acknowledgments are always necessary in preparing a textbook because so many people and organizations are involved. First, we want to thank The National Science Foundation for their support. Their funding was important, but more critical than money was our need for credibility. Without the Engineering Education Coalition Program, the need for curriculum reform would not have received adequate attention from the college administrators and our colleagues. The NSF basically called for a reform of the curriculum and with generous funding gave it the required status.

Second, we need to recognize the contributions of many of our colleagues at the University of Maryland. In 1990 the author taught the pilot offering of the course with the assistance of Dr. Guangming Zhang. Guangming attended every class and prepared an excellent set of notes that were essential in the second series of pilot courses that were taught by Tom Regan and Isabel Lloyd. Guangming also taught the course, with slight modifications, to several classes of high school women as part of an early entrance summer program at the University of Maryland. His success in teaching engineering design to 16 year old women clearly showed his superior ability as a teacher and the robust nature of the course material that we had developed. We want to thank Ms. Jane Fines who maintained contact with many of our students in several offerings of the course. She gave us very valuable feedback about the reactions of the students as we modified the course over the years. Thanks are also due to Ms. Janet Yowell from the University of Colorado for helpful editorial comments on several chapters that she had the opportunity to review.

As always, thanks are due to the administrators who encouraged and supported the development of this course. Dr. George Dieter and Dr. William Destler, the former and current Deans of Engineering, authorized the small class size essential for effectively teaching this course and committed to significant long term expenditures necessary to support the faculty involved. They also publicly supported the efforts of the small group of instructors during the early years as we were institutionalizing the course. In fact, Dr. William Destler, first as a Department Chair and now as a Dean, takes time from his busy schedule to lead a section of this course.

James W. Dally
Knoxville, TN

DEDICATION
ABOUT THE AUTHORS
CODE
PREFACE

CONTENTS

PART I POSTAL SCALES

CHAPTER 1 ENGINEERING DESIGN AND PRODUCT DEVELOPMENT

CHAPTER 2 POSTAL SCALES AS WEIGHING MACHINES

CHAPTER 3 MECHANICAL SCALES

CHAPTER 4 ELECTRONIC SCALES

PART II ENGINEERING GRAPHICS

CHAPTER 5 THREE-VIEW DRAWINGS

CHAPTER 6 PICTORIAL DRAWING

CHAPTER 7 TABLES AND GRAPHS

PART III SOFTWARE APPLICATIONS

CHAPTER 8 Pro/ENGINEER

CHAPTER 9 MICROSOFT EXCEL–97

CHAPTER 10 MICROSOFT PowerPoint

PART IV PRODUCT DEVELOPMENT PROCESSES

CHAPTER 11 DEVELOPMENT TEAMS

CHAPTER 12 A PRODUCT DEVELOPMENT PROCESS

PART V COMMUNICATIONS

CHAPTER 13 TECHNICAL REPORTS AND INFORMATION GATHERING

CHAPTER 14 DESIGN BRIEFINGS

PART VI ENGINEERING AND SOCIETY

CHAPTER 15 ENGINEERING AND SOCIETY

CHAPTER 16 SAFETY, RISK AND PERFORMANCE

CHAPTER 17 ETHICS, CHARACTER AND ENGINEERING

PART I

POSTAL
SCALES

CHAPTER 1

ENGINEERING DESIGN AND PRODUCT DEVELOPMENT

1.1 THE IMPORTANCE OF PRODUCT DEVELOPMENT

We use hundreds of products every day. We are surrounded by products. This morning I prepared breakfast by toasting bread, frying eggs, and making coffee. How many products did I use in this simple task? The toaster (Black & Decker), the frying pan (T-FAL), the coffee maker (Krups) and the gas cooktop (Kitchen Aid) are all products that are designed, manufactured and sold to customers both in the U. S. and abroad. Some products are relatively simple, such as the frying pan with only a few parts, yet some are much more complex, such as your automobile with several thousand parts.

Corporations world-wide continuously develop their product lines with minor improvements introduced every year or two with more major improvements every four or five years. Product development involves many engineering disciplines and is a major responsibility of engineers entering the work place, as illustrated in Table 1.1. An examination of Table 1.1, shows that designing, manufacturing, and product sales represent more than 80% of the responsibilities of mechanical engineers in their initial position in industry. This distribution of activities is typical of many of the engineering disciplines.

The viability of many corporations depends on the introduction of a steady stream of successful products to the marketplace. A successful product must satisfy the customer by providing robust and reliable service at a competitive price that completely meets the customer needs.

Table 1.1

Responsibilities of Mechanical Engineers in Their First Position [1]

ASSIGNMENT	PERCENT of TIME
Design Engineering	40
Product Design	24
Systems Design	9
Equipment Design	7
Plant Engineering / Operations / Maintenance	13
Quality Control / Reliability / Standards	12
Production Engineering	12
Sales Engineering	5
Management	4
Engineering	3
Corporate	1
Computer Applications / Systems Analysis	4
Basic Research and Development	3
Other Activities	7

1.1.1 Development Teams

Products are developed by interdisciplinary teams with representation from engineering, marketing, manufacturing, production, purchasing, etc. A typical organization chart for a team developing a relatively simple electro-mechanical component is shown in Fig. 1.1. The team leader coordinates and directs the activities of the designers (mechanical, electronic, industrial, etc.) and is supported by a marketing specialist to ensure that the evolving product meets the needs of the customer. Financial, sales and legal assistance is usually provided on an as-needed basis by corporate staff or outside contractors.

The team leader is also supported by a purchasing agent who establishes close relations with the suppliers who provide materials and component parts used in the final assembly of the product. In the past decade, the relationship between suppliers and end-product producers has changed significantly. Suppliers become part of the development team; they often provide a significant level of engineering support that is external to the internal (core) development team. In some developments, these external (supplier funded) development teams—when totaled—are larger than the internal development team. For example, the internal development team for the Boeing 777 airliner involved 6,800 employees, and the external teams were estimated to include about 10,000 employees [2].

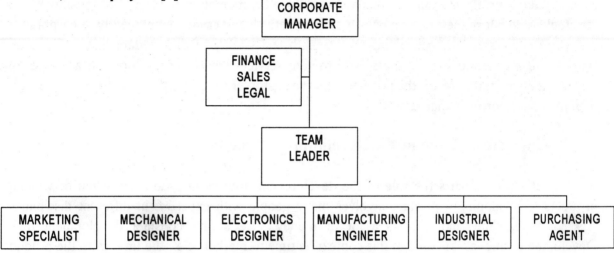

Fig. 1.1 Organization of a product development team for a relatively simple product.

1.2 DEVELOPING WINNING PRODUCTS

There are two primary aspects involved in developing successful products that win in the competitive marketplace. The first is the quality, performance and price of the product. The second is the time and the cost of the development, and also the manufacturing cost of the product in production.

1.2.1 Quality, Performance and Price

Let's discuss the product first. Is it attractive and easy to use? Is it durable and reliable? Is it effective and does it meet the needs of the customer? Is it better than the products now available in the marketplace? If the answer to all of these questions is an unqualified **YES**, then the customer may want to buy the product if the price is right. Next, you need to understand what is implied by product cost and its relation to the price actually paid for the product. Cost and price are distinctly different quantities. Product cost clearly includes the cost of materials, components, manufacturing and assembly. The accountants also include less obvious costs such as the prorated costs of capital equipment (the plant and its machinery), the price of the tooling, the development cost, and even the expense of maintaining the inventory in establishing the total cost of producing the product.

Price is the amount of money that a customer pays to buy the product. The difference between the price and the cost to produce a product is profit, which is usually expressed on a per unit basis.

Profit = Product Price - Product Cost

This equation is the most important relation in engineering or in any business. If a corporation cannot make a profit, it soon is forced into bankruptcy, its employees lose their positions, and the stockholders lose their investment. It is this profit that everyone employed by a corporation seeks to maximize while maintaining the strength and vitality of the product lines. The same statement can be made for a business that provides services instead of products. If a business is to make a profit and prosper, the price paid by the customer for a specified service must be more than the cost to provide that service.

1.2.2 The Role of Time in the Development Process

Let's now discuss the role of the development process in producing a line of winning products. Developing a product involves many people with expertise in different disciplines. Also, it takes time, and it costs a lot of money. Let's first consider development time. Development time, as it is used in this context, is time to market, i.e., the time from the product development kickoff to the introduction of the product to the market. This is a very important target for a development team because of the many significant benefits that follow from being first to market. Many competitive advantages are realized by a corporation with a fast development capability. First, the product's market life is extended. For each month cut from the development schedule, a month is added to the life of the product in the marketplace with an additional month of sales revenue and profit. We show the benefits of being first to market on sales revenue in Fig. 1.2. The shaded area between the two curves to the left side of the graph is the enhanced revenue due to the longer sales.

A second benefit of early product release is increased market share. The first product to market has 100% of market share in the absence of a competing product. For products with periodic development of "new models," it is generally recognized that the earlier a product is introduced to compete with older models—without sacrificing quality and reliability—the better chance it has for acquiring and retaining a large share of the market. The effect of gaining a larger market share on sales revenue is also illustrated in Fig. 1.2. The shaded area between the two curves at the top of the graph shows the enhanced sales revenue due to increased market share.

A third advantage of a short development cycle is higher profit margins. If a new product is introduced prior to availability of competitive products, the corporation is able to command a higher price for the product which enhances the profit. With time, competitive products will be introduced forcing price reductions. However, in many instances, relatively large profit margins can still be maintained because the company that is first to market has added time to reduce their manufacturing costs. They learn better methods for producing components and reduce the time needed to assemble the product. The advantage of being first to market, with a product where a manufacturing learning curve exists, is shown graphically in Fig. 1.3. The manufacturing learning curve reflects the reduced cost of manufacturing and assembling a product with time in production. These cost reductions are due to many innovations introduced by the workers after mass production begins. With time and manufacturing experience, it is possible to drive down production costs.

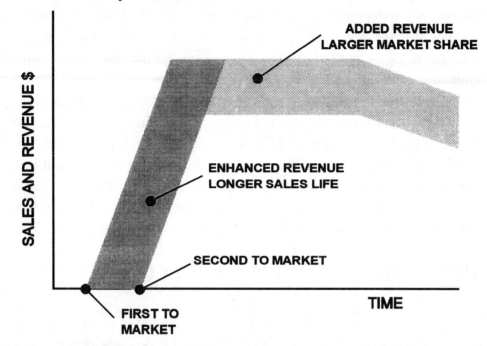

Fig. 1.2 Increased sales revenue due to extended market life and larger market share.

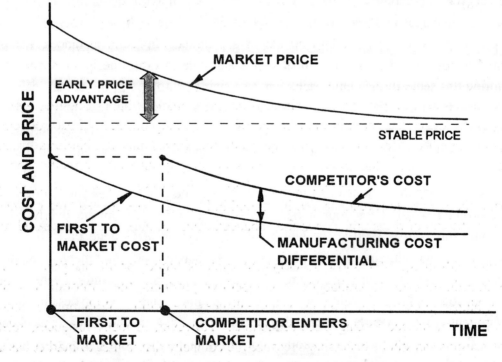

Fig. 1.3 The development team bringing the product to market first enjoys a higher initial price and subsequent cost advantages from manufacturing efficiencies.

Let's next consider development costs which represent a very important investment for the companies involved. Development costs include the salaries of the members of the development team, money paid to subcontractors, costs of pre-production tooling, expense of supplies and materials, etc. These development costs can be significant, and most companies must limit the number of developments in which they invest. The size of the investment may be appreciated by noting that the development cost of a new automobile is estimated [2] at $1 billion, with an additional expenditure of $500 to $700 million for the new tooling required for high-volume production.

We have included this discussion on time and cost of product development to help you begin to appreciate some of the business aspects of developing winning products. Any company involved in the sale of products depends completely on their ability to continuously introduce winning products in the marketplace in a timely manner. To win, the development team must bring a quality product to market that meets the needs of the customer. The development costs must be minimized while maintaining a product development schedule that permits an early (preferably first-to- market) introduction of the product.

1.3 LEARNING ABOUT PRODUCT DEVELOPMENT

The best way to learn about product design is to work on a development team and develop a prototype. For this reason, we have selected a postal scale for your first development effort. The degree of complexity of the postal scale will be defined in detail in Chapter 2. You are to work as a member in a team of five or six students to develop a prototype of a postal scale. A prototype is the first working model of a product. In some instances, companies develop three or four prototypes before finalizing a particular product design. However, time available during this semester will limit each team to a single prototype of the postal scale.

We understand that the development of a scale is a tremendous task particularly when the assignment is given so early in your engineering program. However, our experience with several thousand students beginning to study engineering indicates that you will gain immensely from your efforts. We know from previous classes that you are all very creative, that you are cooperative, work well within a team structure and design and build successful prototypes. Of course, the product selected for your initial development project cannot be overly complex. Since beginning this course 1990 at the University of Maryland at College Park, the students participating in this introductory course have developed playground equipment, windmills for the generation of electricity, furniture, human-powered water pumps, wind-powered vehicles, digital and analog bathroom scales, a solar-cooking unit, and a solar-desalination unit. Extensive student surveys clearly indicate that you will spend many hours each week on the project, have fun doing so, learn a lot about engineering and appreciate the lessons learned in working as a member of a development team.

1.4 PROTOTYPE DEVELOPMENT

We often divide the prototype development process into three major phases which include:

1. Designing the prototype.
2. Preparing the assembly kit.
3. Assembling and evaluating the prototype.

We will begin the process by providing you with a typical description, initially in general terms, of methods to measure weight and to compute postage. Also, you will be provided with design requirements pertaining to the input: the capacity of the scales and the accuracy of the measurement of the weight and the calculation of the postage needed for a letter or package. The description will also include the size and location of the display to insure that it can be read by the individual using the scale. We suggest that you make full use of the library to search for literature and patent files. There is absolutely nothing to be gained by reinventing the wheel (or the scale in this instance).

After you understand in general terms the design requirements for the postal scale, we will introduce the product specification [3]. The product specification, written relatively early in the development process, is usually prepared in tabular format. It tersely states design requirements and design limits or constraints. Remember we will always design with constraints and/or limits. The specification also includes all of the performance criteria and provides design targets on weight, size, power, cost, etc.

We will place explicit design constraints which limit your flexibility in designing the prototype. First, there is a maximum limit of $25.00 per team member for the cost of materials, supplies, and components purchased for the scales and the displays. The idea is to keep your out-of-pocket expenses at an affordable level, while giving the team adequate funds to build a good prototype. We will also limit the size of the scales, so that the laboratory facilities and workshops available in the college can handle the large number of scales that will be designed and constructed this semester.

1.5 DESIGN CONCEPTS

After you and each member of your development team understand the design specification and the design limitations, you are ready to generate design concepts. (A much more complete description of the design requirements, constraints, etc. is given in Chapter 2 together with some initial design ideas for you to consider). Design concepts are simply ideas for performing each function involved in the operation of the scale which will help you meet the product specifications. To illustrate design concepts, let's examine the functions (processes) involved in weighing a package on either a mechanical or an electronic postal scale, as depicted in the block diagrams presented in Fig. 1.4.

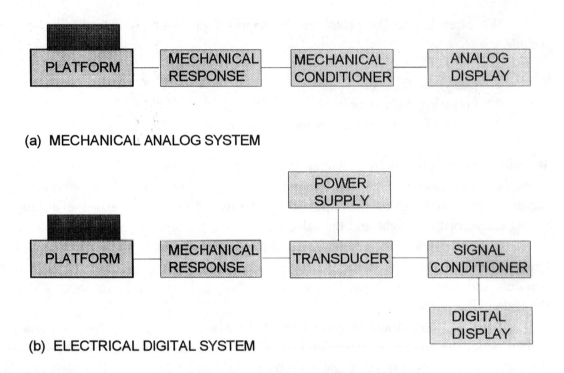

Fig. 1.4 Functions involved in the weighing a package on a postal scale.

1.5.1 The Design Concept—Design Function Relationship

In a typical scale, we have some sort of a platform to accept the item that is to be weighed. When you place a parcel or letter on the platform, it must be large enough to accommodate the item to be weighed. The platform must also be stable enough to avoid wobble or tilt. The customer would be very unhappy if he or she had to replace the package because the platform tilted so much that the parcel shifted and slid off the scale.

No problem—a sheet of wood or a thin plate of aluminum would serve as a nice platform. It is solid and we can make it large enough to hold a package of a size consistent with the capacity of the scale. In addition to supporting the parcel, the platform must help us in the weighing process. It must somehow respond to the weight placed on its surface. The platform must be connected with some sort of mechanism that responds to the weight. But, what kind of a response would be helpful?

Recall some scales that you have seen. In the produce section of the local supermarket, you have seen the scales that permit shoppers to check the weight of their fruit or veggies. These are spring scales. When we place our apples in the pan (platform), a spring deflects with the applied weight, and a pointer attached to the spring indicates (displays) the weight. The platform response in this case is a displacement. The deflection of the spring is converted into a reading of the weight by measuring its displacement. The pointer rotation increases with the spring displacement and indicates the weight on a calibrated dial—an analog display. Unless you spend a lot of money, your bathroom scale at home is probably a spring scale, but the mechanism is a bit more complex than the scale for weighing produce.

Examine Fig. 1.4a as it pertains to a spring-type scale. The mechanical response in Fig. 1.4a is the displacement of the spring. We condition this mechanical output with a mechanism to convert linear motion (the stretch of the spring) to rotary motion (the turning of the shaft of the pointer). The analog is the calibrated dial. The pointer is aligned with the calibrated dial markings to provide a measurement of the applied weight, as shown in Fig. 1.5.

Fig. 1.5 A simple spring scale for weighing produce.

Let's go back to the supermarket and move on to the delicatessen (deli) counter. On top of the counter, you will spot another scale. This one is a lot fancier than the spring scale which you noticed in the produce section of the store. When your sliced salami is placed on the platform, you can read the weight in red-lighted numbers. This is a digital scale. In fact, it is a computer and a printer in addition to a scale. The deli clerk places the sliced salami on the platform and enters its per pound cost. The computer calculates the cost of the quantity purchased and prints a product label with a bar code for use with the automatic check-out procedure.

The spring scales back in the produce department provided an analog display. They were totally mechanical. We converted the displacement of the spring to a pointer movement against the background of a calibrated dial as shown in Fig. 1.5 to give an analog readout of the weight. However, the scales with digital readout have both mechanical and electrical components as illustrated in Fig. 4b. The mechanical components are the platform and some type of a mechanical member that responds with a change in some quantity (displacement, strain, pressure, etc.) in proportion to the imposed weight. Suppose the change we are monitoring is a quantity known as strain. The strain is sensed with strain gages (electrical resistors). The combination of the mechanical member undergoing the strain and the strain gages themselves comprise what is known as a transducer. The strain imposed on the strain gages produces a change in their resistance, which in turn is converted into a voltage output, V_0, with a signal conditioning circuit (Wheatstone bridge). The voltage out of the Wheatstone bridge is amplified, scaled and displayed as weight with a digital voltmeter. Clearly, in developing a scale with a digital display, you will be working with both mechanical and electrical components. An example of a digital scale is illustrated in Fig. 1.6.

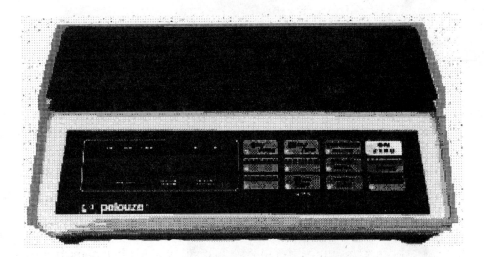

Fig. 1.6 A digital scale which calculates the cost of postage.

We will discuss the history of measuring weight and describe the principles upon which different types of scales are based in Chapter 2. Some initial ideas to consider in the design of the mechanical components are presented in Chapter 3. The background for the digital scale is described in Chapter 4, where information on sensors, signal conditioning circuits and digital displays is provided. Your digital and analog postal scales can have a common platform with interchangeable displays, or if your team prefers, the scales can be completely independent.

You need to develop many different ideas for each subsystem included in the design of your scales. Each concept should be described in considerable detail prior to developing preliminary design proposals incorporating each of the design concepts. The more completely you develop the design proposals, the better prepared you will be to conduct what is termed a design-trade-off analysis, where you compare several design concepts and select the best one.

1.5.2 Design-trade-off Analysis

To perform a design-trade-off analysis, we consider each design proposal individually and list its strengths and weaknesses. Factors usually considered in this analysis include size, weight, cost, ease of manufacturing, performance, appearance, etc. After the strengths and weaknesses of each design proposal have been listed, we can compare and evaluate each design proposal and select the most suitable for our prototype development.

We have briefly discussed design concepts, design proposals and design trade-off analyses for a single function that is involved in the process of weighing an individual. You have several functions to consider:

- supporting the item to be weighed.
- developing the mechanical response.
- mechanical conditioning.
- providing the mechanical display for the analog system.
- developing the transducer.
- designing the signal conditioner.
- providing the display for the digital system.

The procedure in design is to generate ideas for performing each function, expanding these ideas with more detail until they can be treated as design proposals, and finally to perform a design trade-off analysis where the merits and faults of each proposal are evaluated.

When the design proposals are selected for each of the different functions involved in the weighing process, the winning concepts must then be integrated into a seamless system that accurately weighs an individual and displays the weight in an easy-to-read format.

We often refer to the part of a product that performs some function as a subsystem. The collection of all of the subsystems constitutes a complete system, which when manufactured and assembled, becomes the prototype of the scale. The integration of the various subsystems is often difficult, as one subsystem influences the design of the others. The manner in which the subsystems interact is defined as the interface between subsystems. In a mechanical application, the fit of a shaft in a bearing is an example of an interface. Frequently, interfaces between the subsystems are troublesome and difficult to manage. It is vitally important that you control the interfaces to permit seamless integration of all of the subsystems without loss of effectiveness and efficiency.

1.5.3 Preparing the Design Documentation

The final step in designing the prototype is the preparation of the design documentation package. The package, which is an extended engineering document, contains:

- Engineering sketches illustrating the design concepts.
- Engineering assembly drawings that describe how the various components fit together to form the complete system.
- Engineering drawings of all of the component parts in sufficient detail to permit the component to be manufactured by anyone capable of reading an engineering drawing.

We recognize that many of you may need instruction in graphics, and we have included four chapters in this text to help you learn how to prepare three-view and pictorial drawings and to make tables and graphs. One of the four chapters covers a computer aided drafting program, Pro/ENGINEER, that will greatly assistance you in preparing the drawings for your design documentation package.

The design documentation package also contains a parts list that identifies every unique part employed in the assembly of the prototype. The quantity required of each component is also included on the line describing the component. For example, if you are going to use eight No. 6 sheet metal screws to fasten together the joints of your container for the scale, you would:

1. Assign a part number for the screws to identify the need for these screws in the assembly of the prototype.
2. List the quantity required as "8" in the quantity column.
3. Describe the screws in sufficient detail so that another team member can go to a hardware store and procure exactly the type of screw that is needed.

\multicolumn{5}{	l	}{**PARTS LIST FOR A SPRING TYPE POSTAL SCALE**}
PART NC	**NAME OF ITEM**	**DESCRIPTION OR DRAWING NO.**	**QUANTITY**	**PRICE**
1	PLATFORM	SHEET METAL PLATE ---- DWG. NO. 100-01	1	$ 1.20
2	BASE SUPPORT BOX	SHEET METAL BOX -------DWG. NO. 100-02	1	$ 2.55
3	SUPPORT BARS	ALUMINUM FLATS 1/16 X 1/2 X 4 LONG	2	$ 0.75
4	LINKAGE	ALUMINUM BAR ----DWG. 100-03	1	$ 0.45
5	BOTTOM SUPPORT	1/16 x 2 IN. AL. BAR --DWG. NO. 100-04	1	$ 0.40
6	POINTER	1/8 DIA. STEEL WIRE-DWG. NO. 100-05	1	$ 0.60
7	SPRING	CATALOG NO. 9656K26 McMASTER CARR	2	$ 0.55
8	SPRING GUIDE	1/4- DIA.STEEL ROD 3.5 IN. LONG	1	$ 0.35
9	COLLAR	CATALOG NO. 6434K1 McMASTER CARR	2	$ 0.80
10	MARKER BAR	1/16 X 1/4 BRASS BAR---DWG. NO. 100-07	1	$ 0.60
11	TARE SCREW	NO. 8-32 BRASS THUMB SCREW 2 IN. LONG	1	$ 0.85
12	FACE PLATE	WHITE PLASTIC --- DWG. 100-08	1	$ 0.65
13	STEEL SCREWS	# 6 SHT METAL, FT. HD, SLOTTED 1/2 IN. LC	8	$ 0.12

Fig. 1.7 Example of a partially complete parts list for a postal scale.

The description provided in the parts list is brief, but complete and precise. (i.e. No. 6 sheet metal screw, flat headed slotted, steel, 1/2 in. long). In this description, **No. 6** gives the diameter; **sheet metal** refers the type of application for the screw; **flat headed** describes its head; **slotted** indicates that you will use a flat blade screw driver in the installation; **steel** is the material from which the screw is fabricated; **1/2 in**. is the length of the screw.

The parts list contains all of the items that are to be purchased for your prototype. It also contains the components that must be manufactured. If the component is to be manufactured, it is identified on the parts list by its name and a drawing number. We show an example of a parts list in Fig. 1.7.

An engineering report is also included as part of the design package. This report supports the design by describing the key features of your scale. Your report should treat each function involved in the weighing process and describe the design concepts that your team considered. A rationale for each design proposal that was adopted, based on systematic design trade-off studies, is an essential element. Additionally, the design report contains your theoretical analyses used to predict the performance of your prototype. The analysis is conducted before actual testing the prototype, and both the analysis and the test results are useful in assessing the merits of the product development. More information on the preparation of an engineering report is included in a chapter of this textbook on writing technical reports.

1.5.4 Final Design Presentation

The design phase of the product development process is concluded with a final design presentation. This is a formal review of the design of the scale that your team presents to the class (peer review), to the instructor and others in the audience. It is the team's responsibility to describe all of the unique features of the design and to predict the performance of the product. It is the responsibility of the class (peer group) and the instructors to question the feasibility of the design and the accuracy of the predicted performance. If, in your review, you note a shortcoming or a flaw in the design, identify the problem to the team presenting their work. As a peer reviewer of another team's design, be tactful in your critique. Criticism is always difficult to offer to another. Offer constructive suggestions in good faith and in good taste. If you are on the receiving end of criticism, do not be defensive. The individual offering the critique is not attacking your capabilities; he or she is actually trying to give you a suggestion that might help you achieve a better design.

The purpose of the design review is to locate deficiencies and errors. It is better to correct errors in the paper stage of the process, not later in the hardware phase, when it is much more difficult and costly to fix the problems.

1.6 PROTOTYPE ASSEMBLY AND EVALUATION

1.6.1 Preparing the Assembly Kit

The second major phase of the product development process is to prepare what is called an assembly kit. The assembly kit is a collection of all of the parts, in the appropriate quantities, that are required to completely build the prototype. The parts list, as described above, is the essential document that directs our procurement and manufacture of the components needed to build the prototype. The parts list identifies everything that we need and indicates the precise quantity of each item required to assemble a single unit.

An efficient method for collecting the required parts is to divide the list into three groups:

- Those parts to be purchased.
- The materials and supplies to be purchased.
- The components to be manufactured.

One or two team members handle purchasing the necessary components, materials and supplies while the other team members work in the workshop provided by the college to fabricate the parts according to the detailed drawings prepared in the design phase. It is important to recognize that an engineering drawing defines the part that is being manufactured. Always work from the engineering drawing and not from your memory or understanding of the part geometry.

It is suggested that one team member serve as the "inspector," checking the finished parts against the drawings for the parts that have been manufactured. If the parts have been purchased, the "inspector" should check the purchased items against the parts list to insure that the item exactly meets the description and that the correct quantity has been procured.

1.6.2 Prototype Assembly

When the assembly kit is complete and checked against the parts list, we can begin the final phase—prototype evaluation. The first part of the evaluation is to assemble or build the prototype. We often call this step the "first article build," because it is the first time that we have attempted to fit together all of the components needed to assemble the prototype. The "first article build" can go well if: all of the parts are available; they all fit together properly; the tolerances on each part and each feature are correct; and the surface finish on all of the parts is acceptable. Prototype design is often judged against the four Fs—form, fit, finish and function. The "first article build" permits us to assess how well the team performed with regard to the first three of the Fs—form, fit and finish.

If the parts do not fit, modifications of one or more parts are required. These modifications require design changes that are a dreaded and costly process in the real world. In fact, one of the most important criterion used by management to judge the quality of a development team is the number of design changes required both before and after the introduction of a product to the marketplace. Clearly, we want to design each part in a product correctly during our first effort and minimize the number of required design changes.

We anticipate that each team will make a few errors in the preparation of the detail design drawings and that changes will be required. The natural tendency is to take the offending part to the model shop and to correct (modify) it and move on with the "first article build." This behavior is acceptable only if the team revises the drawing of the offending part to reflect the modification made to correct the design deficiency. In the real world, the integrity of the drawing package is much more important than the prototype. The second, and all subsequent assemblies, will be fabricated from the details shown in your engineering drawings and not by examining the prototype. The prototype is often scrapped after it has been built and tested.

1.6.3 Prototype Evaluation

After the assembly of the prototype is complete, it should be carefully inspected to insure it is safe. Fortunately, a scale is a relatively safe product since the parts do not undergo significant motion, and the hazards due to pressures and temperatures should not exist. However, make sure that you have eliminated all sharp edges and points on the prototype, and that you have avoided introducing pinch points. (Pinch points are openings which close due to motion of one or more parts when a product is operated. Fingers, hair and/or clothing caught in pinch points may represent a hazard).

The final step in this development process is to test your prototype to determine if it is functional and accurate. This is a big day and indeed a big hour. Typically a one hour time slot has been scheduled for evaluating each team's prototype in the testing facility available in a college laboratory. Your prototype will be judged based on the following criteria:

1. Performance: accuracy over the entire specified range and readability of the display.
2. Capability: ability to determine required postage for different sizes and types of letters and packages.
3. Cost: minimize.
4. Design innovation: novelty and creativeness.
5. Quality of parts and assembly: workman(women)ship.
6. Appearance: pleasant looking and in keeping with the decor in a typical office.

1.7 TEAMWORK

In this course, you are required—for three reasons—to participate on a development team. First, the project is too ambitious for an individual to complete in the time available. You will need the collective efforts of the entire team to develop a quality postal scale during the semester. We plan on pressing the development teams to complete the project on a prescribed schedule. Second, we want you to begin to learn teamwork skills. Experience has shown us that most students entering the engineering program have not developed these skills. From K-12, the educational process has focused on teaching you to work as an individual often in a setting where you competed against other students in your class. We will demand that you begin functioning as a team member where cooperation, following and listening are more important than individual effort. Leadership is important in a team setting, but following is also a critical element for successful team performance. The final reason is that you will probably find yourself on a development team early in your career if you take a position in industry. A recent study [4] by the American Society for Mechanical Engineers (ASME), the results of which are shown in Table 2, ranked teamwork as the most important skill to develop in an engineering program. Teamwork was also the first skill, in a list of 20, considered important by managers in industry. We are hopeful that this course will be instrumental in exposing you to team working skills so necessary for a successful career.

As you work within your team in the development of a postal scale, you will be introduced to many of the 20 topics listed in Table 1.2. We hope that you will begin to develop an appreciation of the need for these skills and begin to enhance your level of understanding of the design process that requires you to utilize many of these skills.

TABLE 1.2

Skills Considered Important for New Mechanical Engineers
with Bachelor Degrees
Priority Ranking

1. Teams and Teamwork
2. Communication
3. Design for Manufacture
4. CAD Systems
5. Professional Ethics
6. Creative Thinking
7. Design for Performance
8. Design for Reliability
9. Design for Safety
10. Concurrent Engineering
11. Sketching and Drawing
12. Design for Cost
13. Application of Statistics
14. Reliability
15. Geometric Tolerancing
16. Value Engineering
17. Design Reviews
18. Manufacturing Processes
19. Systems Perspective
20. Design for Assembly

1.8 OTHER COURSE OBJECTIVES

While your experience in developing the postal scale (either analog and/or digital) is the primary objective of this course, there are several other related objectives. You will quickly recognize a critical need for graphics as you attempt to describe your design concepts to fellow team members. We have included four chapters on graphics to help you learn the basic skills required at this entry level. These chapters include materials on three view and pictorial drawing, graphs and tables, and instructions for developing entry level skills in Pro/ENGINEER, which is a computer aided drawing program.

You will also be required to become familiar with using three additional software application programs: word processing, spreadsheet, and graphics presentation. Our experience indicates that nearly all of you are already proficient with word processing; therefore, we will not cover this topic here. However, many of you are not familiar with how to use a spreadsheet. Accordingly, we have included a chapter describing Microsoft Excel to help you prepare your parts list, to perform calculations and to prepare graphs. Spreadsheets provide very powerful tools that you will find useful in many different ways for the remainder of your life. We hope that you will take this opportunity to learn how to use this important software. We have also included a description of Microsoft PowerPoint to aid you in the preparation of world-class slides for your design briefings.

As you proceed with the development of the postal scale, we frequently will require you to communicate both orally and in writing. Design reviews before the class give you the opportunity to learn presentation skills such as style, timing and the preparation of visual aids. The design report will give you experience in preparing complete, high-quality engineering drawings and in writing a technical report which contains text, figures, tables and graphs.

The final objective of the course is design analysis. We understand that your engineering analysis skills have yet to be developed. For this reason, we will present key equations that mathematically model the electrical and mechanical components used in the design of many of the subsystems that you employ in the design of the scales. We do not expect you to completely understand all of the theoretical aspects of the relatively complicated subsystems involved. Most of you will take several courses later in the curriculum dealing with these subjects in great detail However, at this stage of your career, we want you to begin to appreciate the relationship between analysis and design. To help you with the analysis, we have included a chapter on the mechanical aspects of weighing machines and another chapter on electrical circuits, digital displays and microprocessors. The coverage is very brief and introductory, but it should give you a direction and a start in understanding both mechanical and electronic systems.

REFERENCES

1. Valenti, M. "Teaching Tomorrow's Engineers," Special Report, Mechanical Engineering, Vol. 118, No. 7, July 1996.
2. Ulrich, K. T., S. D. Eppinger, <u>Product Design and Development</u>, McGraw-Hill, New York, NY, 1995, p. 6.
3. Cross, N., <u>Engineering Design Methods</u>, 2nd Edition, Wiley, New York, NY, 1994, p. 77-81.
4. Anon, "Integrating the Product Realization Process into the Undergraduate Curriculum," ASME report to the National Science Foundation, 1995.
5. Macaulay, D., <u>The Way Things Work</u>, Houghton Mifflin Co., Boston, 1988.
6. Horowitz, P. and W. Hill, <u>The Art of Electronics</u>, 2nd Edition, Cambridge University Press, New York, 1989.
7. Dally, J. W., Riley, W. F. and K. G. McConnell, <u>Instrumentation for Engineering Measurements</u>, 2nd Edition, Wiley, New York, 1993.

EXERCISES

1. List ten products that you have used today and the companies that manufactured and marketed them.
2. Suppose that you do not want to work on product development, manufacturing or sales. What opportunities remain for you in engineering?
3. Write an engineering brief describing the characteristics of a winning product.
4. Write the most important equation in engineering or business.
5. Why is Ford Motor Company reluctant to develop a brand new model of an automobile?
6. What is a design concept? How many concepts will your team generate in developing the postal scale this semester?
7. What are the differences between analog and digital scales? Define the meaning of the word analog and relate it to an analog device. Define the meaning of the word digital and relate it to a digital device.
8. Why is it important to prepare a parts list during a development program?
9. What is an assembly kit and why do we prepare one in the development of the first prototype?
10. Why do we invest scarce company funds in assembling a prototype?
11. What are the safety considerations with which you must be concerned in testing the analog and digital scales?
12. Why do you believe industry representatives ranked teams and teamwork as the most important skill for new engineers?

CHAPTER 2

POSTAL SCALES
AS WEIGHING MACHINES

2.1 DEFINING WEIGHT

A postal scale is the product selected as the theme for Book 3. We intend for you to study the history of measuring weight, to perform a mechanical dissection on one or more scales and then to design, build and evaluate some type of a postal scale. The class will be divided to form development teams each with five or six members. Each team will have the responsibility to develop one of several different postal scales described later in Section 2.4. The level of the complexity of these different postal scales will also be described in Section 2.4. Because you will share responsibility with other team members in developing a postal scale, grading of some assignments will be common among the members of the development team.

Before describing postal scales in detail, let's briefly discuss the meaning of weight, and then consider some of the methods for measuring weight that have been used for thousands of years. Our weight is a force produced by the earth's gravitational pull on our bodies. Our body, or any body for that matter, has a mass which we will define as m_b. The earth has a mass m_e. When two masses are in close proximity, there is an attractive force F which develops between the two bodies. The magnitude of this force is given by:

$$F = C \, m_b \, m_e \, / r^2 \qquad\qquad (2.1)$$

where C is the universal constant of gravitation.

r is the distance between the two masses.

Earth bound bodies, either those of people or objects, are very small compared to the radius of the earth R_e which is equal to 3960 miles or 6.37×10^6 meters. Because of the large size of the earth, we can, without introducing significant error, ignore the size of the earth bound body and set:

$$r = R_e \qquad (2.2)$$

Next, we collect all of the quantities that have anything to do with the earth in Eqs. (2.1) and (2.2), group them together and define a new constant g as:

$$g = Cm_e / R_e^2 \qquad (2.3)$$

We treat g (the gravitation constant for the earth) as a constant equal to 32.17 ft/s^2 or 9.806 m/s^2 depending on our choice of units. Although strictly speaking, g varies a bit because the earth is not a true sphere and R_e does not remain constant when we move from Pikes Peak to Death Valley.

Substituting Eq. (2.3) into Eq. (2.1) and rewriting the result gives:

$$F = m_b g \qquad (2.4)$$

Note, the gravitational force F in Eq. (2.4) is the weight of a body on Earth with a mass m_b.

If you examine the units of g (ft/s^2 or m/s^2), it is evident that it is an acceleration. The force due to the gravitational field of the earth produces an acceleration that acts on all the bodies on or near the surface of the earth. This explains why we fall down a set of stairs or out of a tree.

Let's look at one more relation to help us describe the meaning of weight—Newton's second law of motion:

$$F = ma \qquad (2.5)$$

where a is the acceleration.

m is the mass of some body.

Sir Isaac Newton first described this important relation between force F and acceleration. Let's apply Eq.(2.5) to earth bound bodies. The gravitational field causes an acceleration a = g (where g is the earth's gravitational constant). The force F acting on the body in the gravitational field is the weight W. We can then write:

$$W = F = ma = m_b g \qquad (2.6)$$

If we travel from College Park to Chicago, our weight remains essentially constant. So we get confused and think of our weight as being the constant in Eq. (2.6). Not true! The term that is the constant in Eq. (2.6) is the mass m_b. Prove it! Go jump on the moon. We all know that we weigh much less on the moon—about one sixth as much as here on earth. The reason for the lower gravitational acceleration on the moon (about g/6) is because the mass of the moon is much smaller than the mass of the earth. The mass of our body m_b is the same whether we are on the moon, mars, in space or on earth.

Okay! We understand mass is the constant, and weight is a force due to the earth's gravitational field. Let's next consider the methods that have been developed over time to measure weight.

2.2 SOME HISTORY ABOUT MEASURING WEIGHT

Sometimes we are in such a hurry that we fail to look back and review the accomplishments made centuries ago. This statement is also true in design. As we approach the millennium year 2000, we must remember that we have some records of what people did 8000 years ago. Our ancestors were not lazy or dumb. They invented a lot of very neat products. True, the invention of steam power came late in historical time, electricity even later and microelectronics in the past 35 years, but we had many very important developments long before our generation. Analog weighing machines have a very long history. Electronic (digital) weighing machines were introduced in the market only after the development of suitable microelectronic components; they are so new that we have very little history to review that is not extremely contemporary.

2.2.1 A Simple Beam Balance

The first weighing machines go back to the ancient Egypt and Babylon [1–4]. These weighing machines were beam balances as illustrated in Fig. 2.1. These were very simple devices consisting only of a beam (usually wooden) and three loops of string. The beam was hung from its center loop, and then an unknown weight was placed in the right hand loop. Reference weights were added in the left hand loop until the beam became horizontal (the balance position). With the beam in the balance position, the unknown weight was equal to the reference weight. When the beam was in balance, the measurement of weight of the unknown mass was complete. You could weigh any item that could be supported by the beam and the strings providing you had an acceptable set of reference weights. In the very early days, the reference weights were usually made of stone or glass although later when metals became more common, brass or bronze was sometimes employed.

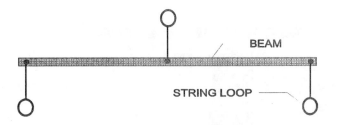

Fig. 2.1 The very early beam balance with string attachments for the support, the unknown mass and the reference weights.

Over the centuries, the beam balances were improved with many design modifications. The first improvements were the additions of pans and hooks to support either the reference or unknown weights as shown in Fig. 2.2. Later, when metal became available for the constructing the beams, pins were placed through the beams to attach the clevis used with the pans or the center support.

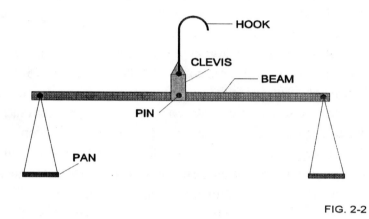

FIG. 2-2

Fig. 2.2 Beam balance with pans for supporting the weights and a pinned central support.

When the accuracy of the weight measurements became more important, there was concern about the effects of friction at the pin associated with the center support. With friction, slight inequalities could exist between the unknown and the reference weights with the beam in apparent balance. To eliminate this concern, the center pin in the beam was replaced with a knife edge support as shown in Fig. 2.3. The knife edge eliminated friction, improving the accuracy of the measurements, but the sharp edge was fragile. Weighing machines were fitted with mechanisms for lifting the beam off of the knife edges to protect the knife edges as weights were added to the pans. When the weights were placed in both pans, the assembly was lowered slowly onto the knife edges to check for balance.

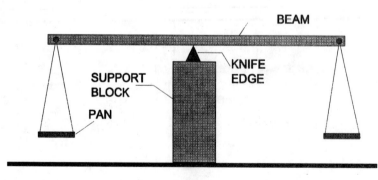

Fig. 2.3 Beam balance with pans and knife edge support to minimize friction.

With time, beam balances became more and more accurate. They were placed into glass cases to prevent air currents from disturbing the balancing operation. Until recently, beam balances could be found in chemistry laboratories. They were equipped with indicators to guide the operator in selecting weights to add to the reference pan. They were also fitted with dampers to reduce the number of oscillations before the beam reached equilibrium.

Beam balances were constructed in many different sizes. Large balances were used to weigh large quantities of grain or other market produce. Medium size balances were employed to measure spices or other more valuable merchandise. Very small balances were used for valuable items like gold or precious stones. It was recognized, even in the time of the Egyptians, that accuracy of a beam balance depended on its capacity. The large beam balances were less accurate than the smaller balances. This fact is still true today for all weighing machines—electronic or mechanical and analog or digital. Accuracy and capacity (range) are inversely related.

2.2.2 The Bismar Balance

The balance, with its many modifications, served well for many centuries[1] before a new type of balance was introduced with unequal arms for the beam. There are two types of unequal arm balances—the bismar and the steel yard. With the bismar balance, illustrated in Fig. 2.4, the beam is equipped with a large fixed weight (called a counterpoise) on the left end. The unknown mass is connected to the hook attached to the right end of the beam. The center support consists of a moveable loop of cord, rope or a metal band. The location of the center support is adjusted by moving it toward one end of the beam or the other until the beam is positioned horizontally. In this horizontal position, the unknown weight of an item can be determined relative to the weight of the counterpoise. The analysis that gives the relation for the unknown weight W_u in terms of the known weight of the counterpoise W_{cp} is based on the equilibrium of moments. To perform this analysis, we start with a free-body diagram of the bismar balance which is illustrated in Fig. 2.5.

Since the beam is horizontal and at rest, it is in equilibrium. We can write the relation for equilibrium of the moments due to all of the forces about point O. The equilibrium relation is written using the Σ notation to indicate that the moments due to each force are to be summed.

$$\Sigma M_O = 0 \qquad\qquad (2.7)$$

The two weights W_u and W_{cp} both produce a moment about point O[2]. The moment is simply the weight times the perpendicular distance from the line of action of this force to the point O:

$$M = W \times d \qquad\qquad (2.7a)$$

Using this simple rule for the magnitude of the moments, we can rewrite Eq. (2.7) as:

---

[1] Equal arm balances are still in use today.

[2] The force F due to the adjustable support does not produce a moment about point O because it passes through this point; consequently, the moment arm d = 0.

$$(W_{cp})d_{cp} - (W_u)d_u = 0$$

Solving this relation for W_u yields:

$$W_u = (d_{cp} / d_u)W_{cp} \qquad (2.8)$$

Since we know the weight of the counterpoise, we can measure W_u by determining the two distances d_u and d_{cp}. Great! We have converted a difficult weight measurement into the simple task of measuring two lengths, d_u and d_{cp}.

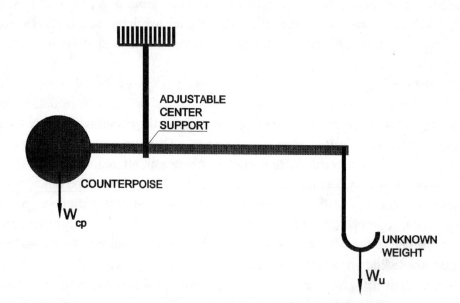

Fig. 2.4 Sketch illustrating the adjustable middle support and fixed weight on a bismar balance.

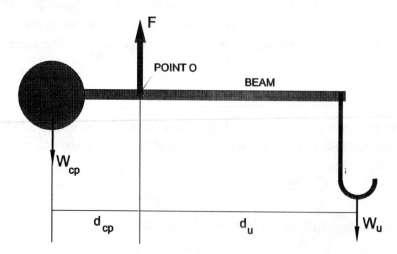

Fig. 2.5 Free-body diagram showing the forces and the distances used in writing the equilibrium relation for a bismar balance.

Weight machines where the weight measurement is somehow transformed into a measurement of lengths or distances are usually called scales. The word scales derives from the instrument used to measure length, namely a scale. For the remainder of this chapter, we will usually use the word scales instead of weighing machines because it is shorter and in more common usage.

2.2.3 The Steelyard Scale

The steelyard scale was another 18th century scale that was very similar to the bismar scale. The design of the steelyard scale fixes the pivot point on the beam at an off-center position as indicated in Fig. 2.6. The position of the pan (or hook) supporting the unknown mass is also fixed at the short end of the beam near the counterpoise. An adjustable weight W_a is the moveable component in the design of a steelyard scale.

To perform an analysis of the steelyard scale, let's draw its free body diagram. Let's cut the vertical support and replace it with a force F. Also, we show the weights due to the counterpoise, the unknown mass and the adjustable weight as W_{cp}, W_u and W_a respectively on the free body diagram presented in Fig. 2.7. The analysis of the free-body diagram of the steelyard scale is identical to that of the bismar scale, and we rewrite Eq. (2.7) and take moments about the point O to give:

$$W_{cp}\, d_{cp} + W_u\, d_u - W_a\, d_a = 0 \qquad (2.9)$$

Simplifying Eq. (2.9) yields:

$$W_u = W_a(d_a/d_u) - W_{cp}(d_{cp}/d_u) \qquad (2.10)$$

However, with the steelyard scale many of the quantities in Eq. (2.10) are fixed and known (i. e. W_{cp}, d_{cp}, W_a and d_u. We can group them together and set them equal to a constants C_1 and C_2:

$$C_1 = W_a/d_u$$
$$C_2 = W_{cp}(d_{cp}/d_u)$$

Substituting these constants C_1 and C_2 into Eq. (2.10) gives a simple linear relation for the weight of the unknown mass that contains only the measurement of a length d_a:

$$W_u = C_1\, d_a - C_2 \qquad (2.11)$$

The steelyard scale is easier to employ than the bismar scale since only one measurement of distance d_a must be measured; a scale for making this measurement is embossed on the beam. With clever design, we have converted a difficult weight measurement into a very simple determination of a single length.

Have you been to your doctor's office lately? The nurse usually weighs you before you have an opportunity to see the doctor. Does the scale in your doctor's office resemble the steelyard scale? Did the nurse move a weight (maybe two)? Are these weights similar to the adjustable weight? The scale in the doctor's office is a platform scale that we will discuss later; however, it is based on the same principle of equilibrium relations as the steelyard and bismar scales.

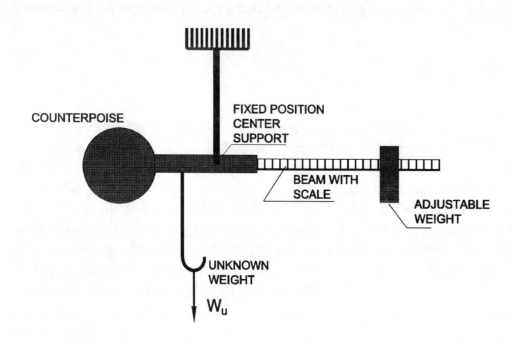

Fig. 2.6 Schematic illustration of a steelyard scale with an adjustable counterweight.

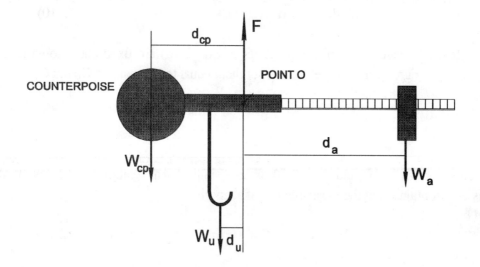

Fig. 2.7 Free body diagram of a steelyard scale.

2.2.4 Spring Scales

Let's consider a different type of scale that does not use a balance beam. One that does not use reference weights or a counterpoise. Do you recognize a helical spring like the one shown in Fig. 2.8?

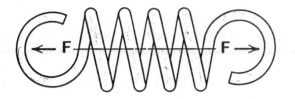

Fig. 2.8 Illustration of a tension type helical spring.

Helical springs are wound from wire to form a helix that serves as the spring element. When springs are subjected to either tension or compression forces, they stretch or compress; when the force is removed, they return to their original shape. Springs are employed for three major tasks in mechanical design. First, they will return some object to its original position like the bolt in a door lock. Second, they store energy in the deflected state which can be returned to the system when the spring resumes its original shape. Finally, springs can be used to measure force. We are interested in springs because we know they may be employed in our weighing machine to convert a force measurement into a length measurement.

Springs have been used in the design of scales for a few hundred years, and they are still utilized in most of the lower cost mechanical scales on the market today. There are three essential components in a spring scale:

- A spring which acts as the sensing element.
- A pointer activated by the deflection of the spring.
- An indicator/scale for displaying the weight.

We will discuss the mechanism of the spring scales in more detail later in Chapter 3. At this time, let's consider the equation describing the response of a helical spring under the action of an applied axial force F.

$$F = k \, d \qquad\qquad (2.12)$$

where d is the axial deflection of the spring in in. or m.

k is the spring constant (or spring rate) in lb/in. or N/m.

Consider a spring with a spring constant of 500 lb/in. This specification for the spring constant (often called the spring rate) means the spring will deflect one inch if we apply a load of 500 lb to it. However, we will learn later that our product specification requires us to measure a weight that varies up to a maximum of 5 to 10 pounds. Can we still use this spring? How much will its free end deflect? Let's analyze the spring before deciding if it might help in the design of a prototype?

Let F = 10 lb which is the required capacity of a scale that your team may decide to design and build. If k = 500 lb/in.[3], determine the deflection d of the spring. Using Eq. (2.12) gives:

$$d = F/k = 10/500 = 0.020 \text{ in.}$$

The spring deflects 0.020 in. when we apply a force of 10 lb to one end and constrained at the other end. This deflection is not large; hence, it is realistic to conclude that a spring with k = 500 lb/in. is too stiff to employ in a low capacity scale.

We note that the spring is a mechanical transducer which converts a force (weight) into a measurement of displacement (length). To make a scale with a spring as a sensor, you need to track and display the spring displacement. A common design is to use a pointer attached to the spring. The position of the pointer is referenced to a scale that is calibrated in terms of weight. We illustrate these concepts schematically in Fig. 2.9.

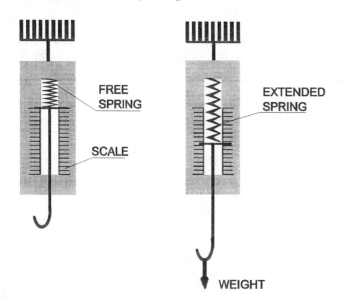

Fig. 2.9 Schematic illustration of a fish scale showing the design features employed in measuring weight with a spring.

2.2.5 The Platform Scale

The platform scale was introduced in the early 18th century when it became necessary to weigh carts and carriages travelling the toll roads in England. Platform scales are essentially complex balances that employ compound levers. The pan for the unknown mass was originally a very large platform which accommodated a cart or a carriage. The force due to the unknown mass is transmitted to a series of compound levers as shown in Fig. 2.10.

---

[3] To give you a sense of size, a spring, wound with round edge flat wire 9/32 by 5/32 in. with a free length of 3 in. and outside and inside diameters of 1.25 in. and 0.625 in. respectively, exhibits a spring constant of 520 lb/in.

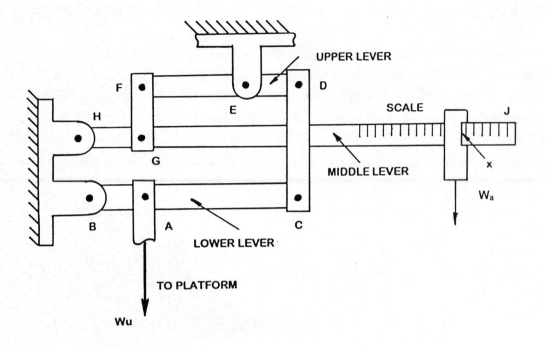

Fig. 2.10 Platform scales employ compound levers to weigh very large masses.

      The platform, shown in Fig. 2.10, is attached to the lower lever at point A. The lower lever arm moves downward at point A, rotating about point B and deflecting downward the pin at point C. This deflection is transmitted from the lower lever to the upper lever with the vertical link which connects points C and D. The downward deflection of point D causes the upper lever to rotate about point E. This clockwise rotation about point E produces an upward deflection of the pin at point F. This upward deflection is transmitted from the upper lever to the middle lever by a second vertical link that connects points F and G. Since point G is moving upward, the middle lever will rotate about the pin at point H and tilt upward. A small weight, on the right side of the middle lever, can be adjusted to bring this lever to a horizontal position. The right side of the middle lever is imprinted with a calibrated scale to give directly the weight of the unknown mass at the balance position of the adjustable weight.

      Whew! Compared to the simple balances introduced earlier, the platform scales are very complex. Why do we introduce this complexity? There are two reasons. First, to limit the vertical movement of the platform. The pins and load application points are arranged to greatly amplify the motion of the middle lever at the end J. This large movement tends to make the scale sensitive while the very limited motion of the platform gives it stability.

      Second, the use of three levers with cleverly arranged pin positions gives a very large leverage and consequently a very small weight W_a can be used to balance a very massive object placed on the platform. The small adjustable weight W_a is a very significant advantage because it moves easily along the scale inscribed middle lever to accurately balance the large weight on the platform.

2.2.6 Indicator Scales

Indicator scales have features similar to the simple balance, bismar and the steelyard scales. The indicator scale pivots about a fixed central support at point O as shown in Fig. 2.11. However, the counterpoise is configured as a pendulum that dominates the right side of the mechanism. The unknown mass is suspended from the hook (or pan) positioned on the left side of the indicator scale. Depending on the weight of the unknown mass the entire system rotates decreasing the dimension d_u and increasing the distance d_a so as to automatically balance the moments about point O.

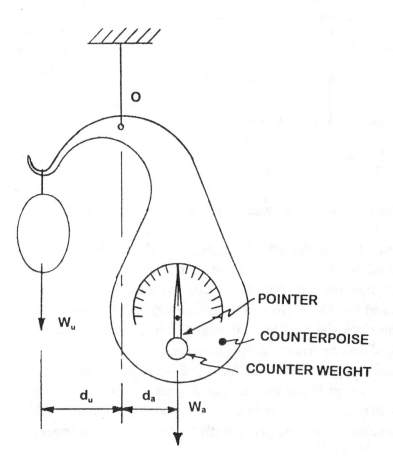

Fig 2.11 Sketch of an indicator scale showing the pendulum weight with the self aligning pointer.

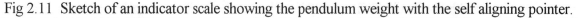

With the indicator scale, unlike the bismar or steelyard scales, it is not necessary to measure the distances d_u and/or d_a to determine the unknown weight. The length measurements are avoided by placing an indicator on the pendulum counterpoise as shown in Fig. 2.11. The indicator is a circular scale and a vertical pointer. The circular scale is inscribed on the pendulum and calibrated to read the weight of the unknown mass directly. A self-aligning pointer (it has a counter weight at its lower end to keep the pointer oriented in the vertical direction) points to the correct marking for the weight providing the circular scale is calibrated. As an unknown mass is added to the indicator scale, the pendulum and the circular scale both rotate below the pointer while the pointer always remains vertical. This is a very clever design for an automatic (analog) measurement of the weight of an unknown mass.

2.2.7 Hydraulic Scales

Let's consider still another concept much different than balances and spring type scales—a weighing machine based on hydraulic principles. Suppose we place our unknown mass on a platform, and connect this platform to a piston and cylinder arrangement as illustrated in Fig. 2.12.

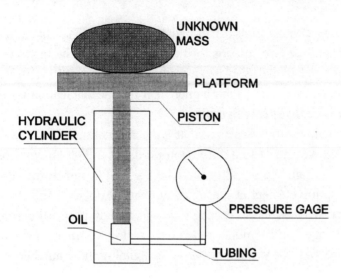

Fig. 2.12   A sketch illustrating a platform-piston system used to develop pressure in a closed hydraulic system.

When the force W_u is applied to the platform, the piston is driven downward into the cylinder. This downward motion is resisted by the oil that is contained in the cavity of the cylinder. A pressure p is developed in the cylinder which is expressed as:

$$p = W_u / A_p \qquad (2.13)$$

Where p is the hydraulic pressure in psi or Pa.
A_p is the cross-sectional area of the piston in in.^{2} or m^2.

Clearly, the weight and the pressure are related in a linear manner as indicated by Eq. (2.13). The piston-cylinder arrangement has changed our problem from measuring weight to one of measuring pressure. Is that an advantage? Yes! To weigh large masses, balances become either extremely cumbersome or complex with many levers and pins. If springs are employed in the design of a scale, they become bulky and costly. We can contain high pressures (up to 3000 psi) with little difficulty, and relatively accurate bourdon tube gages are available to measure the pressure at low cost.

But, how do we convert pressure to weight? It is easy. As you might expect the pressure gage is contained in a circular case with a pointer and a circular scale. The circular scale is calibrated to indicate the pressure. It is a simple matter to replace this pressure scale with one calibrated to provide a read-out in terms of weight.

If you search the historical literature, you will not find a reference to a hydraulic scale (at least I did not in my brief review). We were relatively late (18th century) in discovering pressure and harnessing it to produce work.

2.3 ELECTRONIC SCALES

Electronic scales are much more contemporary than hydraulic scales. They have become common only in the past 25 years. There are two reasons for their late arrival in the market place. First, the technology (stable amplifiers and digital displays) was not available until very recently. Second, the cost of the electronic components was much higher than the mechanical parts (this is still true today although the difference is decreasing with time).

Most electronic scales are related to spring scales in that a mechanical member in the scale deforms under the action of the applied weight. The deformation produces a high strain at one or more points on the spring element. Strain gages, illustrated in Fig. 2.13, which are variable electrical resistors, are bonded to the spring element at these highly strained points. The change in the resistance of the strain gages is converted into a voltage change that is proportional to the weight on the scale. The voltage change can be monitored on an analog voltmeter or a digital voltmeter. If a digital voltmeter is employed, the voltage displayed is digital (lighted numbers).

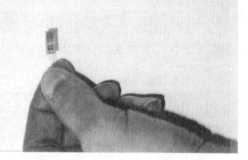

Fig. 2.13 Photograph of a bonded metal-film resistor used to measure strain.

This purpose of this section on electronic scales is only to mention that this "new" scale exists; they are very contemporary in the historical development of weighing machines. We will, in Chapter 4, provide a complete and detailed treatment of the technology used in designing digital scales.

2.4 THE PRODUCT SPECIFICATION

The project described in this book is a postal scale; however a quick trip to an office supply store will show that a wide variety of products are available for the typical mail room. There are mechanical scales with analog display and electronic scales with a digital display. Some scales display the weight, others show the postage and some indicate both weight and postage. We will outline a large number of products which are suitable for use as a postal scale in the home or mail room of a typical office. Your team, acting in cooperation with your instructor, is to select one or more of the products listed below to design build and evaluate during the semester.

1. Mechanical scale with an analog display of both weight and postage.
   - Very low capacity 0 to 12 oz.
   - Low capacity 0 to 4 lb.
   - High capacity 0 to 50 lb.
2. Electrical scale with an analog voltmeter displaying weight and postage.
   - Low capacity 0 to 4 lb.
   - High capacity 0 to 70 lb.
3. Electronic scale with a digital display of voltage.
   - High accuracy.
   - Wide range in capacity.
4. Electronic scale with a digital display of weight.
   - Low capacity 0 to 4 lb.
   - High capacity 0 to 70 lb.
5. Electronic scale with a digital display of postage.
   - Programmable rate computing scale for first and third class postal rates with a capacity of 0 to 4 lbs.

An examination of the many products described shows a wide latitude in complexity. Some products are simple with relatively few part numbers while others involve many parts. Also a few products are mechanical; made from components that we know and understand. Other products are electrical and/or electronic and contain components and devices with which we may not be familiar. In the selection of the product or products that your team is to develop, match the team skills with the talent required to successfully produce the product of your choice.

You will be expected to participate as a member, or possibly the leader, of a student development team in the design, manufacture, assembly and evaluation of the product of your choice. A computer laboratory with software for word processing, computer aided drafting (CAD), spreadsheets, and a graphics presentation program will be important in this development activity. A model shop and/or work shop will also be helpful in fabricating the required components and for fitting and assembly of the postal scales.

The evaluation of the postal scales will be performed by adding calibration weights to the platform or pan that is intended to support the package or letter. These weights are available in different capacities such as ounce, pound, five pound and ten pounds. The calibration weights will be available for your use prior to the final evaluation of your prototype. In the final evaluation to determine the accuracy of the postal scale, You will add at least ten calibration weights, one at a time, recording the measurement after each application of load. A comparison of the calibration weight with the measurement of the weight will indicate the accuracy of the scale over its entire range. We anticipate that you will be able to complete the verification test to determine the accuracy of a scale in five minutes or less.

The visibility of the displays, both analog and digital, will be checked by your instructor. Your instructor, with 20/20 vision, will verify the readings of the weight displayed from a distance of two feet. If you design a high capacity scale, the display must be visible with a package the size of a copy paper box on its platform.

Criteria that will be used to judge the merits of your team's development effort include:

1. Performance
    - Accuracy of 0.5% over the entire range.
    - Resolution of the measurement ($\pm$ 1 count or smallest division on the scale).
    - Visibility of the display.
2. Cost
    - Minimize the cost of materials and components used in the prototype.
    - Minimize cost consistent with maintaining a quality product.
3. Design innovation
    - Simplicity.
    - Novel concepts.
    - Ease in manufacturing and assembly.
    - Ease in adapting to changing postal rates.
4. Quality
    - Reliability of the measurements.
    - Close fitting parts.
    - Minimum friction.
    - Superior workman(women)ship.
    - Alignment.
    - Durability.
5. Appearance
    - Pleasing to the eye.
    - Decor matching a modern office.
6. Package size
    - See extended description below.
7. Safety
    - Avoid all hazards[4].

The criterion—package size—requires more explanation because it is more involved than the other criteria used in determining the adequacy of the postal scale which your team develops. The package size is a design constraint. We are limiting the size of each postal scale by requiring that all the parts fit into a shoe box which normally contains a pair of athletic shoes. (My most recent pair of Nike shoes came in a box 15 in. long by 8 in. deep by 5 in. high.)

You will also be required to assemble your prototypes using simple hand tools such as a drill, screw drivers, hammer, pliers, soldering iron, etc. You will be allotted 50 minutes immediately prior to the class period scheduled for the evaluation test to assemble the prototype. You will also be required to disassemble the scale, placing all of the parts back into the shoe box in the 15 minute period immediately following the evaluation tests. Unless you receive instructions to the contrary from your instructor, you will have only one opportunity to "officially" determine the accuracy of your scales. However, the calibration weights will be available six weeks prior to the scheduled

---

[4] See Chapter 16 on Safety, Risk and Performance.

evaluation test; you are encouraged to check your design concepts well in advance of your final class. When you have completed the evaluation test, pack the parts from the postal scale in the shoe box and dispose of it in an environmentally sensitive manner. If it is functional, perhaps you will use it in your home office or give it as a gift to a friend.

The resolution of the measurement of weight will depend on the capacity of the postal scale. For example, if your scale is intended to weigh letters, the postal rates vary with each ounce[5]. Accordingly, the resolution should be of the order of 1/10 of an ounce. For higher capacity scales intended for weighing packages, the postal rates change with each pound. Hence, the resolution required of a high capacity scale is less than that required in a lower capacity scale intended for a different purpose. (We will discuss the meaning of resolution in more detail in Chapters 3 and 4). The accuracy requirements depend on the weight placed on the scale. From 0 to 13 ounces, the weight measurement should be accurate to ± 0.5 %. From 13 ounces to 80 ounces, the accuracy should be ±2 %. From 5 lb to 50 lb, the accuracy of the weight measurement should be ± 5%.

The postal scale is to be equipped with a control and/or mechanism for zeroing the scale. This zero adjustment is an important feature for all of the systems previously described. The mechanical components in the scale wear or change dimensions with temperature causing the zero position of the indicator to vary from zero. For electrical and/or electronic systems, the output voltage drifts with time and temperature and adjustments are necessary to null out these parasitic voltages.

We anticipate that our customers will purchase these scales to be used on a daily basis in either their home or a mailroom in a typical professional office. Accordingly, the scale should be small to conserve counter space. It should also be attractive with rounded corners, flowing curves and smooth finishes. The colors employed should be consistent with the decor of a modern office.

The platform on the postal scale should be easily accessible and stable. We recognize that the platform of a scale must move, however, packages and/or letters must not shift during this motion. The platform should also be sized in accordance with the capacity of the scale. If it is intended for packages, the platform must be large enough to accommodate them. On the other hand if the scale is intended for letters the platform should not be too large.

The displays must be visible to the individuals using the scales. You should anticipate that the user will stand adjacent to the desk or counter upon which the scale is placed. Reading either the weight or the postage from the display should be possible from the standing position. You may design a large display, magnify the display to enlarge the numerals or you may choose to move the display to a location more suitable for reading.

While 120 V ac power is available in the home or office, we believe that it represents a safety hazard if used in your early prototype. Accordingly, **the use of the 120 Volt ac utility line power supply is strictly prohibited**. We recognize that an electrical supply may be essential for the digital display and/or electronic scale. However, we anticipate that you will employ batteries to provide the necessary power. **Do not employ more than nine volt batteries without permission in writing from your instructor.**

---

[5] Postage for the first ounce for first class mail is $0.32 with a charge of $0.23 for each additional ounce. Since first class postage for a 13 ounce package is $3.08, it is advisable to use priority mail which requires a $3.00 postal charge for all packages weighing 32 ounces or less. Postal rates currently effective in the U. S. are provided in Appendix A.

REFERENCES

1. Kisch, B., <u>Scales and Weights: A Historical Outline</u>, Yale University Press, New Haven, CT, 1965.
2. Klein. H. A., <u>The World of Measurement</u>, Simon and Schuster, New York, 1974.
3. Lieberg, O. S., <u>Wonders of Measurement</u>, Dodd, Mead & Company, New York,1972.
4. Macauly, D., <u>The Way Things Work</u>, Houghton Mifflin Co., Boston, 1988.

LABORATORY EXERCISES

2.1 Using wire from one or more clothes hangers, design and construct a self-indicating postal scale capable of determining the postage for first class letters weighing from 0 to 4 ounces. Adhesives and or solder may be employed to fasten joints in your prototype; however, tape is not permitted. Prepare a report describing your prototype together with a theoretical analysis of the design.

EXERCISES

1. If the gravitational field on the moon produces a gravitational constant 1/6ᵗʰ as large as the gravitational constant on earth, what would you weigh on the moon? What is your mass on earth? What is your mass on the moon?
2. Discuss the advantages and disadvantages of the simple beam balance.
3. Describe the differences between the simple beam balance, the bismar scale and the steelyard scale. Also, describe the similarities between these three weighing machines.
4. We have described the helical spring in this chapter. Define two other types of structural elements that are sometimes employed as springs.
5. What are the advantages of the platform scale? What is its most significant disadvantage?
6. While the indicator scale is very clever in design, why is it rarely employed while the scales based on helical springs are common.
7. We have cited the advantage of the hydraulic weighing machine as it ability to determine the weight of very large masses. What are its two most important disadvantages?
8. The change of resistance of a strain gage (which is a variable resistor) is extremely small. What well known electrical circuit is used to convert this small resistance change into a voltage change?
9. Determine the error band for a scale with a capacity of 5 lb. Show a graph of measured weight as a function of calibration weight with the error band superimposed.
10. Describe the difference between resolution and accuracy.
11. Define friction. What is the equation used to determine the friction forces? What are the factors that affect the magnitude of the friction force? Is friction important in the design of a postal scale? What can you do as a designer to minimize friction?

CHAPTER 3

MECHANICAL SCALES

3.1 INTRODUCTION

All types of scales, mechanical or electronic—analog or digital, incorporate mechanical elements in their design. Even the electronic scales with their digital displays of the weight or the amount of postage begin with mechanical components. The conversion from the initial mechanical response to the weight placed on the scale, to an electrical signal and then to a digital output is an important part of designing a digital postal scale.

Before we become too deeply involved in the mechanical aspects of weighing machines, let's worry about the meanings of four words that we have used in the above paragraph:

- Mechanical
- Electronic
- Analog
- Digital

The word mechanical refers to components that we will be using in our scales such as platforms, springs, pins, levers, shafts, bearings, pointers, bars, scales, bolts, nuts, etc. Electronic refers to components such as resistors, capacitors, inductors, chips, circuits and circuit boards. In performing mechanical design, we will be concerned with motion, equilibrium, strength and friction. With electronic design we are concerned with voltage, current, signal amplification, analog to digital conversion and electrical noise.

The words analog and digital both refer to the way that we display the weight measurement. To illustrate the difference, examine the two voltmeters illustrated in Fig. 3.1. The meter to the left is an analog multi-meter; it measures voltage, resistance and current. The measurement is displayed with a pointer that sweeps over several calibrated scales. The analog multi-meter provides a continuous reading of the quantity being measured as indicated in Fig. 3.2. If the input voltage increases by a very small amount, the voltage displayed on the meter (the output quantity) will also increase by this amount.

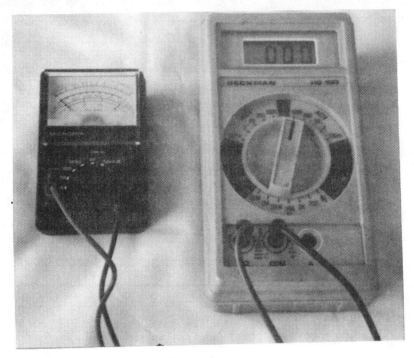

Fig. 3.1 Photographs of analog (left side) and digital (right side) multi-meters.

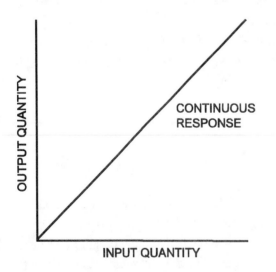

Fig. 3.2 An analog response is a continuous function.

The output of a digital display is not continuous. Inspection of Fig. 3.3 shows that the output voltage, corresponding to a linear ramp input voltage, resembles a staircase. As we slowly increase the input voltage to the digital voltmeter, the output voltage displayed remains constant until we reach a certain threshold voltage. At this threshold voltage, the digital voltage displayed increases as a step function and actually becomes larger that the input voltage. From Fig. 3.3, it is clear that the digital voltmeter is in error almost all of the time. Not to worry! We keep the error small (within allowable bounds) by maintaining very small steps in the staircase response when the input signal is digitized.

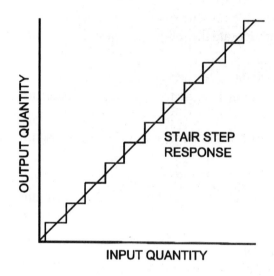

Fig. 3.3 The digital response for a linear ramp input voltage resembles a staircase.

We will discuss digital displays and the electronic aspects of our postal scale in much more detail in Chapter 4. In this chapter, let's closely examine postal scales that are based on the three following principles:

- Equilibrium or $\sum M_O = 0$.
- Springs that convert force into displacement.
- Strain gage based transducers.

We have selected these three topics because they all are viable approaches that your team may wish to consider in designing a prototype postal scale. In this Chapter, we will introduce the analysis methods that you will need to evaluate the merits of the different design concepts and to size the components incorporated in your design. We will also try to illustrate some of the design ideas and/or features that have been employed to develop scales currently available on the market.

3.2 EQUILIBRIUM BASED SCALES

The simple balance, shown in Fig. 2.3 and described in Chapter 2, is an equilibrium based weighing machine. The weight of the reference mass in the left hand pan produces a moment about the center support that is exactly canceled out by the weight of the unknown mass in the right hand pan. The beam of the balance is horizontal when the moments due to the weights in the two pans are equal. When the moments are not in balance, the beam tilts and the motion and/or position of the beam clearly indicates that the selection of the reference mass was not correct.

If the left and right sides of the balance are exactly identical, the friction at the pivot point is negligible and the beam is perfectly horizontal, then the weight of the unknown mass can be equated to the weight of the reference mass. The previous sentence is a long one with three qualifying statements. These qualifying statements are important to you as a designer; they illustrate that the **simple** balance you plan to develop must incorporate the following features:

- Perfectly symmetrical so the left and the right sides are identical.
- A pivot point with a knife edge or a low friction bearing to eliminate the frictional moment.
- A level indicator to insure that the beam is in a perfectly horizontal position when balance is achieved.

When we perform an analysis of a balance, we use the equilibrium relation for moments about the pivot point O:

$$\sum M_O = 0 \qquad\qquad (3.1)$$

We then explore the conditions required for balance to be achieved. However, when we perform the analysis, some assumptions are made which lead to the qualifications listed above. It is vital that you recognize these qualifications because they dictate many of the features that you must incorporate in the design of your balance.

Let's continue with our discussion of the simple balance. It is certainly a viable concept for a scale. In fact, it is still in use today to measure the weight of small items; however, it is not commonly employed to measure the weight of larger masses. The reason is very simple. We become tired of moving the reference masses around in the weighing process when they become large. It takes too long and requires too much energy. The simple balance is too cumbersome and too delicate to be used with objects weighing more than a few pounds.

The simple balance appears to be a viable approach to develop a low capacity scale for weighing letters and small packages. For larger packages, the equilibrium concept remains viable; however, the simple scale with equal length arms on the balance beam is no longer practical. To accommodate the larger packages with higher weight, we employ a balance with arms of different lengths. An example of such a balance is presented in Fig. 3.4. This scale is commercially available and is widely used because of its relatively low cost and high accuracy. The balance shown in Fig. 3.4 has a capacity of 2610 grams and a resolution of one gram. Is a resolution of 1 part in 2610 or 0.0383% considered good?

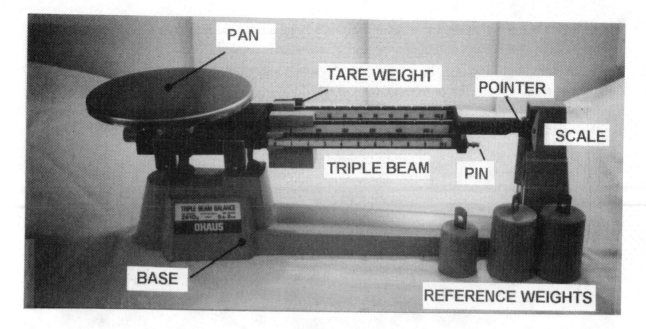

Fig. 3.4 Photograph of a commercial beam balance showing many design features.

Let's explore the features incorporated into this commercial balance to learn how to achieve accuracy, resolution and capacity (range) in a scale. In Fig. 3.4, we have identified several of the components including:

- Pan that supports the unknown mass.
- Base that supports the entire balance.
- Triple beam that supports the reference, adjustable and tare weights.
- Three different size adjustable weights that slide along separate tracks on the triple beam.
- Reference weights that may hang from the triple beam.
- Tare weight that slides along a fourth track on the balance beam.
- Pointer to indicate balance.
- Scale of reference for the pointer.

A close-up view of this scale showing mainly the triple beam is presented in Fig 3.5. In this view, we note the scales more clearly along each track on the balance beam. The central track, with its large adjustable weight, is notched at 100 gram intervals providing a capacity of 500 grams. The rear track, with its intermediate size adjustable weight, is notched at 10 gram increments and provides a capacity of 100 grams. The front track, with a small sliding weight, is marked at one gram increments with 100 divisions for a total capacity of 10 grams. The smallest division, one tenth of a gram on the front track, represents the resolution of the scale.

On the right end of the triple beam, two pins protrude. These pins are notched to accommodate the reference weights (equivalent to 500 and 1000 grams). For the higher capacity applications, either one or two reference weights are hung from the pins.

On the left end of the triple beam, we find a small adjustable weight which is used for initial balance of the scale. With nothing on the pan, this weight is rotated to bring the pointer to the zero mark on the reference scale. The scale is now in balance and ready for use. If a container is placed on the pan to hold a powder or a liquid, the weight of this container must be eliminated prior to any measurement. The adjustable (tare) weight on the fourth track is moved to balance out the weight of the container.

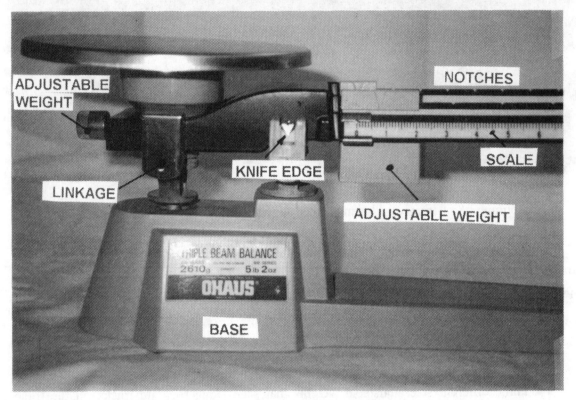

Fig. 3.5 Close-up photograph showing the scales, support pins and knife edges.

The balance incorporates four knife edges one of which is shown in Fig. 3.5. Each knife edge is fabricated from hardened steel. The seat for the knife edge is made from ceramic to minimize wear. Care must be exercised in handling the scale to prevent damage to the knife edges or the seats.

A linkage mechanism, not evident in either Figs. 3.4 or 3.5, is employed in the design as illustrated in Fig. 3.6. The moveable support and the fixed support represent two parallel bars that remain parallel as the triple beam of the balance moves. The important aspect of this motion is that the pan remains level. In addition, the position of the unknown weight does not affect the accuracy of the measurement.

When you build your prototype scale, there are two approaches that you can follow in preparing a calibrated scale for the middle lever. You can perform an equilibrium analysis similar to those described in Chapter 2, and use the results to calculate the position for the marks indicating weights of say 1, 2, 3 and 4 lb. These distances are marked on each of the tracks that you employ in the design of the balance beam.

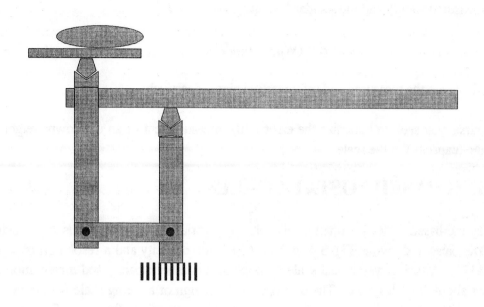

Fig. 3.6 Parallel bar linkage mechanism maintains a level pan and eliminates the requirement to center the unknown weight.

Another approach for preparing the calibrated scale is experimental. You add calibrated weights to the platform in small increments marking the position on the appropriate track where balance is achieved. A better approach is to use both methods for laying out the scale on each track. Use the theoretical approach for the initial markings followed by the calibration experiment to check the accuracy of your solution.

The scale should be linear with equal distance between major marks indicating say 25 gram increments as shown in Fig. 3.7. Minor marks are made between the major marks dividing the scale into say 5 gram increments. These minor marks determine the resolution of the weight measurement. The resolution of the scale is equal to the incremental weight between the minor marks on the calibrated scale.

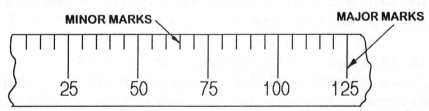

Fig. 3.7 A calibrated scale showing major markings and minor markings—uniform spacing (linear).

Resolution should not be confused with accuracy. You can improve the resolution by making many minor marks on the scale and reducing the increment between the marks to a fraction of a gram. However, if the knife edges in the balancing mechanism exhibit friction or if the scale is not initially balanced, the accuracy will be poor regardless of the resolution of the scale. The error

E is determine by placing reference (standard) weights on your scale and noting the difference between the measured weight and the standard weight. The error E is:

$$E = (W_m - W_R)/W_R \qquad (3.2)$$

where W_m and W_R are the measured and reference weights respectively.

Of course, you seek to minimize the error in the measurement of an unknown weight over the entire range (capacity) of the scale.

3.3 SPRING BASED POSTAL SCALES

The spring based scales are often smaller than equilibrium scales and are usually much less expensive. The balance shown in Fig. 3.4 with a 2660 gram. capacity and a resolution of 1 gram. costs about $125. A spring type postal scale with about the same capacity and a resolution of ½ ounce sells for about 1/4th this price. The concept for the design of a spring scale is to convert the force measurement into a measurement of length—the deflection of the free end of a spring. Of course, the deflection must be large enough to be scaled with sufficient resolution and it must be easily discernable from a suitable distance.

The most common type of spring used in mechanical design is a helical spring as depicted in Fig. 3.8. The coil is formed by winding a small diameter wire about a mandrel to form a larger diameter helix. There are many forms of helical springs including, compression, tension, conical and torsion springs all of which are illustrated in Fig. 3.8.

The relation between the axial force F applied to the spring and the deflection d of the free end of the spring is:

$$F = kd \qquad (3.3)$$

where k is the spring rate or spring constant specified in lb/in. or N/m.

The spring constant depends on the material used in fabrication, the diameter of the wire from which the spring was wound, the diameter of the helix and the number of its coils. Designers do not normally fabricate springs because they are commercially available, competitively priced and marketed in all sizes shapes and forms by a number of different companies.

You have all seen spring scales. They are hanging at many locations in the produce section of the typical supermarket. Low cost postal scales, similar to the one presented in Fig. 3.9, employ springs for weighing envelopes and small packages. The spring scales all have several common features which include:

- A platform, pan or scoop to support the unknown mass.
- A link to connect the platform to the free end of the spring.
- A support for the fixed end of the spring.
- A pointer to indicate the deflection of the spring.

- A calibrated scale to convert the deflection into a measurement of weight.
- An adjusting screw to located the pointer at the zero mark on the scale.

We have located these features for you on a photograph of the mechanism of an antique postal scale in Fig. 3.10.

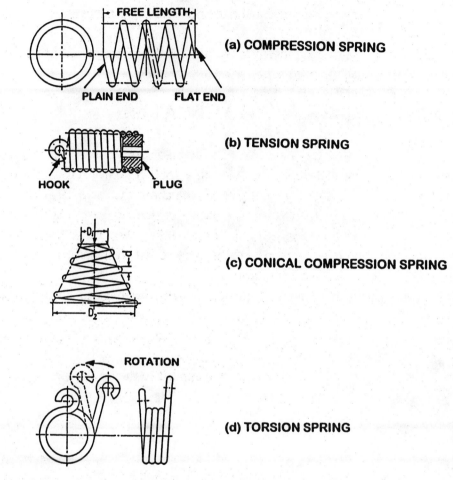

Fig. 3.8 Examples of helical springs.

As we examine the mechanism for the postal scale in Fig. 3.10, we are studying a design that already exists. In design, it is okay to copy if we do not infringe on a patent. In fact, we have coined a word for copying when it comes to design. It is called reverse engineering. If we are designing a new model of a product, we purchase the latest model of all of the competitive products in the market. We test them (this is called benchmarking). We also completely disassemble competitive products, and thoroughly inspect the mechanism to gain a complete understanding of all of the features incorporated in them. If you are worried about a patent infringement, conduct a patent search. Remember, patents protect a novel idea, but the protection is limited to only 17 years. Spring scales were introduced in the 19th century and most of the basic patents expired a century ago.

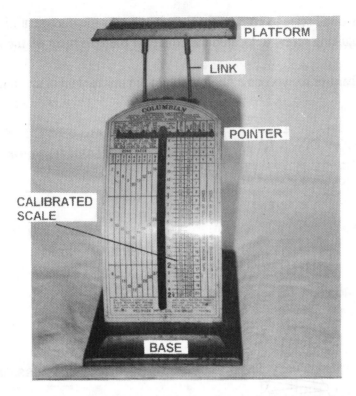

Fig. 3.9 Postal scale with a soft spring giving a large displacement to indicate the required postage[1].

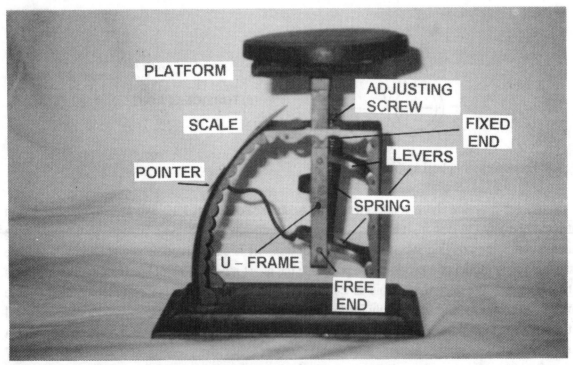

Fig. 3.10 Mechanism of the spring based postal sale.

---

[1] If you note the postage tabulated on the scale in Fig. 3.9, you will recognize that is an antique. The 3 cent charge for a one ounce letter dates back to the 1940s.

Do we expect that you will somehow manage to acquire a scale, take it apart and inspect the mechanism? Dissection of mechanisms is a well established educational technique. We will be very disappointed in you if you miss this opportunity to learn. In fact, we have written Laboratory Exercise #1 which requires your team to perform a mechanical dissection. Hopefully, your instructor will assign it to be performed.

Let's assume that you have procured an inexpensive postal scale. OfficeMax® lists one for $11.99, and you may be able to find one at a discount store for much less. Let's proceed with the mechanical dissection of the postal scale by removing its case. We open the antique postal scale, for example, by loosening the two small screws on each side plate. Removing the side plates reveals the mechanism shown in Fig. 3.10. Next, we remove the platform by pulling it upward disengaging it from the U shaped support structure. This completes the easy part of the disassembly because all of the remaining components on my antique scale are riveted together or are fastened by bending sheet metal tabs. Fortunately, the mechanism of postal scale is visible and a thorough inspection is possible.

The U shaped support structure is central to the mechanism. It contains the spring and serves to anchor the free end of the spring. As weight is applied to the platform, the U shaped structure moves downward stretching the helical spring. The top of the spring (its fixed end) is attached to the case of the scale with a small thumb screw. This screw serves as the zero adjustment since it is rotated to pretension the spring; thus, zeroing the pointer on the scale.

The capacity of the antique scale is 2-1/2 lb. The spring is protected from overload by the U shaped member which stops its downward deflection by contacting the bottom of the case. Loads on the scale in excess of 2-1/2 lb are transmitted by the U shaped member directly to the case—not to the more fragile spring. The stability of the U shaped member is insured by curved bar which connects the two vertical bars to prevent them from separating under high loads.

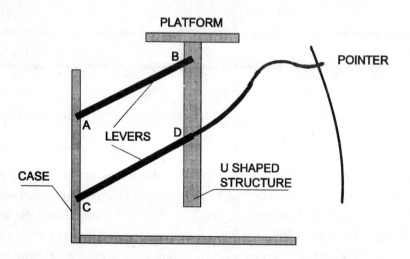

Fig. 3.11 Parallel linkage in scales maintains the platform horizontal and eliminates error due to the position of the weigh on the platform.

The parallel linkage that we discussed in Section 3.2 is also employed in the antique spring scale. This linkage is demonstrated in Fig. 3.11. Levers A-B and C-D constrain the motion of the

U shaped structure so that it moves parallel to the vertical wall of the case. This feature maintains the platform in a horizontal orientation, and eliminates from consideration the position of the unknown weight on the platform[2].

The parallel linkage, shown in Fig. 3.11, also lends itself to mechanical magnification. When a weight of 2-1/2 lb is placed on the platform, the spring stretches and the U shaped member moves downward by 1.25 in. However, the pointer sweeps through an arc of 4.00 in. The length of the levers and the pointer were arranged to provide a mechanical magnification of 3.2. This magnification enhances the resolution of the postal scale.

3.4 STRAIN GAGE BASED SCALES

In the past decade or so, many electronic scales have been introduced to the market. These are premium scales in that they are accurate, respond nearly instantaneously and usually incorporate a digital display. However, they are relatively expensive when compared to mechanical scales. Most of these electronic scales are based on the measurement of a quantity called strain with a strain gage sensor.

Let's introduce the concept of strain, describe how it is measured and finally indicate a method for incorporating a strain gage in a scale. Let's consider some solid body like the potato shown in Fig. 3.12. Suppose we place two well defined marks on the potato at points A and B a distance L_0 apart. Now apply some system of forces to the potato and watch it deform. As the potato deforms, both points A and B move and the distance between them changes to L_f. Applying the forces to the potato has caused it to deform and to be subjected to strain. The magnitude of that strain ε is given by:

$$\varepsilon = (L_f - L_0)/L_0 = \Delta L/L_0 \qquad (3.4)$$

Strain is a geometric quantity. It is simply the change in length of a line segment (with load) over the original length of that same line segment. As such, strain is a quantity without dimensions. In components fabricated from metals, the strains developed, even under very large loads, are very small. We usually refer to strains on a micro scale with magnitudes usually ranging from about 100×10^{-6} to 3000×10^{-6}.

We usually measure strain with a metal foil resistor called a strain gage. The pattern of the resistor for a typical strain gage is shown in Fig. 3.13. A grid configuration is photoetched from the metallic foil to increase the resistance of the gage and to give the gage a high directional sensitivity. Large tabs at each end of the metal foil conductors are provided to accommodate the lead wire connections. The foil is very thin (about 200 μ in.) and fragile; hence, it requires support to avoid damage in normal handling. This support is provided by a thin plastic carrier (illustrated in Fig. 2.13) which facilitates handling the gage.

---

[2] The concept of parallel linkage in scales was invented in 1669 by Roberval a French mathematician.

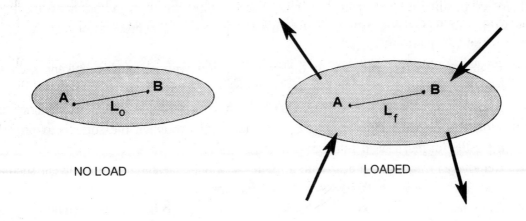

Fig. 3.12 A solid body deforms under the action of applied loads.

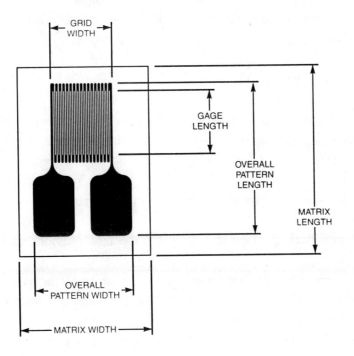

Fig. 3.13 The grid pattern for a typical strain gage. Courtesy of Micro-Measurements Division,
Measurements Group, Inc.

To employ a strain gage, we bond it to a mechanical member with an adhesive. The adhesive serves a vital function in the strain measuring system since it transmits the strain from the mechanical component to the gage sensing element (the foil) without loss. The singularly unimpressive feat of adhesively bonding a strain gage to a mechanical component is one of the most critical steps in the entire process of measuring strain with a bonded resistance strain gage.

We have included Laboratory Exercise (#2) at the end of this Chapter that will be helpful in developing the skills required to mount strain gages. We describe the procedure in the following listing:

1. When bonding a strain gage to a mechanical component, it is important to carefully prepare the surface of the object to which the gage is to be mounted. The preparation includes:

- Degrease the area surrounding the gage location with a suitable solvent such as isopropyl alcohol.
- Abrade (sand) the surface using 220 to 320 grit silicon-carbide papers. Sometimes the surface is etched with a mild acid. The idea is to roughen the surface to provide asperities for mechanical attachment of the adhesive.
- Remove all traces of the acid with a tissue. Continue cleaning with a solvent such as isopropyl alcohol until the tissue is free of discoloration.
- Draw lines on the surface with a hard pencil (4 to 6 H) to be used in aligning the gage at an exact location. Avoid touching the gage location with your fingers as you draw these lines.
- Treat the strain gage location with a neutralizing solution to provide the proper chemical affinity for the adhesive. Dry the location careful with a single pass of a gauze sponge. Inspect the location to insure that it is completely dry.

2. Prepare the strain gage for placement.

- Remove the strain gage from its container with a pair of tweezers. We are to avoid touching the gage during the installation procedure. The natural oils on our fingers may contaminate the bonding surface of the gage.
- Place the gage on a clean glass plate with the grid pattern showing.
- Capture the strain gage by placing a piece of cellophane tape over the grid pattern as illustrated in Fig. 3.14.
- Carefully peel back the tape and note that the tape provides an excellent means for handling the gage.
- Position the gage over the previously prepared location on the specimen.
- Align the centering marks on the gage with the alignment lines previously drawn to mark the exact location of the gage.
- Attach the tape to the specimen to fix the gage at this location.
- Peel away one end of the tape exposing the back side of the gage.

3. Bonding the strain gage with a cyanoacrylate adhesive.

- Apply an extremely thin coat of a catalyst to the back of the gage and permit the catalyst to thoroughly dry before proceeding.
- Apply one or two drops of the cyanoacrylate adhesive to the specimen at the fold of the cellophane tape as shown in Fig. 3.14. The initial point of application of the adhesive should be about 0.2 to 0.3 in. removed from the gage location to avoid the detrimental effects of local polymerization.
- Immediately rotate the tape so that the gage is positioned over its location.

- Apply a slight tension to the tape while wiping across the strain gage with sufficient pressure to squeeze out all of the excess adhesive. The bond line between the gage and the specimen should be thin and of uniform thickness.
- Immediately place your thumb over the gage and apply a firm pressure. The combination of the pressure and body heat will usually polymerize the adhesive in about one minute.
- Remove the tape from the strain gage installation when you are prepared to wire the gage. If the adhesive is completely polymerized over the entire area of the gage, the tape can be lifted up without damaging the gage. However, if the adhesive is not cured or if it does not cover the complete bonding area, all or part of the gage will lift off with the tape. The tape method provides a fail-save system for strain gage installation.

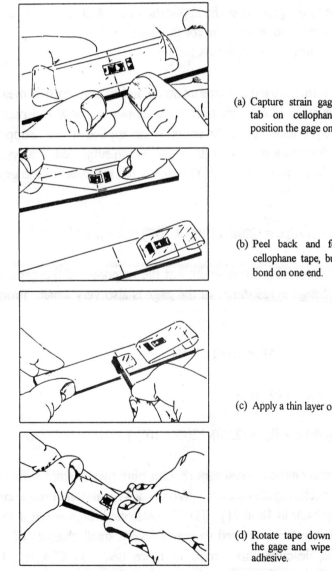

(a) Capture strain gage and solder tab on cellophane tape and position the gage on the beam.

(b) Peel back and fold over the cellophane tape, but maintain its bond on one end.

(c) Apply a thin layer of adhesive.

(d) Rotate tape down to reposition the gage and wipe out excessive adhesive.

Fig. 3.14 Recommended procedures for adhesively bonding electrical resistance strain gages (Courtesy of Micro Measurements).

The next phase of the strain gage installation process is the wiring of the lead wires to the tabs on the gage. This is a tedious task that requires great care; otherwise, the gage will be damaged by the application of excessive heat to the delicate wiring tabs. The metal foil of the gage is only 200 μ in. thick; hence, the entire gage is fragile and must be protected during and after installation. We describe the wiring procedure for the gage in Laboratory Exercise #3.

A strain gage acts as a variable resistor when subjected to a surface strain in the direction of its grid elements. The change in resistance ΔR is with the strain ε is given by:

$$\Delta R/R_0 = (R_f - R_0)/R_0 = (GF) \varepsilon \qquad (3.5)$$

where R_0 is the initial resistance of the gage usually 120 or 350 Ω (ohms).
R_f is the final resistance of the gage after the application of a strain ε.
GF is the gage factor or calibration constant for the gage.
ε is the strain in the direction of the gage axis.

A strain gage is a sensor that is used to measure strain and to convert that measurement into a change in the electrical resistance. Let's consider an example to show the magnitude of the resistance change that the gage produces under the action of an applied strain. Suppose we have a mechanical component that undergoes a strain of 1200×10^{-6} when subjected to a system of applied forces. If the strain gage has a gage factor GF = 2.00, what is the response of the gage in terms of $\Delta R/R_0$? We use Eq. (3.5) to determine:

$$\Delta R/R_0 = (GF) \varepsilon = (2.00) (1200 \times 10^{-6}) = 2.4 \times 10^{-3}$$

The strain of 1200×10^{-6} is a reasonably large strain, but the response of the gage $\Delta R/R_0$ is very small (only 2.4×10^{-3}). The change in resistance of the gage is also very small. From Eq. (3.5), it is evident that:

$$\Delta R = (GF) \varepsilon \times R_0 \qquad (3.6)$$

If $R_0 = 120 \ \Omega$, substituting into Eq. (3.6) gives:

$$\Delta R = (GF) \varepsilon \times R_0 = (2.00)(1200 \times 10^{-6}) \times 120 = 0.288 \ \Omega$$

How do we measure such a small change in resistance? Our ohm meter is not sufficiently sensitive to measure with any accuracy such a small quantity. We use a Wheatstone bridge circuit that was developed by Lord Kelvin way back in 1856 [1]. The Wheatstone bridge is not a structure as the name implies, but a simple electrical circuit used to convert very small changes of resistance into changes in voltage. We will describe this circuit in more detail in Chapter 4 when some introductory principles of electrical circuits are introduced. We will also assign a laboratory exercise in Chapter 4 which is designed to acquaint you with this important circuit.

3.5 STRAIN GAGE BASED TRANSDUCERS

Let's next explore techniques for using strain gages to produce transducers. Transducers are mechanical components that incorporate a sensor. The transducer measures a mechanical quantity like force, torque, pressure, velocity, acceleration, etc., and converts that quantity into an electrical output. A strain gage based transducer incorporates a bonded electrical resistance strain gage as the sensor. Let's study the design of a strain gage transducer for measuring weight. Of course, the transducer converts the weight into an electrical signal that is displayed to provide the output.

A complete transducer system contains five principle components as indicated in Fig. 3.15. We begin with a mechanical input of some quantity (in our case, it is the weight of a letter or perhaps a package). The first component is a mechanical element that deforms due to the weight. In the process, the mechanical element is strained in proportion to the applied weight. The second component is a sensor, such as a strain gage, that converts the strain into an electrical quantity—$\Delta R/R_0$. The third component is a signal conditioning circuit, usually a Wheatstone bridge, that converts the resistance change ΔR into a voltage change ΔV. The fourth component, if necessary, is an amplifier to increase the magnitude of the voltage output from the bridge so that it can be measured on a suitable instrument. The final component is the display which is usually a voltmeter—either digital or analog.

In this section, we will describe techniques for the design of the first two components of the transducer system shown in Fig. 3.15. The design and/or procurement of the last three components will be discussed in Chapter 4.

Let's consider the design of a small postal scale similar in size and shape to the spring type scale depicted in Fig. 3.9. We have a platform that moves down to stretch a tension spring. Suppose we replaced this spring and its more complicated lever mechanism with a support rod and a sensing beam as illustrated in Fig. 3.16. The new concept has fewer parts which include:

- A platform.
- A support rod.
- Low friction guides or flex bars.
- A sensing beam.
- Beam supports.

Of course, we have not shown the case or the electronic components in Fig. 3.16. However, the essential mechanical elements for the scale are introduced and the concept of a scale incorporating a strain gage transducer is apparent.

Let's examine the sensor beam in much more detail to determine if it is a suitable mechanical element for our postal scale. Consider the simply supported beam as depicted in Fig. 3.17. The beam has a span S with a cross section h units high and b units wide. The unknown weight W is applied at the center of the beam. A strain gage sensor is bonded to the underside of the beam directly beneath the loading point.

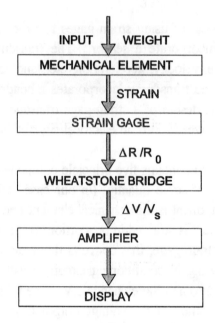

Fig. 3.15 Block diagram showing the five primary components of a transducer system.

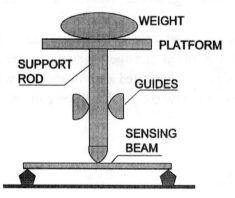

Fig. 3.16 A simply supported beam with a strain gage sensor bonded to its underside.

The weight W produces a bending stress σ in the beam at the position of the strain gage that is given by:

$$\sigma = (3\ W\ S)/(2\ b\ h^2) \qquad\qquad (3.7)$$

From Eq. (3.7), we can determine the bending stresses σ. However, we are more interested in the strain at the location of the gage. How do we determine the strain if we know the stress? We use Hooke's law which states for a uniaxial member such as a beam that:

$$\sigma = E\,\varepsilon \tag{3.8}$$

Now, if we combine the results of Eqs. (3.7) and (3.8), we obtain:

$$\varepsilon = \sigma / E = (3\,W\,S)/(2\,b\,h^2\,E) \tag{3.9}$$

where E is the elastic (Young's) modulus of the material from which the beam is fabricated. For aluminum $E = 10.6 \times 10^6$ psi or 73 GPa.

Okay! We now have a relation for strain in terms of the geometry of the beam (S, b and h), the beam material (E) and the load W applied to the center point of the beam. As a designer, how do you make use of this interesting expression? You must make initial selections pertaining to the geometry and material of the beam. We know the capacity of the postal scale. (Consider 5 lb capacity in this example). In addition, we know that the strain should be large to produce a large strain gage output $\Delta R/R_0$. However, the strain should not be so large as to cause the beam to fail by yielding.

Let's make some initial estimates (guesses) for the dimensions of our beam, and use Eq. (3.9) to determine if the size of the beam selected is feasible. Select S = 4 in. for the span of the beam. Consider dimensions of b = ¼ in. and h = 1/8 in. Fabricate the beam from a 2024 T4 aluminum alloy with $E = 10.6 \times 10^6$ psi. Recall that the maximum weight we will impose on the postal scale is 5 lb. Substitute these design parameters into Eq. (3.9) to obtain:

$$\varepsilon = (3)(5)(4)(4)(64)/(2)(10.6)(10^6) = 725 \times 10^{-6}$$

Let's interpret this result for the strain. We certainly wanted a very large strain, but is 725×10^{-6} too much strain for the beam to withstand. Let's check to determine the stress in the beam by using Hooke's law given in Eq. (3.8).

$$\sigma = E\,\varepsilon = 10.6 \times 10^6 \times 725 \times 10^{-6} = 7685 \text{ psi}$$

Checking on the strength of our aluminum alloy 2024T4 in a materials handbook, we find that it will yield when the bending stress exceeds 47,000 psi. Our initial selection of the beam's dimensions gives a mechanical strain element with a safety factor SF of:

$$SF = S/\sigma = 47,000/7,685 = 6.11 \tag{3.10}$$

where S is the strength of the beam's material (psi)
σ is the applied stress (psi)

A safety factor of 6.11 is very conservative. It indicates that we could place about six times the capacity load, or 30 lb, on the scale before the sensor beam would fail by yielding. We should reduce the dimensions on the beam to increase both the stress and the strain—reducing the safety factor to about 2. The beam can be protected from overload by placing stops beneath it to limit its deflection. An exercise is given at the end of the Chapter that assigns you the task of resizing the beam to produce a strain of 3×10^{-3} when loaded to the scale's capacity of 5 lb.

Suppose a strain gage is placed on the lower surface of the beam directly under the point of the application of the load, and we will record a strain of 3×10^{-3} when the platform is loaded to the maximum capacity of 5 lb. What is the electrical signal output, in terms of resistance, from this strain gage? From Eq. (3.5), we determine:

$$\Delta R /R_0 = (GF)\ \varepsilon = (2.00)\ 3000 \text{ x } 10^{-6} = 6.00 \text{ x } 10^{-3}$$

Suppose the strain gage we install on the sensor beam has an initial resistance of 350 Ω. Then the total change in resistance of the gage is:

$$\Delta R = R_0 \times 6.00 \times 10^{-3} = 350 \times 6.00 \times 10^{-3} = 2.1\ \Omega$$

Clearly, the change in resistance of 2.1 is very small when compared to the original resistance of 350 Ω. Indeed, it is too small to be measured accurately with a typical ohmmeter. A special circuit called a Wheatstone bridge is employed to convert these small changes in resistance to a voltage that can be measured with a suitable voltmeter. We will return to this result in the next chapter in a section describing the Wheatstone bridge circuit. We will describe the procedure for determining the voltage output from the bridge as a function of the bridge configuration, the supply voltage V_s and $\Delta R/ R_0$.

REFERENCES

1. Thomson, W. (Lord Kelvin): "On the Electrodynamic Qualities of Metals," Proceedings. Royal Society, 1856.
2. Beer, F. P. and E. R. Johnston, Jr., Mechanics of Materials, 2nd Edition, McGraw-Hill, NY, 1992.
3. Dally, J. W., Riley, W. F. and K. G. McConnell, Instrumentation for Engineering Measurements, 2nd Edition, John Wiley, New York, 1993.
4. Dally, J. W. and W. F. Riley, Experimental Stress Analysis, 3rd Edition, McGraw-Hill, New York, 1991.

LABORATORY EXERCISES

3.1. As a team, procure a scale of some type; we suggest either a bathroom or a postal scale. The scale should be functional before you begin this laboratory exercise and functional after you have reassembled the scale.

- Calibrate the scale and prepare a graph showing input (calibration weights) as a function of the output (recorded weights) over the entire range of the scale.

- Perform a mechanical dissection of the scale. Remove all of the components that are not permanently fastened together.
- Examine each component and establish the process used to manufacture it.
- Determine the purpose of each component in the functioning of the scale.
- Count the unique part numbers and the total number of components.
- Why is part number count and total component count important?
- Lay out all of the components on a sheet of white paper and photograph them.
- Record your observations during the mechanical dissection.
- Reassemble the scale.
- Recalibrate the scale to ensure that it is functional. If the scale is not functional or not in calibration, disassemble and then reassemble it to correct the problem or problems.
- Prepare an engineering report describing the results of the mechanical dissection of the scale.

3.2. Your development team will be provided with a strain gage installation kit and an aluminum cantilever beam specimen. The strain gage installation kit contains:

- A solvent isopropyl alcohol
- Silicon carbide paper
- A mild phosphoric acid etching solution.
- An ammonia based neutralizing solution.
- Gauze patches for cleaning.
- Cotton swabs for cleaning
- Cellophane tape.
- Practice strain gages.
- A detailed instruction booklet.

The aluminum beam is 1/8 in. thick by ¾ in. wide by 8 in. long. Install the practice strain gage 2.5 in. from one of the ends along the centerline of the beam. The axis of the strain gage should be coincident with the longitudinal axis of the beam.

When the installation is complete, measure the resistance of the strain gage with an ohmmeter. Also, measure the resistance of the gage to ground (the aluminum specimen).

Retain the beam with the gage installed for use in Laboratory Exercise #3.

3.3. Attach lead wires to the strain gage installed on the aluminum beam in Laboratory Exercise #2. A soldering station has been prepared for you to facilitate this assignment which includes the following items of equipment and supplies:

- A power-controlled soldering station that provides low voltage and adjustable temperatures.
- A soldering pencil with an iron tip.
- Supplies of eutectic solder (63/37 tin-lead alloy) in the form of small diameter wire.
- A small bottle of liquid activated-rosin flux with brush.

- A small bottle of rosin solvent with brush.
- A small bottle of urethane varnish with brush.
- Masking tape.
- Gauze patches.
- Three conductor color-coded lead wire lead—gage no. 26 stranded-tinned-copper-wire with vinyl insulation.
- A pair of diagonal cutters.
- A pair of wire strippers.

1. Prepare the soldering iron.

- Adjust the power level on the soldering station to give a temperature of about 390 °F which is sufficient to melt the eutectic solder which has a solidus/liquidus temperature of 361 °F.
- When the tip of the soldering pencil reaches temperature, clean it by removing all of the excess solder and oxidation products with a thick gauze wipe.
- Tin the tip by applying a generous amount solder.
- Wipe the excess solder from the tip with a patch of gauze.
- The soldering iron tip should be cleaned and tinned frequently to insure strong high conductivity connections.

2. Prepare the lead wires.

- Cut a four foot length of triple conductor lead wire.
- Strip back the insulation of all the conductors on both ends of the lead wire by about ½ in.
- On one end of the lead wire, twist the individual strands of each of the conductors to form a round wire.
- Tin these three ends by applying a small amount of solder with the tip of the pencil soldering iron.
- On the other end of the lead wire, separate the red conductor from the black and white conductors.
- Twist the strands from the white and black lead wires together, but separate one single strand from the bundle as shown in the figure below:
- Tin the leads and clip the bundles near the insulation. Later, you will solder the single strand of wire to the strain gage tab.
- Repeat this tinning process with the strands of wire from the red conductor as shown in the figure below.

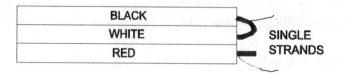

3. Preparing the strain gage.

    - Place a piece of masking tape over the grid of the strain gage to protect it from any flying solder or abuse.
    - Apply solder flux to the tabs.
    - Apply a very small solder dot to the tabs. Avoid excess solder in this step. The dots should be rounded at the top without a point.
    - Tape the lead wires to the surface of the aluminum beam so they are firmly fixed. The single strands from the red and the black+white leads should extend over the tabs.

4. Solder the lead wires to the strain gage tabs.

    - Solder the lead wires to the tabs simply by touching the tinned strands to the solder dot with the soldering pencil. No addition solder should be necessary to provide a strong high conductivity connection. The solder over the single strand of wire should be rounded (without a point) and bright (not dull).
    - Trim the excess length of lead wire from both conductors.
    - Clean the rosin flux from the gage installation.
    - Check the resistance of the gage and the lead wires.
    - Check the resistance to ground to insure that you have not created a short.
    - Coat the gage with urethane varnish or wax to waterproof the installation.
    - Apply a second tape bond to firmly fasten the lead wire to the beam.

EXERCISES

1. Describe the difference between analog and digital displays.
2. Give several reasons why a simple balance is a good choice as a design concept for a postal scale?
3. Give several reasons why a simple balance is not a good choice as a design concept for a postal scale?
4. Draw a calibrated scale 300 mm long with a range from 0 to 1000 N. Use 100 N for the major marks on the scale and 10 N for minor marks on the scale. What is the resolution of the scale?
5. You place a reference weight of 12.5 oz on a postal scale, and measure a weight of 12.63 oz. Determine the error of the measurement.
6. You design a postal scale with a helical spring with a capacity of 40 lb. If the spring has a spring constant of 18 lb/in., what is its deflection?
7. A pinion gear, which drives the pointer on a circular dial, has a 10 teeth. If the teeth on the mating rack gear have a pitch of 1 mm, determine the arc of rotation of the pinion gear if the rack moves through a distance of 6 mm. Is the use of a rack and pinion a suitable mechanism for mechanical amplification?
8. For the rack, pinion dial mechanism of exercise 7, determine the mechanical amplification of the movement of the rack gear if the dial has diameter of 150 mm.

9. A gage length $L_0 = 20$ mm is scribed on a component fabricated from an aluminum alloy 2024 T4. If this component is subjected to a system of forces and the new gage length becomes $L_f = 20.006$ mm, find the strain and the stress at the location of the scribed line.

10. Will the component described in exercise 9 fail by yielding?

11. What assumption did you make in exercise 10 when you converted strain to stress by using Hooke's law?

12. An electrical resistance strain gage with a resistance of 350.00 Ω, and gage factor of 2.06 is subject to a strain of 0.000900. Determine new resistance of the strain gage.

13. Describe the five components usually found in a transducer measuring system. Relate these components to a measuring system on your automobile

14. For the beam shown in Fig. 3.17, determine b, h and S so that a strain of 3000×10^{-6} is developed when the beam is centrally loaded with a 5 lb weight. Assume that the beam is fabricated from an aluminum alloy 2024 T4.

15. Why are people willing to pay more for an electronic scale than a mechanical scale of the same capacity?

CHAPTER 4

ELECTRONIC SCALES

4.1 INTRODUCTION

In Chapter 3, we learned that a spring could be used to convert an applied force (weight) into a displacement. The conversion was beneficial because we measured the displacement with a scale and eliminated the requirement for reference weights used with a balance. In this chapter, let's continue to employ a spring to convert force into displacement. However, instead of measuring the displacement with a scale, we convert the displacement to a voltage, and then measure it with a voltmeter.

The double conversion system from force to displacement to voltage is depicted in Fig. 4.1. The system begins with a mechanical component (the spring), but the second component (electrical) changes the technology of the system from mechanical to electrical. Because the system in Fig. 4.1 has both mechanical and electrical components, it is called a hybrid. We will describe this hybrid system in more detail later in this chapter, and suggest some ideas to aid in your efforts to design the prototype for a postal scale.

Another mechanical-electronic hybrid system incorporates a strain gage as the electrical component to convert strain (a mechanical quantity) into a change in resistance (an electrical quantity). We have already discussed this topic in Chapter 3, but we did not resolve the difficulty presented by a resistance change, $\Delta R/R_0$, that is too small to measure with a common ohmmeter. To measure a very small $\Delta R/R_0$, we employ a signal conditioning circuit called a Wheatstone bridge. This simple electrical circuit is very useful in converting small resistance changes into a voltage. We will describe the Wheatstone bridge and show the relation between $\Delta R/R_0$ and the voltage ratio $\Delta V/V_s$.

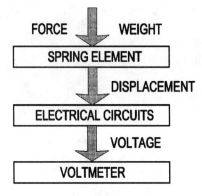

Fig. 4.1 Block diagram of a hybrid system containing both mechanical and electrical components.

As we describe these electrical circuits, it will become necessary to introduce you to some basic concepts including voltage sources, voltage, and current and to discuss passive electrical components. We will deal only with direct current (dc) in describing the simple laws governing current flow in circuits.

We will also introduce some very elementary concepts in digital electronics by covering the binary code and the digitization process. We will discuss the analog to digital converter and describe simple digital displays. Finally, elementary aspects of microprocessors will be described.

4.2 BASIC ELECTRICAL CONCEPTS

Electricity involves charge carriers—either electrons, which are negative charge carriers, or holes which are positive charge carriers. Since we cannot see, smell or touch an electron, electricity has always been somewhat mysterious to most of us. Holes, which are atoms with an electron missing from its outer ring, are even more obscure. Nevertheless, we will try to introduce many of the basic electrical concepts in a simple straight forward manner and investigate the mystery.

4.2.1 The Battery

Let's start with a source of electricity—a battery which is shown in Fig. 4.2. This figure illustrates a wet cell which consists of two metal bars submerged in a liquid electrolyte. The electrolyte is an acid ($H_2 SO_4$) which attacks both metal bars. The zinc gives off positive ions (Zn^{++}) to the electrolyte which leaves the metal electrode with a negative charge. The copper gives off negative ions (Cu^-) to the electrolyte which in turn leaves it with a positive charge. If we attach a voltmeter[1] across the two metal bars (electrodes), we measure a voltage that corresponds to the

---

[1]The voltmeter must have a high input impedance to limit the current flow during the measurement otherwise the voltage measurement will be in error.

difference in the charges that have developed on the Zn and the Cu electrodes. The battery voltage V_s is given by:

$$V = EP_{CATHODE} - EP_{ANODE} \qquad (4.1)$$

where EP is the electrode potential of the materials used for the anode and cathode.

Since the electrode potential of Cu (the cathode) is 0.34 volts and that of Zn (the anode) −0.76 volts, the voltage of a Cu-Zn battery is:

$$0.34 - (-0.76) = 1.10 \text{ V (volt).}$$

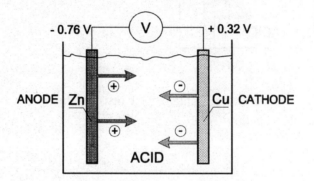

Fig. 4.2 A Cu-Zn wet cell (battery) that serves as a voltage source.

The battery is frequently employed as a voltage source. You have a battery in your automobile and probably in your wrist-watch. A battery converts chemical energy into electrical energy and produces a relatively low but essentially constant (dc) voltage. We generate higher voltages by connecting several individual battery cells in a series arrangement. We can also select certain elements with higher electrode potentials in the electromotive series in the fabricating the unit cells to generate slightly higher voltages.

Let's change the connections on the battery shown in Fig. 4.2, and add a resistor R in the circuit from the Zn electrode to the Cu electrode as shown in Fig. 4.3. What happens? We have completed the circuit. The battery in Fig. 4.2, with an open circuit, acts as a voltage source; however, with a closed circuit, as illustrated in Fig. 4.3, the battery becomes a current source. An electric current (I) flows from the positively charged Cu electrode through the resistor to the negatively charged Zn electrode. The electrons (charge carriers), shown in the insert of Fig. 4.3, flow in the opposite direction from the negative charge to the positive charge.

Okay! We understand that a battery is a voltage source which supplies current when connected to a closed circuit. Let's now examine the relation between the voltage provided by the battery and the current which flows in the circuit. To develop this relation, refer to the circuit diagram presented in Fig. 4.4. Note the symbols used for the voltage source (the battery) and the resistor (R). We depict the current (I) flowing through the circuit with the arrowhead and the voltage provided by the battery with the symbol (V_s).

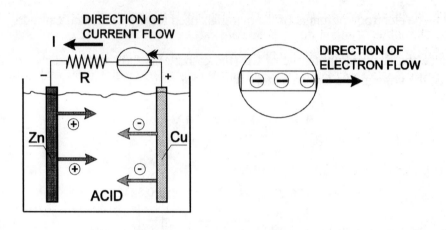

Fig. 4.3 A battery with a closed circuit containing a resistor shows the opposing directions of current flow and electron flow.

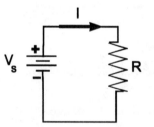

Fig. 4.4 A simple circuit loop containing a voltage source (battery) and a resistor.

The relation between the applied voltage V_s, the resistance R, and the current flow I is given by:

$$V_s = I\,R \tag{4.2a}$$

$$I = V_s / R \tag{4.2b}$$

Equations (4.2a) and (4.2b) are both expressions of Ohm's law. When V_s is expressed in volts and I in amperes, R is given in ohms Ω.

The resistance R of a conductor is given by:

$$R = \rho L / A \tag{4.3}$$

where ρ is the specific resistance of the metal conductor (Ω-cm).

L is the length of the conductor (cm).

A is the cross-sectional area of the conductor (cm^2).

We can make resistors by winding very-small-diameter, high-resistance wire on coils or by evaporating thin films of metal on ceramic substrates. The relation in Eq. (4.3) allows us to select the metal and to size the conductor to provide a specified resistance.

A resistor is a dissipative component. When a current flows through a resistor, power is lost and heat is generated. The power dissipated by the resistor is given by:

$$P = V_d \, I \qquad\qquad (4.4)$$

where P is the power expressed in Watts (W).

V_d is the voltage drop across the resistor.

Substituting Eqs. (4.2a) and (4.2b) into Eq. (4.4), we obtain:

$$P = I^2 R = V_d^2 / R \qquad\qquad (4.5)$$

We have used subscripts with the voltage in Eqs. (4.2a) and (4.2b) and the subscript d with the voltage in Eq. (4.5). For the simple circuit shown in Fig 4.4, the supply voltage V_s is equal to the voltage drop across the resistor V_d. In this case, the subscripts can be interchanged. Clearly, it is easy for us to determine the power lost if we know the current flowing through a resistor or the voltage drop across the resistor.

Let's consider another passive electrical component, namely a capacitor as shown in Fig. 4.5. The capacitor consists of two flat, parallel-plate electrodes separated by a dielectric which also serves as an electrical insulator. The capacitance C of a flat plate capacitor is given in units of picofarad and is determined by:

$$C = k \, \varepsilon \, A/h \qquad\qquad (4.6)$$

where ε is the dielectric constant.

A is the area of the plates.

h is the distance between the two plates as shown in Fig. 4.5.

k is a proportionality constant; 0.00885 for dimensions in millimeters.

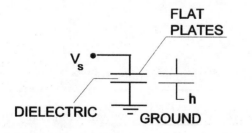

Fig. 4.5 A flat parallel plate capacitor.

When a voltage V is applied across the capacitor, it stores a charge q that is given by:

$$q = C\, V \qquad\qquad (4.7)$$

where C is the capacitance given in farads denoted with the symbol F.
    q is the charge on the capacitor given in coulombs.

When the capacitor is charged, it becomes a storage device. The electrical energy stored $\mathcal{E}$ is determined from:

$$\mathcal{E} = C\, V^2\, /2 \qquad\qquad (4.8)$$

where $\mathcal{E}$ is the stored energy in Joules when C is expressed in farads.

When a voltage is first applied to the capacitor, current flows and the capacitor becomes charged. However, when the capacitor is fully charged and maintained at a constant voltage, no current flows through it. The capacitor conducts current only when the voltage changes with respect to time.
    The final passive component, important in electrical systems, is the inductor shown in Fig. 4.6. We produce an inductor by winding a coil of small-diameter, insulated wire about a core. The inductance L is given by:

$$L = 4\pi N^2\, A\mu/\, \ell \times 10^{-9} \qquad\qquad (4.9)$$

where   L is the inductance in Henrys
        N is the number of turns of wire in the coil.
        μ is the permeability of the media within the coil.
        A is the cross sectional area of the coil.
        ℓ is the length of the coil.

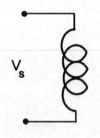

Fig. 4.6 An inductor coil across an alternating voltage source.

A solenoid, which is a coil with an inductance, is commonly used to develop a magnetic field. The solenoid produces a magnetic force on a metal plunger that extends into the core of the coil. A solenoid engages a relay which switches your battery across the starter motor .

In dc circuits, an inductor coil acts like a small resistor and it has little effect on the current flowing in the circuit. However, when the coil is exposed to a current that fluctuates with respect to time (an alternating current) there is a significant voltage drop across the coil which is given by:

$$V_d = L \, (dI/dt) \tag{4.10}$$

where (dI/dt) is the rate of change of the current with respect to time.

4.2.2 Kirchhoff's Laws

We introduced Ohm's law with a simple circuit loop containing a voltage source and a resistor as shown in Fig. 4.4. This is a good beginning; however what happens when we begin to insert additional components in the circuit? We need to develop a few more tools useful in analyzing slightly more complex circuits. Let's begin by adding one additional resistor to the simple circuit loop as shown in Fig. 4.7a. Then determine the relation for the current flowing in this slightly more complex circuit.

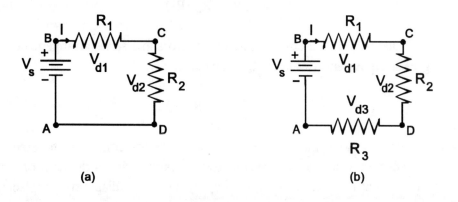

(a)	(b)

Fig. 4.7 Circuit loops containing two or three resistors.

To approach this problem, we need to use **Kirchhoff's voltage law** that states:

The sum of the voltages changes about a circuit loop is zero.

If we refer to the drawing of the circuit shown in Fig. 4.7a, we can write Kirchhoff's law in an equation format as:

$$+ V_{A-B} - V_{B-C} - V_{C-D} - V_{D-A} = 0 \tag{4.11}$$

In the interpretation of Eq. (4.11), we move in the direction of the current flow (clockwise in this case). As we move from node A to B, we pass through the battery which increases the voltage at B relative to A by an amount V_s. This is the reason for the + sign shown on the term V_{A-B}. Continuing around the circuit loop, we proceed from node B to C and pass through resistor R_1. There is a voltage drop V_{d1} as the current passes through this resistor equal to $I \, R_1$. The fact that the voltage drops across R_1 is the reason for the minus sign on term V_{B-C}. Moving onto the next leg of the circuit between node C and D, we encounter a second voltage drop $V_{d2} = I\,R_2$. Because the voltage decreases as we move from node C to D, a minus sign is assigned to the term V_{C-D}. Finally, we consider the final leg of the circuit between nodes D - A. We have a perfect conductor between these two points with no loss or gain of voltage. For this reason, $V_{D-A} = 0$. Substituting these four voltage changes into Eq. (4.11) gives:

$$V_s - I\,R_1 - IR_2 - 0 = 0$$

$$I = V_s /(R_1 + R_2) \tag{4.12}$$

If we apply the same analysis techniques to the circuit with the three resistors shown in Fig. 4.7b, it is easy to add another term to the relation shown above and write:

$$V_s - I\,R_1 - IR_2 - IR_3 = 0$$

$$I = V_s /(R_1 + R_2 + R_3) \tag{4.13}$$

Examine Eqs. (4.12) and (4.13), and observe that the resistors in a series arrangement in the circuit add together to give an equivalent resistance R_e given by:

$$R_e = R_1 + R_2 + R_3 = \Sigma R \tag{4.14}$$

Equation (4.14) leads to a well know resistor rule for resistors connected together in a series arrangement. The effective resistance of several resistances in series is the sum of the individual resistances. The equivalent resistance is used to simplify the circuit diagram by replacing the three resistor circuit with a single equivalent resistor as shown in Fig. 4.8.

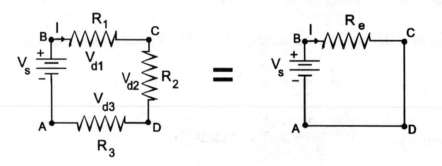

Fig. 4.8 An equivalent circuit with an equivalent resistor replaces a more complex circuit.

What is the effect if the resistors are not all connected in a series arrangement? What if we have a parallel arrangement of the resistors such as shown in Figs. 4.9a and 4.9b? To analyze this type of a circuit, we need to introduce **Kirchhoff's current law**. This simple law states:

   The sum of the currents flowing into a node must be equal to the sum of the currents flowing from that node.

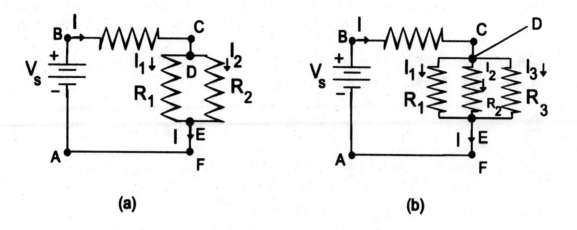

     (a)   **(b)**

Fig. 4.9 Circuits with parallel arrangements of resistors.

Let's refer to the node X presented in Fig. 4.10 with currents flowing into and out of this node. Kirchhoff's current law permits us to write:

$$I_1 + I_2 - I_3 - I_4 = 0 \qquad\qquad (4.15)$$

   To deal with the resistors in parallel, as shown in Fig. 4.9a, we will apply Eq. (4.15) to the current flow at node D and write:

$$I - I_1 - I_2 = 0 \qquad\qquad \text{(a)}$$

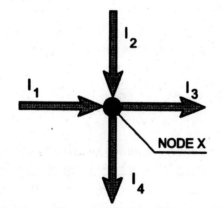

Fig. 4.10 Node X with current flows in and out.

Next, consider Kirchhoff's voltage law Eq. (4.11), and apply it to the circuit in Fig. 4.9. It is clear then that the voltage gain due to the battery is offset by the voltage drop across the two parallel resistors between points D - E. We write this fact in an equation:

$$V_s - V_{D-E} = 0 \qquad\qquad (b)$$

We know that the current splits at node D with I_1 flowing through R_1 and I_2 through R_2 to produce the voltage drop V_{D-E}. This knowledge permits us to write:

$$V_{D-E} = I_1 R_1 = I_2 R_2 = V_s \qquad\qquad (c)$$

Equation (c) gives relations for both I_1 and I_2:

$$I_1 = V_s / R_1 \qquad \text{and} \qquad I_2 = V_s / R_2 \qquad\qquad (d)$$

Following the form of Eqs. (d), we write:

$$I = V_s / R_e \qquad\qquad (e)$$

where R_e is the equivalent resistor that replaces the two parallel resistors in Fig. 4.9a to give the same equivalent circuit as shown in Fig. 4.8.

Let's substitute Eqs. (d) and (e) into Eq. (a) to obtain:

$$V_s/R_e = V_s [(1/R_1) + (1/R_2)] \qquad\qquad (f)$$

Eliminating V_s from Eq. (f) gives:

$$1/R_e = (1/R_1) + (1/R_2) \qquad\qquad (4.16a)$$

or
$$R_e = (R_1 R_2) / (R_1 + R_2) \qquad\qquad (4.16b)$$

If we find three resistors in parallel, we can follow the same procedure in the analysis of the circuit and show that the equivalent resistance for the three parallel resistors is:

$$1/R_e = (1/R_1) + (1/R_2) + (1/R_3) \qquad\qquad (4.17)$$

While resistors connected in series were summed to give the equivalent resistance, resistors in parallel follow a different rule. For parallel resistors, the reciprocals are summed and set equal to the reciprocal of the equivalent resistance. This fact implies that the equivalent resistance for a parallel arrangement of resistors is less than that of the smallest individual resistance.

Okay! Hopefully, you recognize that Ohm's law and Kirchhoff's two laws can be employed to analyze circuits. Also, you should be able to take a complex circuit and use equivalent resistances to reduce the circuit to its most elementary form. In the next section, we will use these analytical tools together with a few passive electrical components to develop a circuit that can be used to convert a displacement into a voltage.

4.3 THE POTENTIOMETER

A potentiometer is a variable resistor. The simplest way of making a potentiometer is to take a length ℓ of a very high resistance wire and attach it across a voltage source such as a battery. We provide a wiper (an electrical contact) which slides along the length of the wire as shown in Fig. 4.11. The wiper divides the slide wire and its total resistance R_w into two parts. The resistance R_x, defined in Fig. 4.11, is a function of the position x of the wiper as shown below:

$$R_x = (x/\ell)\ R_w \tag{4.18}$$

The current flowing through the potentiometer wire is determined from Eq. (4.2b) as:

$$I = V_s\ /R_w \tag{a}$$

The voltage drop V_x across the resistance R_x is obtained from Eqs. (4.18), (4.2a) and (a) as:

$$V_d = I \times R_x = (V_s/R_w) \times (x/\ell)R_w = C_p\ x \tag{4.19}$$

where $C_p = V_s/\ell$ is a constant depending on the slide wire length and the supply voltage V_s.

Examination of Eq.(4.19) shows that we have another approach for converting a displacement x into a voltage. This interesting fact may have an application in the development of your postal scale. Suppose you develop a spring element with a spring rate k that undergoes a displacement d so that:

$$W = kd \tag{4.20}$$

where W is the weight (force) producing the displacement d of the spring.

If you join together the free end of the spring to the wiper of the slide wire resistor with a linkage, a relationship exists between the displacement of the spring d and the position of the wiper x. Let's design the linkage so that:

$$x = K_L\ d \tag{4.21}$$

where K_L is the amplification factor for the displacement provided by the linkage mechanism.

We have all the elements in Eqs. (4.19), (4.20) and (4.21) to derive the relation between the voltage drop, the amplification factor, the constant C_p, the spring rate and weight:

$$V_d = (C_p K_L /k)W \tag{4.22}$$

Equation (4.22) is a design relationship. It indicates that you can generate a voltage V_d that is linearly proportional to the weight W on the platform of the postal scale. The factors C_p, K_L, and k, which occur in the constant of proportionality ($C_p K_L /k$), are determined by your selections of components employed in the design of the prototype . You have a lot of decisions to make in the process of designing a scale based on a potentiometer sensor; however, Eq. (4.22) provides an excellent guide in selecting individual components and their sizes.

Suppose the potentiometer provides a voltage V_d that is proportional to the weight. How do you convert this voltage to provide a measurement of weight? A voltmeter will measure the voltage, but not the weight . If you employ an analog voltmeter, all you do is:

- Remove the cover.
- Modify the scale to display weight instead of voltage.

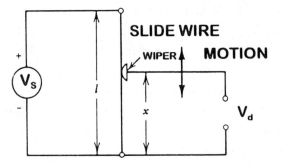

Fig. 4.11 A slide-wire resistance potentiometer.

The simple slide wire resistor, shown in Fig. 4.11, is easy to understand. Unfortunately it is rarely used because the specific resistance of common metals is too low. For instance, if we used a constantan alloy wire (45% Ni and 55% Cu) with a diameter of 0.010 in. for the potentiometer, the resistance per unit length of the wire would only be 0.25 Ω/in. A potentiometer made with a single strand of this wire would have a resistance much too low to be useful. When the resistance applied across the terminals of the battery is too low, the current drawn from the battery becomes excessively high and the life of the battery is prohibitively short. To alleviate this difficulty, high-resistance, wire-wound potentiometers are manufactured by winding the resistance wire around an insulating core as shown in Fig. 4.12. The potentiometer illustrated in Fig. 4.12a may be used for linear displacement measurements.

Cylindrically shaped potentiometers, similar to the one illustrated in Fig. 4.12b, may be used for angular measurements. Of course, cylindrical potentiometers are used to measure linear displacements by converting their rotary motion to a linear motion. An example of a conversion method is illustrated in Fig. 4.13 where a pulley is attached to the shaft of the potentiometer. A small diameter wire, wrapped around the pulley, leads to a plate that is fixed to the free end of a helical compression spring. As the weight W deflects the spring, its linear motion is transmitted to the pulley; thus, rotating the potentiometer.

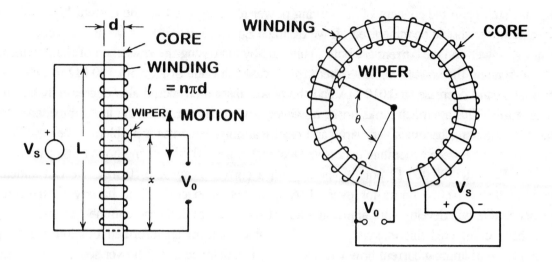

Fig. 4.12 Wire wrapped resistance potentiometers with (a) linear core and (b) cylindrical core.

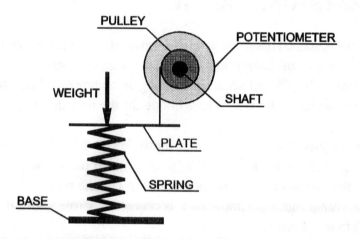

Fig. 4.13 A mechanism for converting a cylindrical potentiometer into a linear motion sensor.

The resistance of commercially available potentiometers ranges from 10 to about 10^6 Ω depending on the selection of the wire material, its diameter and the length of the coil. If we are to employ a potentiometer in the design of our postal scale, the resistance of the potentiometer (its input impedance) should be relatively high to reduce the current drain on the battery. For example, suppose you have a potentiometer with a total resistance of R_p =100,000 Ω and a battery with a voltage of 10V. The current flowing through the potentiometer is determined from Ohms law [Eq. (4.2b)] as:

$$I = V_s / R_p = 10/100,000 = 0.10 \text{ mA}$$

The result of 0.10 mA must be interpreted before it has meaning. Is 0.10 mA satisfactory or is it a disaster? We know that quantities such as work, energy and power should be conserved. It is an easy step from the concept of conserving power to recognize that the current flow through the potentiometer should be minimized. If we examine Eq. (4.5), $P = I^2 R_p = V^2 /R_p$, it becomes evident that the power and the current are both minimized by increasing the resistance of the potentiometer R_p while fixing the voltage V. If we were to increase the value of R_p from 100 kΩ to one MΩ, the current would decrease to 0.010 mA and the power dissipated would also decrease by an order of magnitude. This approach looks good; however, if we carry it too far the input impedance (R_p) of the potentiometer becomes very large and electrical noise becomes a problem. We usually seek a compromise between generating excessive electrical noise and minimizing power dissipation.

For our 100,000 Ω potentiometer with a current flow of 0.10 mA, we can anticipate a battery life of 10,000 hours for every 1 Ampere-hour capacity of the battery. An exercise is provided that will develop your appreciation for battery life of common commercial batteries. After you have completed this exercise, you will be able to form an informed opinion regarding the importance of limiting current flow when a battery is employed as the power source. However, it is clear from inspection that selecting a high resistance potentiometer clearly enhances the battery life.

4.4 THE WHEATSTONE BRIDGE

In Chapter 3, we described the strain gage as a sensor which converts strain into a resistance change. We also indicated that the change in resistance of a metal foil strain gage was so small that it could not be measured with sufficient accuracy by using an ohmmeter. This problem was solved by Lord Kelvin when he invented the Wheatstone bridge for the measurement of small changes of resistance ΔR.

Let's examine the Wheatstone bridge circuit shown in Fig. 4.14. The bridge consists of four resistors arranged in the shape of a diamond. We have identified these resistors as R_1, R_2, R_3 and R_4. A dc voltage supply is connected across node A and C with the positive terminal connected to point A. A voltmeter, with a high input impedance, is connected across nodes B and D to measure the output voltage V_0 from the bridge.

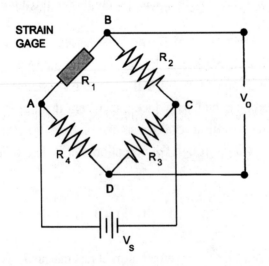

Fig. 4.14 A Wheatstone bridge circuit with a single active strain gage in arm R_1.

Selecting the four resistances for the Wheatstone bridge circuit such that:

$$R_1 R_3 = R_2 R_4 \qquad (4.23)$$

gives a null output voltage:

$$V_0 = 0$$

With the output voltage $V_0 = 0$, the bridge is in perfect balance. Equation (4.23) describes the constraint on the values of the resistors necessary to achieve this balance. This is an important relation because we will always initially balance the bridge to achieve a zero output voltage prior to applying a strain to the strain gage or a weight on the postal scale.

Next, let's place a strain gage in arm R_1 of the bridge between nodes at A and B so that:

$$R_1 = R_g \qquad (4.24)$$

where R_g is the resistance of the strain gage[2].

Suppose we apply a load to a mechanical member upon which a strain gage is bonded. The mechanical member and the gage both are subjected to the same strain ε. The strain gage responds by changing its resistance by an amount ΔR_g given by:

$$\Delta R_g / R_g = (GF)\varepsilon \qquad (4.25)$$

where GF is the gage factor or calibration constant for the strain gage (approximately 2.0 for most metal foil gages).

As the strain gage changes its resistance, the bridge becomes unbalanced and an output voltage V_0 occurs across the nodes B -- D. This output voltage is given by:

$$V_0 = \frac{R_g R_2}{(R_1 + R_2)^2} \frac{\Delta R_g}{R_g} V_s \qquad (4.26)$$

Let's simplify the design of the bridge by selecting all four of the resistors with the same value:

$$R_1 = R_g = R_2 = R_3 = R_4 \qquad (4.27)$$

Substituting Eq. (4.27) into Eq. (4.26) yields:

---

[2] Most commercially available strain gages are specified with initial resistance of 120 or 350 Ω although a few types are available with higher resistance—500 and 1000 Ω.

$$V_s = \frac{V_s}{4} \frac{\Delta R_g}{R_g}$$ (4.28)

Finally, by substituting Eq. (4.25) into Eq. (4.26) we obtain a relation for the voltage output from the bridge in terms of the strain applied to the gage.

$$V_0 = (V_s/4)(GF)\varepsilon$$ (4.29)

Okay! The Wheatstone bridge converted the resistance change from the strain gage to a voltage V_0. Let's consider an example involving the determination of the output voltage from the bridge. Suppose we select a strain gage with a gage factor (calibration constant) equal to 2.00, and apply the gage to a beam. When loaded with a weight of 10 pounds both the beam and the gage are subjected to a strain of 0.003333. Moreover, we use a six volt battery to power the bridge. What output voltage will be measured when we attach a high impedance voltmeter across the terminals at nodes B - D of Fig. 4.14?

The solution is given by Eq. (4.28) as:

$$V = (6)(2.00)(0.003333)/4 = 10 \text{ mV}$$

The value of 10 millivolts for the full load applied to our postal scale may be too small to provide an accurate measurement of the 10 pound weight. The adequacy of the output voltage depends on the sensitivity of the voltmeter. If the voltmeter can provide a measurement accurate to ± 0.1 mV, then an output voltage of 10 mV gives 100 distinct readings. For a 10 lb weight, the increment between readings is $(10 \text{ lb})/100 = 0.1$ lb. The accuracy of ± 0.1 mV in the voltage measurement corresponds to an accuracy of ± 0.1 lb in measuring the weight. In terms of percent, the accuracy $\mathscr{A}$ in the weight measurement is given by:

$$\mathscr{A} = (100)(W_m - W_T)/W_T$$ (4.30)

where W_m is the measured weight.

W_T is the true weight.

Applying Eq. (4.30) to the results shown above, establishes the accuracy of the weight measurement as:

$$\mathscr{A} = (100)(\pm 0.1)/(10) = \pm 1.00\%$$

If we are required to measure the weight with more accuracy, we either increase the output voltage from the bridge or increase the sensitivity of the voltmeter. We can increase the sensitivity and the accuracy of the voltmeter, but higher sensitivity more accurate voltmeters cost more. We can increase the output from the bridge by increasing the battery voltage, increasing the number of active strain gages or increasing the strain on the gage. However, the increases which can be achieved by using one or more of these alternatives are usually relatively small. One must expect an output of less than 100 mV from strain gage bridges even under maximum load conditions with four active gages and a high supply voltage. To increase the output voltage substantially, requires the use of a suitable voltage amplifier. We will discuss this approach in more detail in the next section.

4.5 VOLTMETERS

4.5.1 Analog Voltmeters

There are two markedly different types of voltmeters. The first is the analog meter with its calibrated scale and a pointer. The second is the digital meter with a digital readout of the voltage. Both of these meters are available commercially as multi-meters providing a means for measuring current and resistance in addition to ac and dc voltages.

Let's consider an analog meter which incorporates a D'Arsonval galvanometer as the sensor element as shown in Fig. 4.15. The galvanometer has a coil supported in a magnetic field with either jeweled bearings (low friction) or torsion springs. When a current flows through the coil, a magnetically induced torque causes it to rotate until it is restrained by springs that are part of the suspension system. The coil reaches an equilibrium position and the pointer indicates the reading of the current or some other quantity being measured.

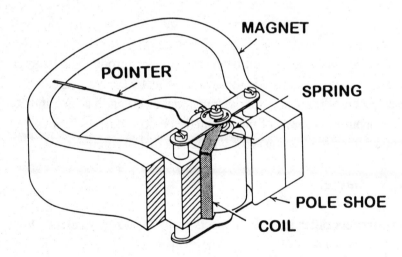

Fig. 4.15 The D'Arsonval galvanometer used as the sensing movement in analog voltmeters.

The galvanometer is converted into a voltmeter by placing a resistor in series with the coil to control the current flow. The range and sensitivity of the voltmeter is varied by changing the value of this series resistance. Typical analog voltmeters have several scales with full-scale readings varying from 100 mV to 1000 V. These are low cost instruments ($10 to $50) with accuracies of ± 2 to 3% of full scale. When operated on the lower voltage ranges, the input impedance of the voltmeter is relatively low and loading errors can occur because of the current that passes through the instrument. The input impedance for a typical analog voltmeter (without amplifier) in terms of ohms per volt full scale varies from about 2000 to 20,000 Ω/V.

What is a loading error? Do we have to deal with it? You do when using an analog meter in an attempt to measure small (tens of millivolts) voltages. When analyzing a circuit like the Wheatstone bridge and developing equations for the output voltage, we assume that the voltmeter used to measure V_0 does not draw current from the bridge. Unfortunately, analog voltmeters with D'Arsonval galvanometers require a significant current to activate the movement.

The error E due to the current draw through the meter is given by:

$$E = \frac{R_s / R_{Imp}}{1 + R_s / R_{Imp}} \qquad (4.31)$$

where R_{Imp} is the input impedance for the meter.

R_s is the impedance of the voltage source.

Let's consider an example to illustrate loading error. Suppose that the Wheatstone bridge is comprised of four resistors each 350 Ω. Also, we attempt to measure its output with a voltmeter with an input impedance of 3500 Ω. Since the bridge exhibits a source impedance $R_s = 350 \Omega$, the ratio of the source resistance to the input impedance resistance of the voltmeter is $R_s/R_{Imp} = 350/3500 = 0.1$. From Eq. (4.31), we determine the loading error due to the voltmeter as:

$$E = 0.1/(1 + 0.1) = 0.091 = 9.1\,\%$$

This example shows that the voltmeter in this case introduces a large error—too large for our project. How do we reduce the loading error due to the voltmeter? In our design of the circuits and the selection of the voltmeter, we must make certain that the ratio $R_s / R_{Imp} < 1/100$. In other words, if the input impedance of the voltmeter is more than 100 times the output impedance of the voltage source, then the loading error will be less that one percent.

4.5.2 Digital Voltmeters

Digital voltmeters differ in many respects from analog voltmeters. First, a low resistance galvanometer is not involved; consequently, the input impedance of the digital voltmeters is very high. The digital voltmeter shown in Fig. 4.16, purchased at Radio Shack for $24.95 (plus tax for the governor), has an input impedance of 10 MΩ. This impedance is so large (10 MΩ is common) that loading errors are not an issue with modern digital instrumentation.

Digital voltmeters (DVM) are relatively new; they have largely replaced the analog voltmeters because they are more accurate, exhibit extremely small loading errors, have better resolution and are easier to read. The voltage is displayed with lighted numerals as shown in Fig. 4.16, rather than a pointer on a continuous scale as is the case with analog voltmeters.

The range of a DVM is determined by the number of full digits on the display. For example, a three digit DVM records a count of 999. If the full scale or the DVM is set at 1 V, the count of 999 would be recorded as a voltage of 0.999V. Many DVMs are equipped with a partial digit to extend the range. The (1/2) partial digit displays numbers 0 and 1, and the (3/4) digit displays 0, 1, 2, 3 and 4. The value of the partial digit in extending the range of a DVM is apparent in the following example. Suppose you have a 3 digit DVM and attempt to measure a voltage of 16.66 V. Your 3 digit voltmeter would record this measurement as 16.7 V. If you made the same measurement using a 3–½ digit DVM, the reading would be 16.66 V. Since the 3–½ digit DVM provides a count to 1999, you would not lose the last digit in making the measurement.

Fig. 4.16 Photograph of an inexpensive pocket size digital multi-meter.

The resolution of a DVM is determined by the maximum count displayed. For example, a 3–½ digit DVM with a maximum count of 1,999 has a resolution of 1 part in 2,000[3]. The sensitivity of a DVM is the smallest increment of voltage that can be detected. It is determined by multiplying the lowest full scale range by the resolution. For example, the 3–½ DVM with 100 mV as the lowest (full scale) range has a sensitivity of 100 mV x 1/2000 = 0.05 mV.

The accuracy of a DVM is usually expressed as $\pm$ x% of full scale $\pm$ N digits. The specifications for the inexpensive Radio Shack DVM, with 3–¾ digits, indicates an accuracy $\pm$ 0.2% of full scale plus $\pm$ 1 in the last digit. If we are operating this DVM on the 100 mV scale, the accuracy is $\pm (0.002 \times 100 \text{ mV} + 0.1 \text{ mV}) = \pm 0.3$ mV.

The input to a multi-meter may be voltage dc or ac, current or resistance. In all instances, the input is immediately converted to a dc voltage as the signal enters the instrument. The voltage is amplified and then input to an analog to digital (A/D) converter. The A/D converter changes the dc voltage input to a clock count by using what is called the dual slope integration method illustrated in Fig. 4.17.

There are three different time controlled operations involved with the dual-slope integration technique.

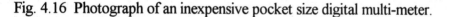

[3] A 3 ½ DVM has a total count of 2000 because, in digital representation, we use one count for zero and 1999 for the remaining numbers shown on the display.

1. **Auto zero**—where the output voltage from the integrator is zeroed for a fixed time (of the order of 100 ms).
2. **Run-up**—the dc voltage input is integrated for a fixed time (about 100 ms). The output of the integrator circuit is a voltage which increases with time to give a ramp function as illustrated in Fig. 4.17.
3. **Run-down**—the voltage on the integrator is discharged at a constant rate with respect to time to produce a linear decreasing voltage (the run-down ramp also shown in Fig. 4.17).

Because both the ramp-up and ramp-down voltages exhibit slopes on the graph of voltage versus time, this A/D conversion technique is known as dual-slope integration.

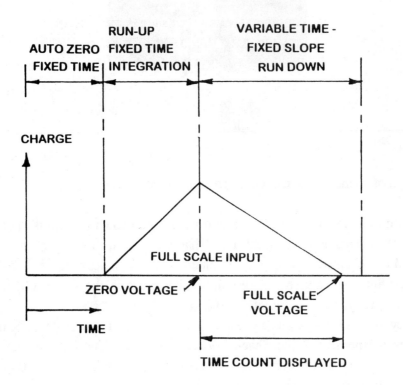

Fig. 4.17 Voltage time trace for a dual slope integrating DVM.

A counter (clock), which is initiated at the beginning of the ramp-down, runs until the voltage from the integrator goes to zero. It stops at this instant. Since this time interval is proportional to the dc voltage first applied to the integrator, the dual slope technique effectively converts voltage to a time count. The time count is presented on the digital display as the dc voltage. Counting is relatively easy to accomplish in digital electronics because extremely accurate quartz crystal oscillators are available at low cost. For example, a 10 MHz counter, capable of providing a train of well defined voltage pulses at 0.1 µs intervals, is available at prices ranging from $1 to $3 depending on packaging and quantity ordered.

4.6 AMPLIFIERS

Amplifiers are active devices used in conditioning electrical signals. An amplifier is employed in nearly every instrument system to increase small voltage outputs from a transducer to a level sufficiently large to be measured with a suitable voltmeter. The symbol for an amplifier is shown in Fig. 4.18. The input to the amplifier is denoted as V_I and the output as V_0. The ratio V_0/V_I is defined as the gain G of the amplifier. When the input voltage is increased, the output voltage also increases. In the linear range of the amplifier, the relation between the voltages is given by:

$$V_0 = GV_I \qquad\qquad (4.32)$$

Amplifiers, with gains of 10 to 1000 or more, are common and relatively inexpensive. However, an instrument amplifier with a gain of your choice probably will not be available at your local electronics supply store. However, they are available from more specialized suppliers[4]. Operational amplifiers, which differ from instrument amplifiers, are available at most electronic supply outlets for very modest costs. They can be used in simple circuits to build an instrument amplifier with stable gains that will meet your specifications.

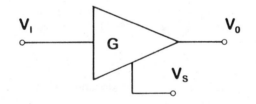

Fig. 4.18 Symbol representing an amplifier.

An operational amplifier (op-amp) is an integrated circuit with its components placed on a small silicon chip to form a complete amplifier circuit. The gain of an op-amp is extremely high (10^5 to 10^6); unfortunately, the gain is not stable. Also, with such a large gain even a small input voltage would saturate the amplifier and drive it into its non-linear region. Neither of these characteristics is desirable if you attempt to use the op-amp directly as an instrument amplifier. Its gain is too high and it is unstable.

To avoid these difficulties, it is common practice to use the op-amp as a circuit element in an instrument amplifier. We can build many different electronic instruments with op-amps as the central element including, instrument amplifiers, differential amplifiers, inverting and non-inverting amplifiers, integrating amplifiers, filters, etc. Let's examine two of these amplifier circuits that you may find useful in developing the prototype of a digital postal scale.

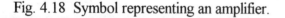

[4] Procuring electronic components at local retail outlets is usually difficult. A reliable supplier, with excellent service, is Digi-Key. Their phone number is 1-800-344-4539 and Internet address is www.digikey.com.

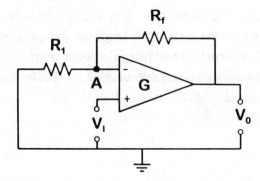

Fig. 4.19 Circuit diagram for a non-inverting instrument amplifier that utilizes an op-amp as the central electronic component.

The circuit diagram shown in Fig. 4.19 is for a non-inverting amplifier. Non-inverting means that the output voltage V_0 is the same sign as the input voltage V_I. Two resistors are employed with this simple circuit. Resistor R_1 is on the lead to the negative pin of the op-amp, and R_f is a feedback resistor which returns current to the node A. An analysis of this circuit for the non-inverting amplifier indicates that the output voltage is given by:

$$V_0 = \left(1 + \frac{R_f}{R_1}\right) V_I = G_c V_I \qquad (4.33)$$

Note, the gain G_c of the amplifier circuit is given by the selection of the resistors R_1 and R_f as:

$$G_c = [1 + (R_f / R_1)] \qquad (4.34)$$

Suppose you select $R_1 = 0.1$ M and $R_f = 1$ M, then the gain $G_c = 11$. The gain of the non-inverting amplifier is stable even if the gain of the op-amp fluctuates with time. The circuit (amplifier) gain depends only on the ratio of the two resistors and is essentially independent of the gain of the op-amp.

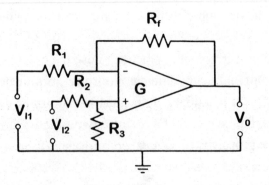

Fig. 4.20 Circuit diagram for a differential amplifier using both inputs to the op-amp.

A differential amplifier, presented in Fig. 4.20, is also useful for making engineering measurements. In fact, it is the preferred amplifier to employ with the Wheatstone bridge circuit. It is called a differential amplifier because the instrument amplifies the difference between the two input voltages V_{I1}, and V_{I2}. The differential amplifier is slightly more complex than the non-inverting amplifier because it requires four resistors instead of two. The output voltage V_0 is a function of the values selected for these four resistors as indicated by:

$$V_0 = \frac{R_3}{R_2}\left(\frac{1+\dfrac{R_f}{R_1}}{1+\dfrac{R_3}{R_2}}\right)V_{I2} - \left(\frac{R_f}{R_1}\right)V_{I1} \qquad (4.35)$$

If we select $R_f/R_1 = R_3/R_2$, Eq. (4.35) reduces to:

$$V_0 = (R_f/R_1)(V_{I2} - V_{I1}) \qquad (4.36)$$

where the circuit gain G_c is again given by the ratio of R_f/R_1.

In connecting the Wheatstone bridge to a differential amplifier, the circuit diagram shown in Fig. 4.21 should be used. Gains of 10 to 100 are usually sufficient to increase the output voltage to a value easily measured with relatively inexpensive voltmeters. The input impedance of an op-amp is so high (several MΩ) that loading errors imposed on the bridge circuit are negligible.

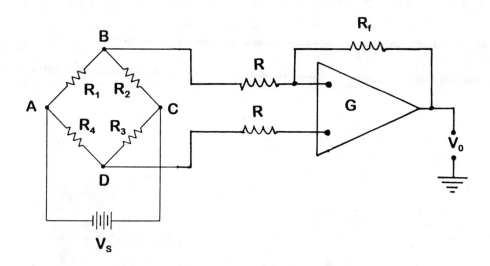

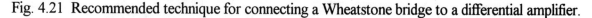

Fig. 4.21 Recommended technique for connecting a Wheatstone bridge to a differential amplifier.

If you find a well stocked electronic supply store, there will be many different op-amps available and most will be inexpensive (less than a dollar each). Which one should you buy? The widest selection of op-amps includes the following categories:

1. General purpose (lowest cost).
2. FET input, low bias current.
3. Electrometers (lowest bias current).
4. High accuracy, low differential drift.
5. Chopper amplifiers (lowest drift).
6. Fast with wideband.
7. Isolated.

In this project categories 1, 2 and 4 are probably the most suitable. Since drift is always a problem with op-amp circuits, we recommend the low differential drift op-amp.

It is a lot of fun to build an amplifier particularly if it works well and provides the gain that is required. Try building a low cost op-amp to increase the voltage from the Wheatstone bridge or to reduce the error due to loading the measuring circuit with an analog voltmeter.

4.7 DISSECTION AND ANALYSIS OF AN ELECTRONIC SCALE

In Chapter 3, we demonstrated the procedure for dissecting a mechanical postal scale with a helical spring. Let's perform a similar dissection with an electronic bathroom scale. We recognize that a bathroom scale is different from a postal scale; however, there are certainly enough similarities to make it a worthwhile endeavor. The bathroom scale employed in this dissection is rectangular in shape 10.7 in. wide by 12 in. long by 1.2 in. deep. The top cover of the scale is fabricated from a white plastic, as shown in Fig. 4.22, with a red plastic insert sufficiently large for a four digit display of the weight. The underside of the top cover, also presented in Fig. 4.22, shows a waffle-like construction. This waffle-like construction increases the rigidity of the top cover which must not deflect excessively with an applied weight of up to 300 lb. Additional rigidity is obtained by inserting two steel bars across the width of the cover plate. The very complex shape of the cover did not pose a problem in manufacturing because the cover was produced by injection molding.

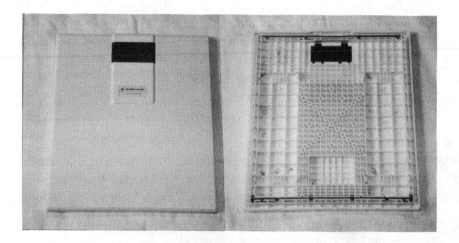

Fig. 4.22 Photographs of the top cover of the electronic bathroom scale.

After removing the top cover, we can inspect the inner workings of the scale as illustrated in Fig. 4.23. The workings of the scale are housed in the base which is also an injected molded component similar to the top cover. The sensor beams to the left and right sides of the scale, with center mounted strain gages, are evident after removing the force transmission plate. The force transmission plates carry the weight from the top cover to the sensor beams. Knife edges are mounted on the transmission plates to apply the forces to the sensor beams at precise locations. A closer examination of the sensor beams indicates that they are subjected to a constant moment between the load application points. An analytical model, illustrated in Fig 4.24, shows the loads applied to the sensor beam and the strain gage locations.

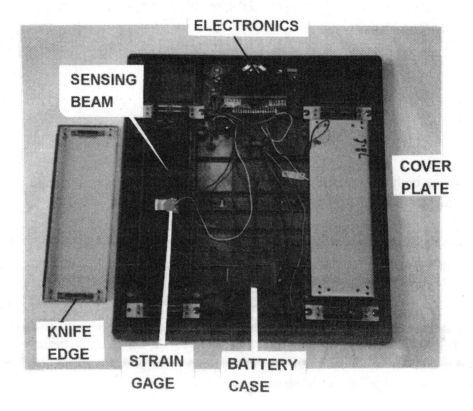

Fig. 4.23 Inner mechanisms and electronic components in the bathroom scale.

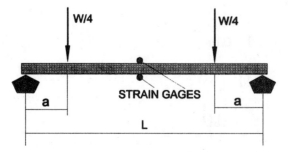

Fig. 4.24 Analytical model of the sensor beam.

To analyze this beam, we construct the free body diagram presented in Fig. 4.25. In this diagram, we have replaced the supports with forces and have cut away the right side of the beam at the location of the strain gage. To maintain the beam in equilibrium an internal moment, due to bending stresses, is generated. The magnitude of this moment is:

$$M = Wa/4 \qquad (4.37)$$

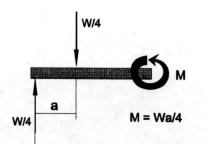

Fig. 4.25   Free body diagram of the sensor beam permitting the determination of the internal moment M.

If we recall the well known flexure formula for beams in bending, we may determine the stresses on the sensor beams from:

$$\sigma = Mc/I \qquad (4.38)$$

where   M is the internal moment acting at some point along the length of the beam.
c is the distance from the neutral axis of the beam to the outside surface.
I is the moment of inertia (the second moment of the area).

For a beam with a rectangular cross section:

$$c = h/2$$
$$I = bh^3/12$$
$$c/I = 6/bh^2 \qquad (4.39)$$

Substituting Eq. (4.39) into Eq. (4.38) yields:

$$\sigma = 6M/bh^2 \qquad (4.40)$$

Finally, we employ Hooke's law, $\sigma = E\,\varepsilon$, to obtain the expression for the strain at the gage location as:

$$\varepsilon = 6M/bh^2E \tag{4.41}$$

where b and h are the height and width of the beam respectively, and E is the modulus of elasticity.

Let's use these result and determine the stress and strain in the sensor beams found in the bathroom scale. Measuring the dimensions of the two identical sensor beams gives the following results:

$$h = 0.094 \text{ in.}$$
$$b = 1.5 \text{ in.}$$
$$a = 0.75 \text{ in.}$$

We note also that the beams are fabricated from a hot rolled steel with a modulus of elasticity E = 30×10^6. They are then cold formed to provide indentations in which the knife edges seat. Let's also consider a 240 lb person weighing himself. Substituting the design parameters for the beam into Eqs. (4.37) and (4.41) gives:

$$\varepsilon = \frac{6Wa}{4bh^2E} = \frac{6 \times 240 \times 0.75 \times 10^{-6}}{4 \times 1.5 \times (0.094)^2 \times 30} = 679 \times 10^{-6}$$

From this result, it is clear that each of the four strain gages employed on the sensor beams is subjected to a strain of 679×10^{-6} when a person weighing 240 lb steps on the scale. The strain gages on the bottom surface of the sensor beams record a tensile strain and those on the top surface of the beams respond to a compressive strain.

The four active strain gages are connected in a Wheatstone bridge circuit as indicated in Fig. 4.26. With four active strain gages in the Wheatstone bridge circuit, $R_1 = R_2 = R_3 = R_4$, the bridge is in balance and Eq. 4.26 is modified to account for the four active strain gages.

$$V_o = \frac{1}{4}\left(\frac{\Delta R_1}{R_1} - \frac{\Delta R_2}{R_2} + \frac{\Delta R_3}{R_3} - \frac{\Delta R_4}{R_4}\right)V_s \tag{4.42}$$

With the strain gages placed so that the tension gages are in arms 1 and 3 and the compression gages in arms 2 and 4, the Wheatstone bridge circuit acts to add together the four strain gage signals. Note:

$$\Delta R_1/R_1 = \Delta R_3/R_3 = (GF)\varepsilon = 2.0 \times 679 \times 10^{-6}$$

$$\Delta R_2/R_2 = \Delta R_4/R_4 = GF\varepsilon = 2.0 \times (-679 \times 10^{-6})$$

Substituting these results into Eq. (4.41) yields:

$$V_o = (1/4)[679 - (-679) + 679 - (-679)]\, 2.0 \times V_s = 1358 \times 10^{-6}\, V_s$$

If the supply voltage on the Wheatstone bridge is set at 8 V, then output voltage $V_o = 10.86$ mV. When normalized relative to the weight applied to the scale the output voltage becomes:

$$V_o/W = 10.86 \text{ mV}/240 \text{ lb} = 45.25 \text{ } \mu V/lb$$

This output voltage is very small; it will require amplification before converting from analog to digital form.

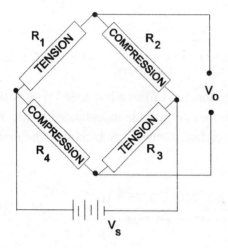

Fig. 4.26 Wheatstone bridge arrangement with four active strain gages to enhance signal output.

Let's return to the dissection of the electronic scale and focus on the electronics incorporated in the design. Perhaps the most obvious feature is the line of four seven-segment displays which serve as the digital readout. The seven-segment displays are illustrated in Fig. 4.27. To light one or more segments, we place a voltage (about 2 to 5 V) on a pin with a current of about 3 to 10 mA. The number 8 is formed by lighting all seven segments; however, the number 7 can be displayed by lighting only 3 segments. The total current supplied to the display depends on the number formed in each of the four digits.

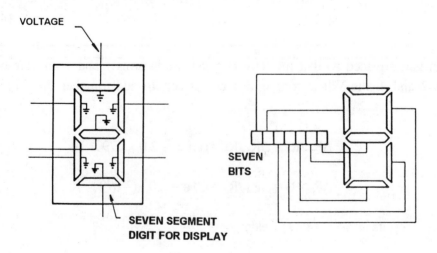

Fig. 4.27 Seven segment display digit.

We have removed the seven-segment digital display from the printed wiring board as shown in Fig. 4.28. The printed wiring board (PWB) supports the electronic components on one side while providing a plane of wiring connections on the back side. The pins of the electronic components are inserted into the PWB and soldered to a small copper pad located on the back side of the board. The solder provides both the electrical connection as well as the mechanical fastening for the component. The wiring from one pin to another is completed with traces of copper photoetched on the back side of the board. Component locations are marked on the front side of the board with printed outlines and designations such as R_1, R_2, C_1, C_2, VR_1, D_1, J_1, Q_1 etc. These symbol stand for resistance, capacitance, variable resistors, diodes, jumpers and transistors.

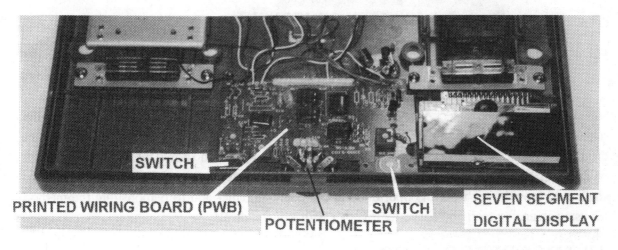

Fig. 4.28 The printed wiring board (PWB) found underneath the seven segment displays.

We will not attempt to describe in detail all of the circuits involved in this scale. Clearly, the battery serves as the power supply. The output from the Wheatstone bridge is amplified and displayed on the seven segment numerals. A switch mount in the top right corner of the PWB converts the reading between lb and kg. This conversion is accomplished by changing the gain of the amplifier by switching from one feedback resistance to another. The potentiometer mounted in the base but attached to the PWB serves to zero the scale prior to making a weight measurement. A dome switch actually incorporated in the PWB board turns on the scale when someone steps on the scale.

Seventeen leads connect the outputs from the PWB to a smaller circuit card containing the four numeral display as illustrated in Fig. 4.29. An unknown component is mounted on this small circuit card (probably an A to D converter). It is adhesively fastened to the board with an opaque compound that also serves as a protective coating. Leads from the mystery component go to small pads etched in the copper foil of the circuit card. There are seven pads for three of the numerals, three for the remaining numeral and one for the decimal point. The circuitry is for a 3–½ digit display capable of 199.9 lb.

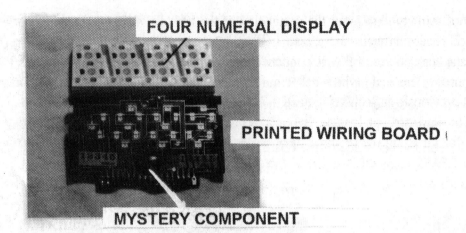

Fig. 4.29 The small circuit card which supports the display and provides capacitive coupling to the numerals.

The electronic dissection has provided us with significant insight regarding the design of the scale based on a pair of strain gage sensor beams. While we have not established the circuit diagram and specified each component in the system[5], the dissection shows the clear necessity for an amplifier and an analog to digital converter. Laboratory exercises are given at the end of this chapter to provide you with an opportunity to build and amplifier and to demonstrate the method for using an analog to digital converter together with a digital display.

4.8 MICROPROCESSORS

In previous sections of Chapters 3 and 4, we have illustrated concepts for you to employ in design postal scales that display the weight of a letter or package in either an analog or digital format. However, we have not provide any background information pertaining to the development of a smart scale—one that not only weighs, but also computes the postal fees required for a specified class of service. The required postage may depend on:

- The weight.
- The distance of the delivery.
- The service requested (the time required for the delivery).

The regulations covering mailing using the U. S. Postal Service are long (10 pages) and complex; we have provided a copy of them in Appendix. Also, shipping instructions and costs for delivery of packages by the United Parcel Service (UPS) are different from those used by the U. S. Postal service. It appears that a smart postal scale must incorporate a scale to measure the weight, a

---

[5] A complete dissection would have included tracing the wiring to each pin of each electronic component to obtain the data necessary for constructing a circuit diagram for the entire system. Then each component would be examined to determine it capacity, ratings and price. We did not pursue this approach because we prefer you to develop a new modular design with greater use of integrated circuits.

keyboard of some type to input the destination of the letter or the package, a memory to store the information regarding rates and a computer to determine the required postage. We show such a smart postal scale in Fig. 4.30. It contains three new subsystems—the memory, the keyboard and the computer. The scale which delivers a signal to the computer indicating the weight has been described previously in Sections 4.6 and 4.7.

The keyboard or keypad is not a problem. Commercially available keypads, which are sufficient for inputting the required data, cost only $11 to $12. Full keyboards with 104 keys are priced at $29.95. Again, we stress there is no need to reinvent the wheel. If a product is commercially available, use it in your design.

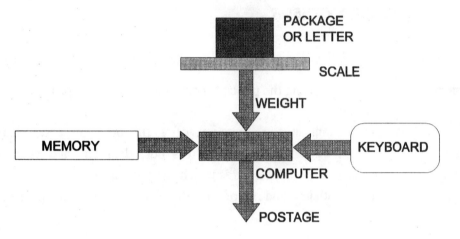

Fig. 4.30 Block diagram showing the functions involved in a smart postal scale.

The computer necessary for establishing the postage is a relatively simple device. There is no need for a Cray supercomputer to calculate the postage of 6 oz letter going by airmail to Portugal. A simple microprocessor has more than enough computing capability.

There are many microprocessors on the market each with different characteristics. It is impossible to even begin to describe the many options available in this text. Instead, we will describe some of the characteristics common to all microprocessors, and occasionally use the Motorola 68008 as an example to illustrate specific features.

Microprocessors are fabricated on a single chip of silicon and contain many hundreds of thousands of transistors, resistors and capacitors. These components are arranged to provide memory [both random access memory (RAM) and read only memory (ROM)] and an arithmetic logic unit (ALU). An oversimplified block diagram of a microprocessor with RAM, ROM and an ALU is shown in Fig. 4.31.

In a typical microprocessor (μP), there are several registers(RAM). The Motorola 68008 μP contains eight data resisters, seven address registers, two stack pointers, a program counter and a status register. The data registers and address registers are general-purpose RAM used to store bytes (8 bits), words (16 bits) or longs (32 bits). The pointers hold memory addresses to be employed in the next cycle. The program counter and the pointers work together to request instructions which are located at a certain memory address.

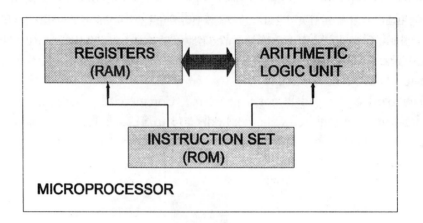

Fig. 4.31 A simple block diagram showing the principle components in a microprocessor.

Communication is important in the operation of a μP both on and off the chip. Communication is accomplished by employing an address bus, data bus and control bus as illustrated in Fig 4.32. A bus is a set of parallel wires which carry signals either out of the μP as is the case with the address bus, or both into and out of the μP as is the case with the data and control buses.

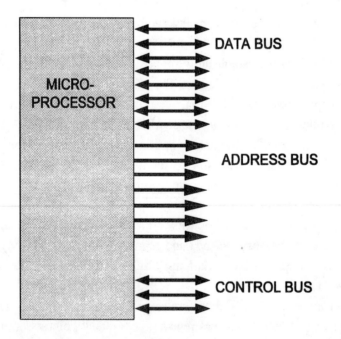

Fig. 4.32 Address, data and control buses provide communication paths for the μP.

The buses connect the μP with ROM, and RAM chips as well as input-output (I/O) devices. While some memory is provided on a μP, it is not sufficient for most applications and more capacity is provided with additional ROM and RAM devises located off of the μP. Information is conveyed between the μP and say the ROM in the manner depicted in Fig. 4.33.

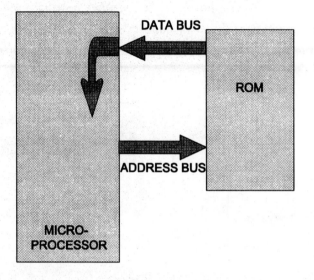

Fig. 4.33 Specific data is transmitted to the μP from ROM by using the address and data buses.

The μP cycles executing a series of programmed instructions. As it is cycling to control some parameter or to measure some quantity, we sense that parameter or quantity. When the μP has achieved the measurement or managed the control, a comparator, shown in Fig. 4.34, senses completion of the task and sends a signal to an I/O chip. The I/O chip, in turn, uses the control bus to send an interrupt signal to the μP stopping the routine (program).

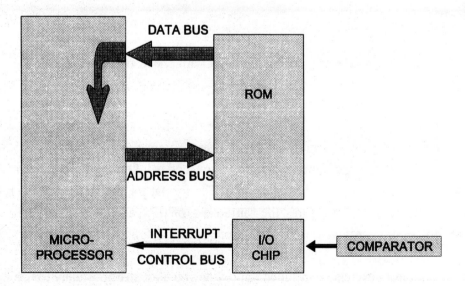

Fig. 4.34 I/O chip and comparator provide an interrupt signal to the μP over the control bus.

The μP is driven by a clock or in some cases depending on the design of the μP by two clocks. The clock signals, are essentially square waves which vary from the low to the high state as shown in Fig. 4.35. The frequency of the clock signal may vary depends on the μP. For example, with the Motorola 68008, you may specify 8, 10 or 12.5 MHz frequencies. With the 10 MHz clock, the time for a single machine cycle is 100 ns. Most instructions in the program can be executed in one machine cycle (100 ns); however, some jump instructions (branch points in the program) will require two or possibly three machine cycles. Nevertheless, we can examine the program and predict with precision the time required to execute a defined sequence. With the speed of available μPs today, the execution time for even long programs is amazingly short.

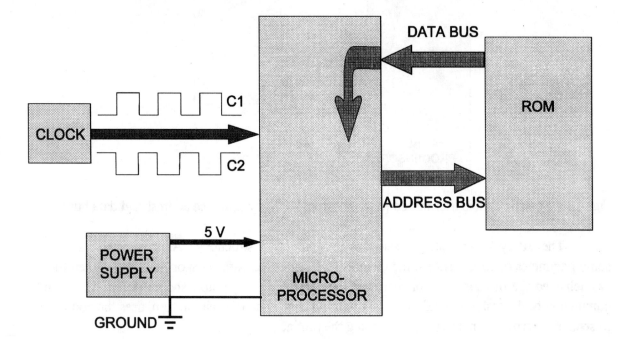

Fig. 4.35 A clock drives the sequencing of the μP with one instruction executed per cycle.

We will make no attempt to provide instructions for programming the μPs. The program depends on the make and model of the μP. For the Motorola 68008 μP, there are a total of 70 different commands in the instruction set. These commands are divided into four categories including—arithmetic, logic, control and condition codes. Other μPs employ a RISC (reduced instruction set chip) with only 35 instructions. The format for the instructions is relatively simple, but unless one frequently programs the same class of μPs, a considerable investment in time is necessary to become proficient. This proficiency soon degrades with time if not employed on a periodic basis.

A laboratory exercise is included to provide you with an opportunity to use a μP to measure temperature and to operate a heater and a fan.

4.9 SUMMARY

In developing a postal scale with a digital display, we have to convert from a mechanical response of a spring or a strain element to an electrical output somewhere in the system. To aid you in dealing with the electrical aspects of your design, we have introduced some of the most elementary electrical concepts.

Batteries (wet cells) have been described to show you the method of converting chemical energy into electrical energy. This treatment also provided an opportunity to acquaint you with voltage, and both electron and current flow.

We have also described Ohm's law, resistance, and the power dissipated by current flow through a resistor. We have also introduced the other common passive components—the capacitor and the inductor. You may not use the capacitor or the inductor in the design of the postal scale, but there are advantages to think of the resistor, capacitor and inductor together as the three passive components found in almost all electrical systems.

Circuit analysis is vitally important for all engineers regardless of their discipline. For this reason, we have introduced Kirchhoff's two laws, and showed two examples of circuit analysis leading to the equations for the equivalent resistance of circuits with series and parallel connected resistors.

A variable resistor known as the potentiometer was introduced as a sensor that could be used to convert motion, either linear or rotational, into a change in voltage. For those of you planning to incorporate a spring in the design of the postal scale, the potentiometer may be the sensor that will prove to be the most suitable choice. In this discussion, we combine the potentiometer with a spring and develop the design relation that gives the output voltage from the potentiometer in terms of the design parameters and the weight on the postal scale.

The Wheatstone bridge circuit is described for those of you intending to use strain gages as the sensor in your weighing machine. We start with the initial balance condition for the bridge and then show the output voltage for a single gage placed in one arm of the bridge. Finally, the equation for the voltage output from the bridge is given as a function of the gage factor, the strain and the supply voltage.

Both analog and digital voltmeters are described in considerable detail. The analog voltmeter with its D'Arsonval galvanometer is inexpensive, but it suffers from the possibility of severe loading errors. We describe the method for determining if loading errors are a factor in your design.

Some of you may find that the signals from your sensors and the signal conditioning circuits are too small. If this is the case, the information provided on instrument amplifiers and operational amplifiers will be helpful. The circuit gain from an instrument amplifier of 10, 100 or even 1000 can enhance the output and eliminate the difficulties associated with a weak signal.

We introduce the digital voltmeter (DVM) and describe its characteristics such as resolution, sensitivity and accuracy. The dual slope integration technique which is used in many of the lower cost DVMs is described. The dual slope technique is interesting because it permits a simple digital clock to be used as a voltmeter.

A digital scale intended for use in a bathroom was dissected and described. This scale employed sensor beams with strain gages to generate and electrical signal from a mechanical response. The sensor beams are modeled and the strains generated in the beams are determined. A four active arm Wheatstone bridge is used two accommodate the strain gages from the two sensor beams. The output from the Wheatstone bridge was so small that the need for an instrument amplifier was apparent.

The electronic dissection showed that the design of the power supply and the instrument amplifier was accomplished with a large number of discrete components assembled on a PWB. The coupling of the seven-segment numerals with an adjacent circuit board was described.

Finally, we introduced a microprocessor (μP) as a devise capable of computing postage for packages and letters requiring different classes of service. While the wide variety of suitable μPs precluded a complete description of a specific device, we described the features common to all μPs. These features include the address and data registers, the stack pointers, and the program counters. the bus structures employed to communicate on and off the chip were also discussed. The cycling of the μP depends on clock pulses at very high frequencies. Instructions listed in the program written for the μP are usually executed in one clock pulse although branch points in the instructions usually require two or more pulses. Programming is accomplished using a set of commands that are unique for a given class of microprocessors. Fortunately the number of commands is usually relatively small.

REFERENCES

1. Dally, J. W., Riley, W. F. and K. G. McConnell, Instrumentation for Engineering Measurements, 2nd edition, John Wiley & Sons, New York, 1993.
2. Sheingold, D. H., editor, Transducer Interfacing Handbook, Analog Devices, Norwood, MA, 1981.
3. Dorf, R. C., Introduction to Electric Circuits, 2nd edition, John Wiley & Sons, New York, 1993.
4. Bennett, W. C. and Evert, C. F., What Every Engineer Should Know About Micro-computers, Motorola Series in Solid State Electronics, Marcel Dekker, New York, 1980.
5. Dally, J. W., Packaging of Electronic Systems: A Mechanical Engineering Approach, McGraw-Hill, New York, 1990.
6. Horowitz, P. and Hill, W., The Art of Electronics, 2nd ED., Cambridge University Press, New York, 1989.
7. TelCom Semiconductor Inc., 1997 Data Book, Mountain View, CA, 1996.

LABORATORY EXERCISES

4.1   The dissection of the electronic scale showed that the manufacturer had employed many discrete components. Moreover, the designer of this scale did not use a modular approach. In more modern design, integrated circuits which contain large numbers of transistors resistors and capacitors on a single chip of silicon of often employed. With integrated circuits fewer discrete components are required and the chip manufacture does much of the wiring on chip. The result is a more stable device with improved performance. To demonstrate amplifiers that you may wish to use with your design we suggest that you use two instrument amplifiers in series to produce an amplifying system for your Wheatstone bridge with a gain of 1000. The first amplifier is a difference amplifier which accepts the two leads from the interfaces and provides an output voltage which is G ($V_5 - V_1$). We are suggesting the Burr-Brown model No. INA — 106KP — ND difference amplifier with a gain of 10 which sells for $7.02. The output from this amplifier is referenced to ground potential (single ended output). The second instrument amplifier in the series is also of Burr-Brown design (model No. INA — 141P — ND) with gains of 10 and 100 established by selection of resistors residing on the integrated circuit. Both the input and output voltages for this amplifier are referenced to ground. Its cost is $8.10 when ordering a single unit. Both amplifiers are housed in an 8 pin dual in-line package (DIP).

Conduct the following calibration experiment for the difference amplifier:

1. Using a reference voltage source and only the difference amplifier, increase the input voltage from zero until the amplifier saturates.
2. Record the input and output voltage at many different levels.
3. Plot these data and establish linearity and the gain of the amplifier.
4. Describe the response of the amplifier when it is saturated.
5. Reduce the voltage from its maximum linear value to zero while again recording the input and output voltages.
6. Prepare a graph of the output voltage versus the input voltage over the operating range of the amplifier.
7. Did you observe any hystresis?

Is it possible for you to increase the gain of this difference amplifier from 10 to some other value? If so change the gain and repeat the calibration experiment described above.

Connect together in a series arrangement the difference amplifier, with a fixed gain of 10, and the single-ended input instrument amplifier. Repeat the calibration experiment with the single-ended input instrument amplifier (INA — 141 — PD) set with a gain of 10.

Set the gain of single-ended input instrument amplifier to 100 and repeat the experiment. How high can you manage to set the gain before the system gain becomes unstable? Describe the instability.

4.2 TelCom Semiconductor, Inc. produces a number of integrated circuits design to act as analog to digital converters which interface well with numerical displays. Below, we show a circuit diagram for a 3–½ digit A/D converter interfacing with a liquid crystal display (LCD). Using the breadboard supplied and the available components construct the A/D converter and the display. Supply a known analog voltage to terminal V_x and check the display to establish the accuracy of the digital conversion process. Slowly increase the analog voltage noting the step increases in the digital output.

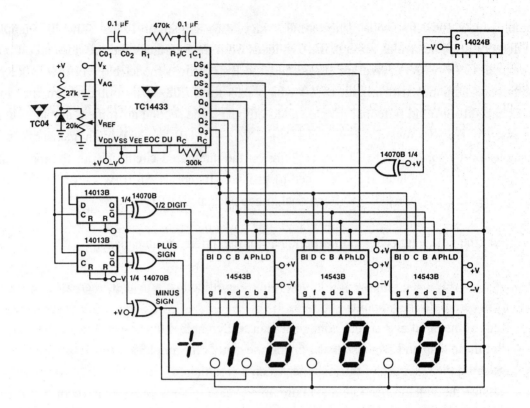

Fig. LX 4.2   A 3 ½ digit voltmeter with a LCD display.   Courtesy of Telecom Inc. 1997 Data
Book.

4.3   A µP has been developed on a breadboard for you to employ in a simple experiment.   A
schematic diagram of the arrangement is shown below:

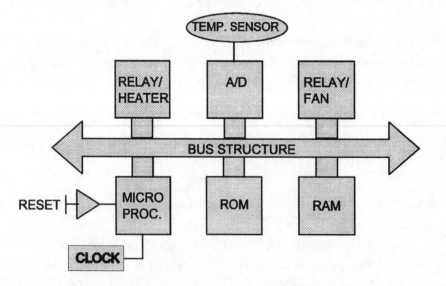

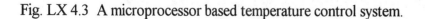

Fig. LX 4.3 A microprocessor based temperature control system.

The microprocessor is to control the temperature of a small quantity of water. The temperature of the water is adjusted by adding heat with a small immersion heater or by cooling the container with a small fan. The temperature of the water is to be automatically controlled by first increasing it from room temperature 72 °F to 95 °F and then decreasing it to 88 °F. The temperature is then maintained at 88 °F for ten minutes before terminating the experiment.

Record your observations during the experiment. Study the program driving the μP.

EXERCISES

1. Write an engineering brief and describe a hybrid system containing both mechanical components and electrical devices. Give an example of a hybrid system based on your own experience.

2. Draw the symbols used on drawings to represent a resistor, a capacitor and an inductor. Also, write the relations for the voltage across these components in terms to the current flow.

3. If you design a circuit with a 9 V battery containing an equivalent resistance of 2,000 Ω, find the current drained from the battery. If the battery has a capacity of 2A-h, determine the life of the battery. Assume that it cannot be recharged. What is the power dissipated by the resistor?

4. A size D, 1.5 V, alkaline battery has a capacity of 1200 hours when the battery is discharged through a 100 Ω resistor. Determine the current flow through this resistor and state the capacity in terms of Ampere-hours (Ah).

5. A 9 V alkaline battery has a capacity of 6 hours when the battery is discharged through a 100 Ω resistor. Determine the current flow through this resistor and state the capacity in terms of Ampere-hours (Ah).

6. Write Kirchhoff's two laws. Use circuit diagrams to illustrate these two laws.

7. Draw a circuit with four series connected resistors and give the expression for the equivalent resistance. If these resistors have values of 1000, 400, 1500 and 10,000 Ω, find the equivalent resistance. Draw a circuit diagram which contains the equivalent resistor which is also equivalent to your initial circuit diagram.

8. Draw a circuit with four parallel connected resistors and give the expression for the equivalent resistance. If these resistors have values of 1000, 400, 1500 and 10,000 Ω, find the equivalent resistance. Draw a circuit diagram which contains the equivalent resistor which is also equivalent to your initial circuit diagram.

9. In adapting a potentiometer to a spring, we employ some type of a linkage which amplifies the small displacement of the free end of the spring. Sketch the design of such a linkage and provide dimensions so that the linkage amplifies the motion by a factor of three.

10. Suppose that your search of the electronic supply stores does not yield a linear motion potentiometer, but you locate several rotary models. Sketch a method showing how to use a rotary potentiometer to measure linear displacement.

11. If you employ a strain gage as a sensor with a gage factor of 2.05, a supply voltage of 9 V on the Wheatstone bridge circuit, determine the output voltage from the bridge when the gage is subjected to a strain of 1800 x 10^{-6}. Comment on the magnitude of this voltage. Is it large enough for you to measure directly with an inexpensive analog voltmeter?

12. Prepare a table showing the loading error as a function of the ratio of the source impedance to the meter impedance. From this table establish a rule to maintain the loading error to less than two percent.

13. Draw a circuit of a non-inverting instrument amplifier with a gain of 100. Use an operational amplifier in this circuit and select the values of the resistors to be employed.

14. Draw a circuit of a differential inverting instrument amplifier with a gain of 1000. Use an operational amplifier in this circuit and select the values of the resistors to be employed.

15. Examine the specifications for a DVM and cite the statement for the accuracy. Convert this specification into a numerical value for the accuracy if you are using the most sensitive scale on the DVM. Perform the same conversion for the least sensitive scale.

16. Consider the sensor beams described in Section 4.7. If the beam thickness h is decreased from 0.094 to 0.062 in., determine the voltage output V_o from the Wheatstone bridge. Recall that a weight of 240 lb was applied to the scale and the bridge was supplied with $V_s = 8V$.

17. Find a printed wiring board (PWB) from a discarded piece of electronic equipment. After conducting an electronic dissection of the PWB, write an engineering brief describing the various uses of the PWB.

18. Sketch a seven-segment numeral used in digital displays. Describe the segments used to display the numbers, 2, 4, 6 and 8.

19. Sketch a seven-segment numeral used in digital displays. Describe the segments used to display the numbers, 3, 5, 7 and 9.

20. Describe the components found on a typical microprocessor.

21. If we have both ROM and RAM on a microprocessor, why do we usually have to communicate with off chip RAM and ROM.

22. What is an address bus?

23. What is a data bus?

24. What is a control bus?

25. What is the purpose of a clock when used with a microprocessor?

26. Is it necessary to program a microprocessor?

27. Describe how a microprocessor can be utilized to control temperature according to some time-temperature profile.

PART II

ENGINEERING GRAPHICS

CHAPTER 5

THREE-VIEW DRAWINGS

5.1 INTRODUCTION

Engineering graphics is a broad term that is used to describe a means of communication. We normally think of communication in terms of writing and speaking because they are more commonly employed in the normal course of our life. However, when trying to communicate design ideas, we find writing and speaking insufficient to express our thoughts. A more visual way to communicate is needed. It more effective to present our ideas by means of drawings, sketches, pictures, and graphs of many different types. Visuals aids, such as drawings, convey our ideas quickly and with remarkable precision. It is easier to transmit and to receive information if it is conveyed in the form of drawings, sketches, or graphs.

The objective of this part of the textbook is to introduce you to visual methods of communication. You should understand the advantages of presenting information in drawings, sketches, and graphs because they are very common techniques used to communicate very complex ideas quickly and with precision. In this part of the textbook, you will begin to learn techniques for preparing orthographic projections, isometric drawing, sketches, and several different types of graphs.

There are two general approaches used in preparing drawings and graphs. The first is a manual approach where we draw the visuals by hand using a few simple drawing instruments. The second is using a computer and suitable software programs which greatly facilitate the preparation of a drawing or graph. In this chapter, we will cover the manual methods for preparing drawings. In a separate chapter, we introduce a computer aided design (CAD) software program used in preparing engineering drawings. In still another chapter, we describe EXCEL, which is a spread sheet program that is used in computation and in preparing several different types of different graphs. Finally, we describe a graphics presentation program, PowerPoint, employed to prepare slides and overhead transparencies for oral presentations.

5.2 VIEW DRAWINGS

Let's begin by considering the block-like object shown in Fig. 5.1. We could describe this object as a rectangular block with a slot cut from its top. However, this written description is vague because we have not conveyed the relative proportions of the block, the precise location of the slot, or the size of its features. In preparing an engineering drawing, it is essential to communicate proportion, exact location, and size of every feature of the object which we are describing. The objective of the drawing is to convey sufficient information for the object to be produced anyone capable of reading a drawing. We use pictorial drawings and/or multi-view drawings to quickly and accurately convey this information to the reader. The drawing shown in Fig. 5.1 is a pictorial, that we will use to aid in describing multi-view drawings.

The pictorial (isometric) drawing in Fig. 5.1 is shown with three arrows. These arrows represent the directions of observation of the object which show the front, top and side views. When considered individually, each view is a two-dimensional rendering of the three-dimensional object. Of course, two-dimensional drawings are incomplete, because they show only the information observed in one of several possible views. Nevertheless, the two-dimensional views are important because we can show objects to a true scale on a sheet of ordinary paper or the screen of a computer monitor. We avoid the problem of incomplete information inherent in two-dimensional drawings by presenting a sufficient number of views to completely locate and size each feature of the object.

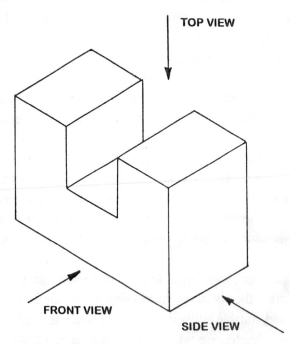

Fig. 5.1 Pictorial drawing showing a rectangular block with a slot. The three viewing directions show the front, top, and side views.

A three-view drawing of the rectangular block presented in Fig. 5.1 is illustrated in Fig. 5.2. The single pictorial drawing of the block is now represented with three different two-dimensional drawings representing the front, top, and side views. These three drawings completely define the proportions of the rectangular block, its size, and the location and size of the slot.

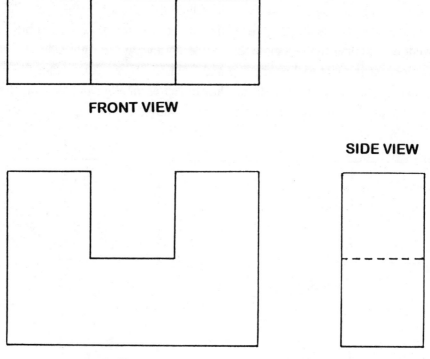

FRONT VIEW

SIDE VIEW

TOP VIEW

Fig. 5.2 A three-view drawing of the object shows the front, top, and side views.

The arrangement of the views is important. The front view is placed in the lower left corner of the paper, the top view is directly above it, and the side view is to the right of the front view. This arrangement is important because it permits us to prepare the three-view drawing using orthographic projection. With orthographic projection, we project dimensions from one view to another. For example, we measure width and height of the block, draw the front view, and project construction lines upward to the top view. The width of the object is identical in both the top and front views. We also project construction lines to the right from the front view to define the height of the side view. The front and side views both show the height dimensions of the block. We can also project from the top view to the side view if we draw a 45° construction line from the right hand corner of the front view as shown in Fig. 5.3. We then project construction lines to the right from the top view until they intersect the 45° construction line. We turn them downward to show the depth of the block on the side view. All of the construction lines used to prepare Fig. 5.2 are illustrated in Fig. 5.3. Orthographic projection is very important because it saves you time in preparing your drawings. It reduces the number of time consuming measurements that you must make in preparing the different views necessary to completely describe any given part. The construction lines also help you in forming visual images of the locations of each feature on the adjacent views.

One more point to consider before we leave Fig. 5.2. Did you notice the dashed line on the side view? Why were all of the lines solid except that one? We follow a convention in preparing engineering drawing for the use of line styles and weight. Solid lines represent edges of features that are observed in a particular view. As we observe the top of the block shown in Fig. 5.1, we can see the four sides of the rectangle and the two edges of the slot. We use solid lines then to represent all six edges visible in the top view. However, when we view the object from the side only the four edges outlining the rectangle are visible. Viewing the right side, we cannot observe the slot. If we look at the pictorial or the top and front views, we know that the slot exists. To show the slot on the side view, we use a dashed line to represent a line hidden from view. Drawings differ from photographs because hidden features can be shown on the appropriate view with dashed lines.

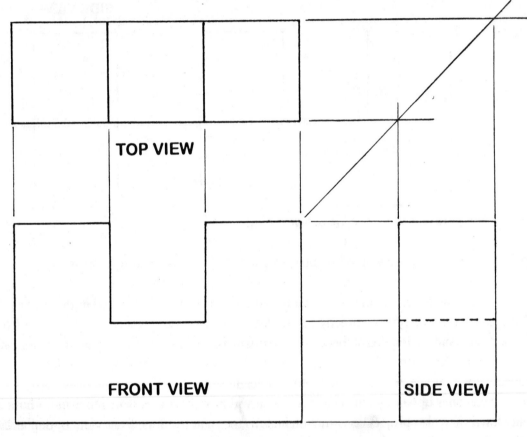

Fig. 5.3 Construction lines used to convey dimensions from one view to another.

5.3 LINE STYLES

We use different line styles and different line weights (thickness and darkness of the lines) in preparing engineering drawings as illustrated in Fig. 5.4. The weight and the thickness of the lines are controlled by the selection of the pencil lead and the sharpness of the point. The degree of hardness is governed by variable amounts of clay that is mixed with graphite to form pencil lead.

Hard leads have a small amount of clay while soft leads have higher clay content. The scale from hard and light to soft and dark is given below:

Hard and Light $\Leftarrow$ 6H–5H–4H–3H–2H–H–F–HB–B–2B–3B–4B–5B–6B $\Rightarrow$ Soft and Dark

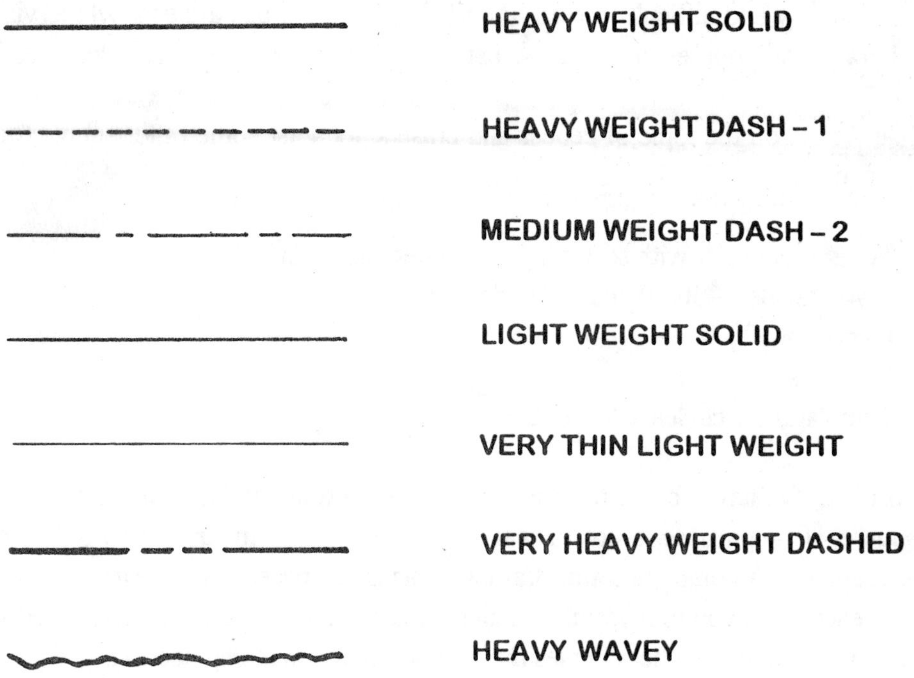

—————————————— **HEAVY WEIGHT SOLID**

— — — — — — — — **HEAVY WEIGHT DASH – 1**

——— — ——— — — **MEDIUM WEIGHT DASH – 2**

———————————— **LIGHT WEIGHT SOLID**

———————————— **VERY THIN LIGHT WEIGHT**

———— —— ———— **VERY HEAVY WEIGHT DASHED**

〰〰〰〰〰〰〰 **HEAVY WAVEY**

Fig. 5.4 Different line styles used in view drawing.
    a.   Heavy-weight, solid-lines representing edges
    b.   Heavy-weight, dashed-lines representing hidden lines.
    c.   Medium-weight, long-dash then short-dash lines for centerlines.
    d.   Light-weight solid-lines for dimensioning.
    e.   Very thin, light-lines for construction.
    f.   Very heavy-weight, dashed-lines for section cuts.
    g.   Heavy wavy-lines to indicate a break in the view.

We use 4H to 6H leads with a very sharp point to draw construction lines. These lines are so light and thin that they will not show when the drawing is reproduced. Light-weight dimension lines are drawn so they will remain visible when the drawing is copied. We use an H or F lead to give the correct relative darkness and a sharp point with a very slightly rounded tip. Medium-weight centerlines are drawn with a HB or B lead with a rounded point. Heavy-weight lines, both solid and dashed, are drawn with B or 2B lead with a rounded point.

The very soft leads—3B to 6B are usually employed by artists to prepare sketchings, but not by engineers in their sketches. Engineers handle a completed drawing much more than an artist. In handling, the soft lead smears degrading the appearance of the drawing and detracting from the quality of the copies usually made from the originals.

We recognize that many of you will be short on pencils, but perhaps team members can share resources. Please do not use ball-point pens for drawing. There are two problems with ball-point pens. First, you have no control over the width of the line because the line width is determined by the diameter of the ball in the pen. Second, you may make errors in preparing the drawing. The lines drawn with a ball-point pen cannot be erased; you eventually complete a drawing that is sloppy and difficult to read. Work in pencil and use a clean, white, vinyl-eraser to eliminate all traces of your errors. Strive hard to produce an accurate, clean, and error-free drawing.

While we are on the topic of pencils and erasers, let's list some other simple tools that you will want to acquire.

1. Transparent ruler with scales in both inches and millimeters.
2. Two triangles—30/60 and 45/45 degrees.
3. Protractor
4. Compass
5. Templates for circles, ellipses, etc.

This is a short list. We have not included a complete set of drawing instruments or a drafting board with a T-square. Our purpose is not to teach drafting, but to introduce you to the essentials of engineering graphics. A complete set of drafting tools is not necessary to learn the early lessons in graphics. We encourage you to invest the small sum needed to acquire the items listed above. You will find them useful in many other courses during your program in engineering.

Finally, a suggestion is made for the paper to use in preparing your drawings. We recommend National Brand engineering paper 8-1/2 by 11 in size. It is light green in color to relieve the strain on your eyes. It also has a grid (five squares to the inch) on the back side of each sheet. The grid shows through to the front side providing guidelines that are helpful in projecting the construction lines needed to prepare three-view drawings. The paper erases well, produces good copies, and is pre-punched for a three-ring binder.

Equipped with the proper pencils, eraser, tools and paper, you are ready to learn techniques for drawing several different features normally encountered in preparing multi-view drawings of engineering components.

5.4 REPRESENTING FEATURES

We will demonstrate the techniques for drawing various features in three-view drawings. The examples selected include a block with a slot and a step, a block with a tapered slot, and a block with a step and a hole.

5.4.1 Block with Slot and Step

The rectangular block in Fig. 5.1 has one feature—a slot. We encounter many different geometric features in depicting parts that we design, including holes, curved boundaries, steps, and tapers. A pictorial of an object with both a slot and a step, presented in Fig. 5.5, serves as our next example. Let's prepare a three-view drawing of this object using the methods of orthographic projection. Examine the pictorial starting with the front view.

As you examine the front view in Fig. 5.5, visualize the outline of the block. Even with the step and the slot, the outline of the block (the outside edges) is a rectangle. Draw the rectangle, representing the outline of the front view, in the lower left hand corner of your quadrille paper. We will not worry about exact dimensions of the rectangle at this stage; however, try to maintain the proportions shown in Fig. 5.5. When drawing the rectangle in the front view, extend the vertical lines upward into the region of the top view and the horizontal lines to the right into the region of the side view. These are construction lines; you should keep them thin and light. Now examine the step in the upper right of the front view of the block. Note the intersection of the planes defining the step with the front plane of the block. These intersections produce two new edges that are visible in the front view. Draw the location of these edges with the horizontal and the vertical lines shown in the upper right hand corner of the rectangle (see Fig. 5.6). Extend the vertical line used to locate the step in the front view upward and the horizontal line to the right. Finally, locate the slot on the left side of the pictorial drawing in Fig. 5.5. Again note the planes defining the slot and observe that they intersect the front plane to form two vertical edges. We draw two vertical lines in the correct location to represent these edges. Project these lines upward into the region of the top view with construction lines. Okay! We have completed the front view and are ready to draw the top view.

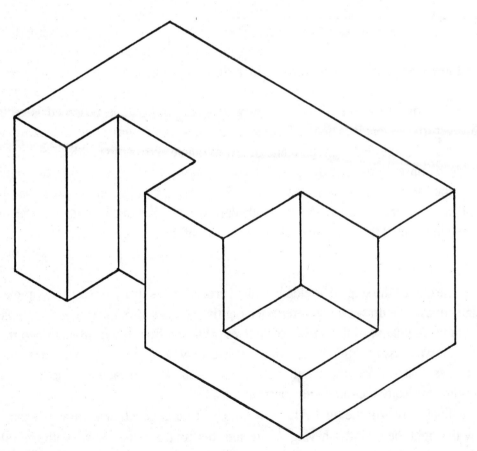

Fig. 5.5 A pictorial drawing of a block with a slot and a step.

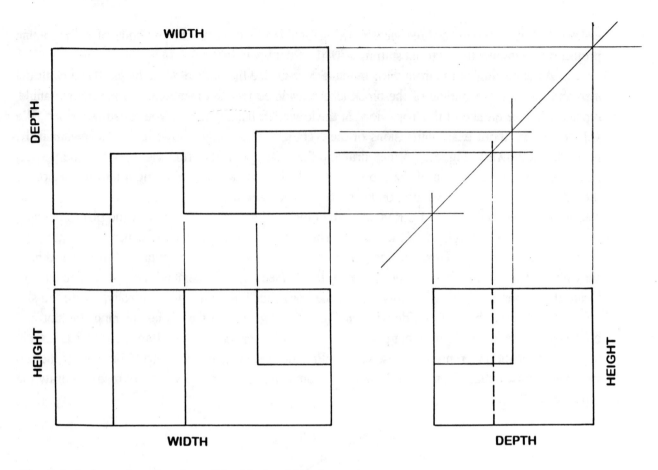

Fig. 5.6 A three-view drawing of the block with a slot and a step.

Look downward at the pictorial drawing. Again, the outline observed is rectangular in shape; however, the rectangle is not perfect since it is interrupted by the slot. Looking downward, we observe that the edge defining the outline of the rectangle across the width of the slot is missing. It is possible to observe the bottom plane at the location of the slot. Draw the outline of the rectangle in the location of the top view of Fig. 5.6. Use the construction lines that exist at the location of the top view to establish the width dimensions. An open section in the rectangle is evident at the location of the slot. Next, show the depth of the notch in the top view by drawing the three lines that represent the edges formed by the three vertical planes of the slot with the top plane of the block.

Next, examine the step. The step has two vertical planes that intersect the top plane to form the defining edges. We draw two lines representing those edges in our drawing to complete the top view. If you made full use of the construction lines projected from the front view into the region of the top view, it was easy to draw the top view. It was only necessary to locate the position of three horizontal lines that defined the depth of the slot, step. and rectangular block. These depth dimensions are carried to the right with construction lines.

The final view is of the right side. It is possible to draw a left side view, but the convention is to draw the right view. After having completed the front and top views, with the orthographic projection of construction lines, the side view is easy to prepare. Draw a construction line at 45° so that it intersects the horizontal construction lines from the top view. Then draw a second set of construction lines downward from these intersection points into the region of the side view. The side view now contains seven construction lines. The three horizontal lines projected from the front view give the height of the step and the rectangular block. The four vertical lines projected from the top view give the depth of the slot, step, and block. We have the location and dimensions of all of the lines in the side view from our orthographic projections. No measurements are necessary. To complete the side view, we darken select portions of the construction lines that represent edges in the side view. Again, examine the side of the pictorial drawing in Fig. 5.5 and observe that the rectangular outline of the block is completed. Also note the edges produced by the intersections of the planes forming the step with the plane of the right side of the block. Draw these two edges in the upper left corner of the side view (see Fig. 5.6). From a visual perspective, our drawing of the right side is complete; however, a multi-view drawing often carries more information than what we can observe directly. We know from the front and top views that a slot exists. Our projection lines from the top view to the side view show the depth of the slot. Even though we cannot "see" the slot in the side view, we represent its presence and its location with a dashed (hidden) line. The three-view drawing of a rectangular block containing a slot and a step is complete as illustrated in Fig. 5.6.

5.4.2 Block with Tapered Slot

Let's consider drawing still another feature; a block incorporating a tapered slot as illustrated pictorially in Fig. 5.7. Begin a three-view drawing with the front view in the lower left corner of your drawing paper as shown in Fig. 5.8. Examine the front view in Fig. 5.7 and observe the edges produced by all the planes intersecting the front plane. Drawing these edges gives us four vertical lines and three horizontal lines. Note, the tapered surface intersects the front plane giving an edge shown as a horizontal line in the front view. The top surface is represented by still another edge visible in the front view. We project all of these lines into the remaining two views with construction lines.

Examine the top view and observe that the rectangular outline of the block remains intact. Observe three edges inside the rectangular outline. The edge due to the intersection of the taper plane with the top surface produces a horizontal line, and the two vertical planes, defining the width of the slot, intersect the top surface to produce two vertical lines in the top view. Project the horizontal lines in the top view to the right so that they intersect the 45° construction line shown in Fig. 5.8. Then project these three lines downward from the intersection points on the 45° line into the side view.

The six construction lines projected into the side view give provide the dimensions needed to complete this view. Clearly, it is evident from Fig. 5.7 that the rectangle outlining the side view is intact. In fact as we view the right side, only a simple rectangle formed by the four edges is visible. The information regarding the tapered surface, evident in the pictorial drawing and the front view, is not visible in the side view. We add information, showing the location of the tapered surface on the side view, by using the dashed (hidden) lines.

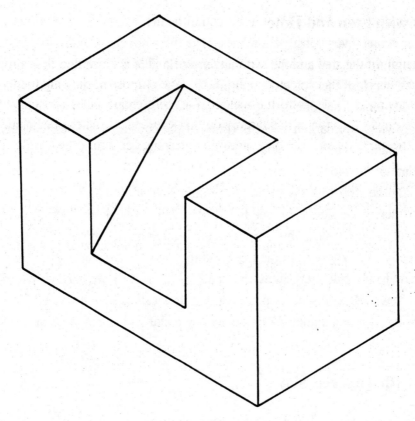

Fig. 5.7 A pictorial drawing of a block with a tapered slot.

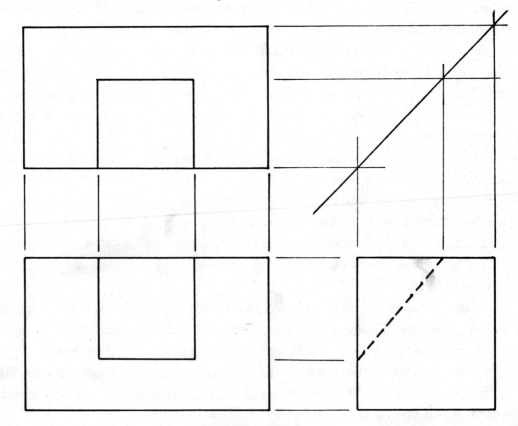

Fig. 5.8 A three-view drawing of the block with a tapered slot.

5.4.3 Block with Step and Hole

A pictorial drawing of a block with a step and a hole is illustrated in Fig. 5.9, and a three-view drawing of this object is presented in Fig. 5.10. The purpose in showing the fourth example of a three-view drawing is to illustrate the method for representing holes and curved boundaries in engineering drawings. Assume in this discussion that you understand the techniques used to draw the outlines of the three-views. The only question remaining concerns drawing the curved boundary that outlines the step and the hole that is drilled through the step.

The circular boundary outlining the step is evident only in the top view (see Fig. 5.10). The front and side views give no evidence of the presence of a curved boundary because the vertical curved surface does not produce intersections (edges) on the front or side planes. We produce the curved surface in the top view by using a compass with its point located in the center of the hole. The radius is set on the compass so that the circular arc is tangent to the horizontal lines forming the outline of the block in the top view. It is advisable to draw the arc with the compass prior to drawing the horizontal lines that are tangent to the ends of the arc.

The compass is also employed to draw the hole in the top view. Note the two orthogonal lines defining the center of the hole. These are center lines that locate the position of the hole relative to other features on the block. The presence of the hole is also depicted in the front and side views even though it is not visible in either of these views. We use dashed (hidden) lines to locate the edges of the holes. We also show the centerline of the hole in the front and side views. Note the long-dash, short-dash line used to denote the centerlines in all three-views. The centerlines are very important when we dimension the drawings because we locate holes with the centerlines and not the their edges.

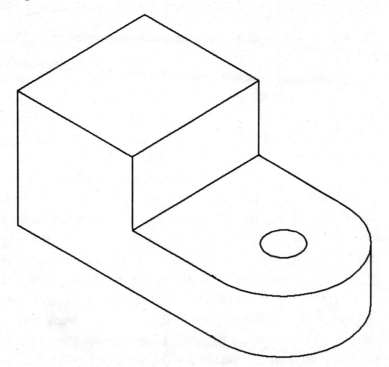

Fig. 5.9 A pictorial drawing of a block with a step and a hole.

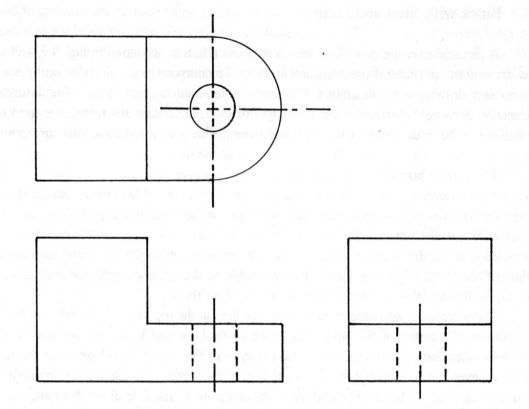

Fig. 5.10 A three-view drawing of the block with a step and a hole.

5.5 DIMENSIONING

In the previous section, we prepared three-view drawings maintaining proportionality, but without concern for the dimensions. This approach was a simplification taken to clarify the discussion of three-view drawings. In actual practice, the dimensions are critical because they control the size of the component and the location of all of its features. Let's examine dimensions from two different points of view. First, as the individual preparing the drawing, you must know (or decide) upon the dimensions required to completely define the component. If you are the designer of a new component, it is necessary to start with a blank sheet of paper and assign the dimensions that you believe will optimize the design of the component. If you are redesigning an existing component, it is possible to start with the drawing of that component and modify existing dimensions to refine the design. The point here is that you (as the person preparing the drawing) must know the dimensions. It is your responsibility to include on your drawing all dimensions necessary for some stranger, perhaps in some other country, to fabricate the component.

Next consider the second point of view—as the person using the drawing. This person may manufacture the part, may assemble the product that incorporates the part, or may repair the product. Engineering drawings serve many purposes and are used by several people not involved in the design. The three-view drawing defines this component without ambiguity. The dimensions give its size precisely. The component as defined by the drawing and its dimension can be made by anyone in the world. The component must be interchangeable with another component manufactured previously by another plant in another country.

Let's begin to learn about dimensioning by adding dimensions to the drawing of the block with a slot shown in Fig. 5.2. We have copied this drawing and added dimensions as indicated in Fig. 5.11. Examine the front view of this drawing noting that we have defined the width of the block as 3. This dimension is inserted in a break in the dimension line. No units are given except for a notation in the drawing block that all dimensions are in inches, mm, ft, etc. The dimension line is terminated by arrowheads that point to the two extension lines. The arrowhead is long (about 1/8 in.) and thin. The extension lines are separated from the view drawings by a small gap (about 1/16 in.).

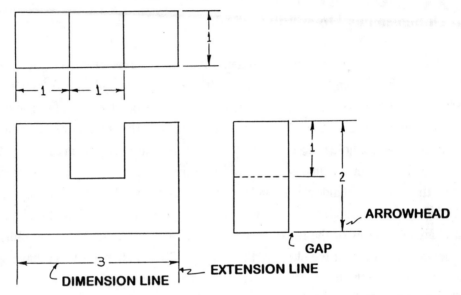

Fig. 5.11 Dimensioning example showing the dimension line, extension line, arrowhead and gap.

In the example presented in Fig. 5.11, we have dimensioned the block using inches as the unit of measure. The fact that we have defined the width as 3, indicates to the person manufacturing the block that the width can be $3 \pm 1/64$ in. The tolerance on the width is implied or is given in the drawing block. If you, as the designer, wish to specify a block manufactured with more accuracy, you would specify the dimension using decimals. For example, if you provided the dimension as 3.00, the person manufacturing the part would understand that the width of the block would have to be between 3.00 ± 0.01 in. If you require still more accuracy, specify the dimension as 3.000; the block must be produced with a width between 3.000 ± 0.005 in.. Remember—there is a great difference between 3, 3.00 and 3.000 in the accuracy of the block and tooling required to manufacture it. There is also a great difference in the cost. As we tighten the tolerances, the cost of manufacturing a component increases significantly.

Let's return to Fig. 5.11 and examine the dimensioning of the top and side views. We have already shown the width on the front view, but we still must define dimensions for the depth and height of the block and the location and size of the slot. Note, the top view is used to give dimensions for the width and depth, and the side view is used to give dimensions for the height and depth. In the top view of Fig. 5.11, we have specified the depth of the rectangular block as 1. We have also specified the dimensions locating the slot and defining its width on this view. The dimensions of the height of the block and the height of the slot are provided on the side view.

We have completed the dimensioning of the block and the slot. Every dimension necessary to manufacture the component has been specified in the drawing. Also the drawing has not been over-dimensioned. (We have not specified the same dimension twice). The choice of where we place the dimensions is somewhat arbitrary. We can dimension the width on either the front or top views, the depth on the top or side views and the height on the front or side views. Two choices for the location of each dimension are possible. We usually select between one view or the other to define a particular dimension to keep the drawing clear and uncluttered..

5.5.1 Dimensioning Holes and Cylinders

To demonstrate techniques for dimensioning either holes or cylindrical boundaries, let's add dimensions to the drawing shown in Fig. 5.10. Again start with the front view and locate the centerline of the hole as 3.20 from the left edge of the block. A hole is always located by a dimension from a suitable edge to its centerline. Use centerlines for dimensioning because the drill employed to make the hole is inserted into the component at the location of the center line.

We choose not to dimension the width in the front view because we prefer to specify the radius of the curved (cylindrical) boundary in the top view. Examine the top view and note the radius of the curved boundary is specified as 1.20 R. The R is added to the dimension to indicate that it is given in terms of the radius and not the diameter. In specifying the radius R, we have in effect defined the width of the block as the sum of 1.20 + 3.20 = 4.40. To have specified the total width of the block in the front view would have been over-dimensioning.

We show the location of the step from the left edge of the block as 2.00 in the top view. The height dimensions of the features are given in the side view. We have indicated the total height of the block as 2.50 and the height of the step as 1.00 from the bottom edge of the block. Note, we have dimensioned the diameter of the hole in the side view as 0.75 DIA. We dimension holes in terms of their diameter because the drill sizes used in manufacturing the holes are given in terms of diameter and not radius.

Have we completed the dimensions as shown in Fig. 5.12? You might question if we have specified the depth of the block since it has not been indicated in the side view. However, we specified the depth with the radius of the curved boundary in the top view. The depth is equal to two times the radius R or 2.40.

Again we have not shown units in the dimensioning. The units are defined in the drawing block along with other information pertaining to the component. The use of two standards for units, the U. S. Customary and the SI systems cause some difficulty in preparing and reading drawings. Some drawings are prepared using U. S. Customary units of in. or ft, and others are prepared using the SI system with units given in mm, cm or m. Sometimes, when the drawing is to be used by many people from different countries, both systems are employed in the dimensioning. An example of dual dimensioning is given in Fig. 5.13. In this case we do not break the dimension line because the U. S. Customary unit is placed above the line and the SI unit is placed below the line and enclosed with parentheses.

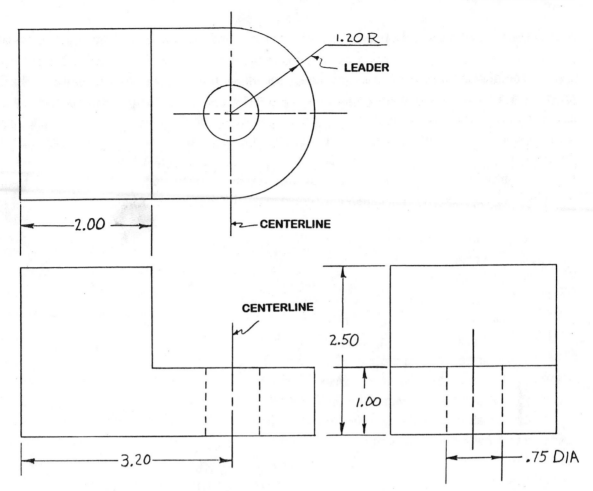

Fig. 5.12 Dimensioning example showing centerlines, leader, and radius and diameter indicators.

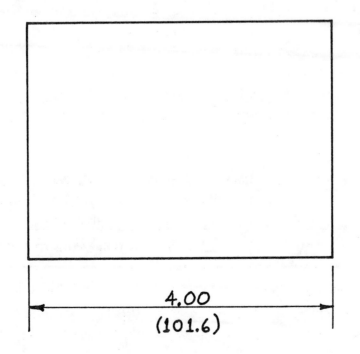

Fig. 5.13 Illustration of a convention used for dual dimensioning.

5.6 DRAWING BLOCKS

The drawing block serves a very important function on an engineering drawing. In the previous section, we referred to the fact that the units for the dimensions were specified in the drawing block. The unit of measurement used in preparing the drawing is only one of the many facts presented in the drawing block. As illustrated in Fig. 5.14, the drawing block is located in the lower right hand corner adjacent to the border of the drawing. A typical drawing block conveys important information pertaining to all drawings produced by a certain company and to the individual drawing. Information commonly shown in the drawing block includes:

1. The name of the company issuing the drawing.
2. The name of the part that the drawing defines.
3. The scale used in preparing the drawing.
4. Tolerances to be employed in manufacturing the part.
5. Date of the completion or the release of the drawing.
6. Material to be used in manufacturing the part.
7. Heat treatment of the part after manufacturing.
8. Units of measurement to be used in manufacturing the part.
9. Initials of the individual preparing the drawing.
10. Initials of the individual checking the drawing.
11. A unique drawing number to identify the drawing.

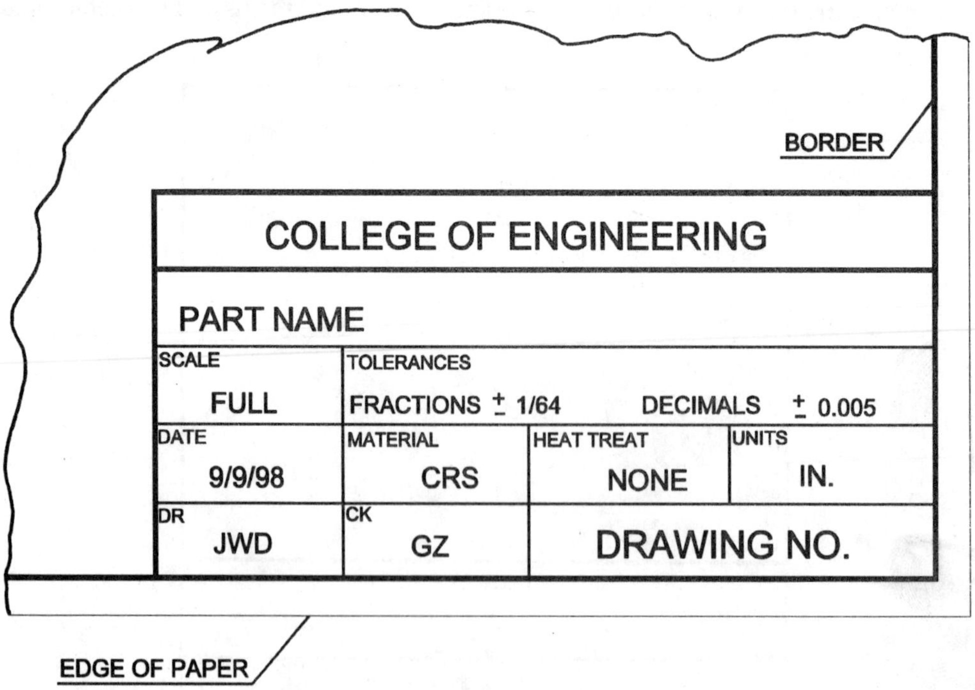

Fig. 5.14 An example of the information presented in a drawing block.

Let's examine each of these items. The name of the company is self evident. It is illustrated in Fig. 5.14 with the College of Engineering. In designing a part, identify it with some name such as bracket, support, shaft, etc. and include this name on your drawing. The part name should be brief (no more than two or three words).

Next you will indicate the scale used in preparing the drawing. The example shown in Fig. 5.14 indicates that we used full scale in drawing the three-views of the part. The scale that we select depends on the size of the part and the size of the paper that is used for the drawing. In a typical college class, we are constrained to standard size copy paper (8-1/2 by 11 inches) because of size limitations imposed by available laser printers. Suppose the part that you are designing is 12 inches wide, 6 inches high and 1 inch deep. Will the three-view drawing of this part fit on this standard size paper? The answer to this question is a clear NO. We need to scale down the part dimensions on the drawing so that it will fit on a single sheet of paper. In this case you would probably use ¼ scale. With ¼ scale the front view of the part would be 3 inches wide by 1 ½ inches high. You would dimension the drawing using the actual sizes of the part (i.e. you would show the width as 12 inches although it is only 3 inches on the scaled down drawing). You indicate to those reading the drawing that you have reduced the size of each view by a scaling factor of ¼.

Tolerances that are to be employed in manufacturing the part are specified in the drawing block. Many companies have standardized their tolerances and these limits are preprinted in drawing blocks that are incorporated on the company's drawing paper. The date of completion of the drawing is also shown. In some instances, the drawing release date is used instead of the completion date. The release date identifies when the design engineers turn over the ownership of the design of a product to the operations (production) function which is responsible for producing the product.

The material from which the part is to be fabricated is often identified in the drawing block. In the example shown, the abbreviation CRS indicates that cold rolled steel will be used in manufacturing the component. Sometimes the heat treatment of the component, if required, is identified in the drawing block. Heat treatment is a multi-step process employed to enhance the strength and hardness of a component. Usually heat treatment follows the machining of a part; we indicate whether or not heat treatment is required to assist those in charge of operations in controlling the flow of parts in the production process.

The units used in the drawing are given in the drawing block. We can use U. S. Customary units, SI units, or dual units. We cannot use mixed units. (Note the difference between mixed and dual units). U. S. Customary units are in terms of inch (in.) or foot (ft). SI units are in terms of millimeter (mm) or meter (m).

Those responsible for the drawing are identified. The individual who has prepared the drawing signs the block with his or her initials. The drawing is checked for completeness and accuracy, and the individual checking the details also initials the drawing block. Finally, the drawing is given a number. In a large company, which may release thousands of drawings each year, the drawing numbers are controlled by the engineering records department. This department maintains a numbering system which insures that each part or component has a unique drawing number. The engineering records department also organizes the numbering system to group together all of the drawings needed to build a specific product.

5.7 ADDITIONAL VIEWS

Some of the parts that we design are complex and addition views may be needed to ensure that the drawings are properly interpreted. We are not restricted to three-view drawings. For very simple parts, we can adequately describe the object with one or two views. There is no need to waste time drawing additional views. For complex parts that are more difficult to visualize, we can provide additional views to clarify the drawing. These extra views can show either external surfaces or internal sections. If it helps to visualize the object, we can add back, bottom and left side views to the more commonly employed front, top and right side views. To draw these additional external views, we simply rearrange the layout of the views on the drawing as shown in Fig. 5.15. This drawing, which shows five views of a multiply tapered block, illustrates the proper arrangement to show additional external views. It also maintains the advantages of orthographic projection. Clearly, the extra views help us visualize the complex shape of this block.

An addition internal (sectional) view is often more useful in clarifying a drawing than an additional external view. The concept of a sectional view requires us to mentally slice the object into two pieces with a cutting plane. We then open the part and view the internal section revealed by the cut. An example of the sectioning technique is illustrated in Fig. 5.16 where Fig. 5.8 has been modified by adding a sectional view. On the front view drawing in Fig. 5.16, a very heavy dashed line is shown to indicate the location of the cut. The cut line is turned through 90° at both ends and arrowheads are drawn to indicate the direction of the view. The arrowheads are each labeled with the letter A. In this case, we are viewing the section cut from the right side. Accordingly, we expect the section view to closely resemble the right side view. We then draw the section that is revealed by the cut.

The section view may be placed at any convenient location on the drawing. We have placed it in the open space in the upper right hand corner of the drawing. The section view is usually encircled with a wavy line to distinguish it from the usual three-views. We also match the section view with the cut by using the letters A-A for identification. In some drawings, we might make several cuts each revealing a different section with each sectional view identified by A-A, B-B, C-C, etc. In the sectional view, we show the view revealed by the cut with cross hatching. The remaining part of the view (outside the cut section) is drawn without cross hatching. Prepare a section view of the component shown in Fig. 5.12. Make the section cut along the centerline through the hole in the top view, and then examine the section from the right side.

5.8 SUMMARY

We have introduced engineering graphics by describing techniques for preparing multi-view drawings. Multi-view (usually three-view) drawings are used to communicate the relative proportions of a component and the exact size of all of its features. The drawing techniques are covered in detail with a few simple examples. We hope that you will invest the time necessary to learn these techniques because learning to draw and read three-view drawings is an essential means of communicating in engineering.

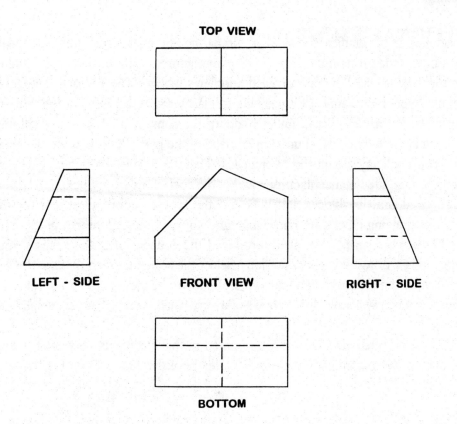

Fig. 5.15 A five view drawing of a complex block.

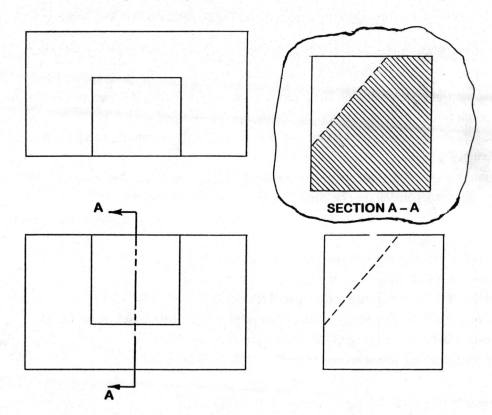

Fig. 5.16 A section cut and the corresponding section view.

Dimensioning is a significant aspect in preparing an error free drawing. All of the dimensions must be included so that the part can be manufactured from the drawing. Often the part is manufactured in another city, or state or even another country. The person manufacturing the part does not know you and cannot ask questions to clarify the ambiguities in the drawing. The drawing and all of its dimensions must completely define the component. An important implication of the dimensioning is the tolerances required in fabricating the part. As you write the numbers for the dimensions, you implicitly assign the tolerances. There is a great difference between 3, and 3.000 in the precision required in manufacturing and in the cost of the component.

We introduced drawing blocks and indicated the type of information that they normally convey. Remember drawing blocks are not unique and you will see differences in the information presented by different companies. We also introduced the concept of scale in this section. The scale used in preparing a drawing is your decision. Select the scale factor so that the three-views fit the paper without crowding. Sometimes you will scale down the view drawings so they will fit on the sheet. Other times you will scale up the views of a very small part so that very small features are clearly represented..

The coverage of three-view drawings presented in this chapter is very brief. If you need additional information, we recommend the complete textbook by James Earle [1] for an in depth discussion.

REFERENCES

1. Earle, J. H., Engineering Design Graphics, 4th edition, Addison Wesley, Reading, MA 1983.

EXERCISES

1. Prepare a three-view drawing similar to the one shown in Fig. 5.2 except change the width of the notch from 1 inch to 1 ½ inch.
2. Take a piece of clay or foam plastic and use a razor knife to manufacture a block with the shape shown in Fig. 5.6.
3. Prepare a three-view drawing of a block with a taper notch like that shown in Fig. 5.8. Select the dimensions yourself, but be consistent from one view to another with these dimensions.
4. Prepare a three-view drawing similar to that shown in Fig. 5.10 except increase the diameter of the hole to 1 ¼ inch.
5. Dimension the drawing that you prepared for exercise 1.
6. Dimension the drawing that you prepared for exercise 3.
7. Dimension the drawing that you prepared for exercise 4.
8. Design a drawing block that your team can employ with their drawings of the scale.
9. Prepare a five view drawing of the object shown in Fig. 5.9.
10. Prepare a drawing with a section view for the block defined in Fig. 5.1.

CHAPTER 6

PICTORIAL DRAWINGS

1.1 INTRODUCTION

Pictorial drawings are three-dimensional illustrations of a machine component, structure, or an object. For a person trying to visualize some object, the pictorial drawing is the most effective means to convey its size and shape. Some people have difficulty placing the three standard (front, top and side) views together to "see" the object. Pictorials drawings assemble the three-views on a single sheet providing a three-dimensional rendering facilitating visualization. Because pictorials are so easy to visualize, they are often used for catalogs, maintenance manuals, and assembly instructions.

Three different types of pictorials are in common usage:

- Isometric
- Oblique
- Perspective

A simple cube with three different types of pictorials is illustrated in Fig. 6.1. The isometric pictorial is drawn with its three axes spaced 120° apart. The term isometric means "equal measurement" indicating that the three sides are all scaled by the same factor relative to their true length. Parallel lines defining edges on the object are also parallel on the isometric drawing. Drawing paper with isometric axes is available in well stocked office and drafting supply stores. We encourage you to use it because it greatly facilitates the preparation of an isometric pictorial.

Oblique pictorials are drawn with the front view in the x-y plane. We project oblique lines, which represent the z axis, at some angle—often 45°. However, the angle used for the oblique lines can vary from 0 to 90°. Parallel lines defining edges on the object are also parallel on the oblique drawing. If the true length of the lines is employed in scaling all three sides, it is known as a cavalier oblique pictorial. The cavalier oblique style is frequently used, but the resulting pictorial is distorted. The distortion is due to the depth dimension which appears to be too long.

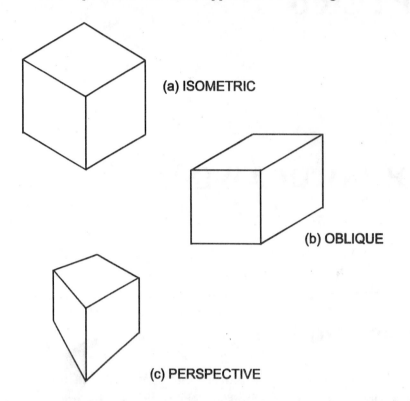

(a) ISOMETRIC

(b) OBLIQUE

(c) PERSPECTIVE

Fig. 6.1 A cube represented with isometric, cavalier oblique, and perspective pictorials.

The perspective is a pictorial drawing that represents what we actually see. Artists usually draw or paint using the perspective style. Engineers sometimes represent their designs in this style; however, this is the most difficult of the three types of pictorials to master. In perspective drawing, we do not have a well defined coordinate system. Parallel lines tend to converge to a vanishing point as they recede from the observer. Scales used on the different axes are different in order to foreshorten the lines located some distance from the picture plane. The use of converging lines instead of parallel lines and the foreshortening of select dimensions give the drawing perspective. The prospective drawing looks like a photograph of an object.

6.2 ISOMETRIC DRAWINGS

In this discussion of the three forms of pictorial drawings, we will introduce the axes used to frame the drawing, the direction of viewing the three-dimensional object, and the dimensions used for the width, height and depth. Let's begin with the isometric pictorial, shown in Fig. 6.2, that employs axes which make 120° with each other. The axes divide the paper into the three zones utilized to present three-views. If the axes form the letter Y with the vertical line oriented downward from the two branches, we are looking downward at the object. From this perspective, we visualize the top view in the region between the branches of the Y as shown in Fig. 6.2. The front view is displayed in the region to the right of the vertical axis, and the left-side view is located in the region to the left of this axis.

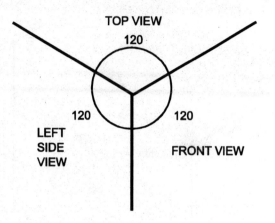

Fig. 6.2 Isometric axes at 120° angles give three regions for the front, top and left-side views.

Okay! We have defined the axes and the direction of viewing the object in Fig. 6.2. Let's discuss the dimensions used on the drawing for the width, height and depth. The word isometric means equal measurement; hence, it is clear that the same scale is used along all three axes. To illustrate the techniques used to prepare isometric drawings, consider a simple rectangular block with width W, height H, and depth D. We have prepared an isometric pictorial of this rectangular block in Fig. 6.3. To draw an isometric pictorial of a rectangular block:

1. Use a 30° triangle to draw the isometric axes identified with the numbers 1, 2, and 3 as shown in Fig. 6.3.
2. Using a scale, measure a length of H down the vertical and establish point A.
3. From point A, draw two additional lines (numbers 4 and 5) parallel to lines number 1 and 2.
4. Along line 5, measure the width W locating point B. Similarly measure the depth D along line 4 to locate point C.
5. From points B and C, draw the vertical lines 6 and 7 that intersect lines 1 and 2, and position points E and F.
6. From point F, draw line 8 parallel to line 2, and from point E draw line 9 parallel to line 1.
7. Lines 8 and 9 intersect at point G to complete the isometric drawing.

The isometric pictorial that we have drawn shows the left-side view, the top view, and the front view because we are viewing the object from above looking from the left to the right. The origin of the isometric coordinates is positioned at the upper left hand corner of the rectangular block.

The procedure for preparing isometric drawings is easy to implement; the lines are all parallel to the isometric axes, and the measurements are all to the same scale.

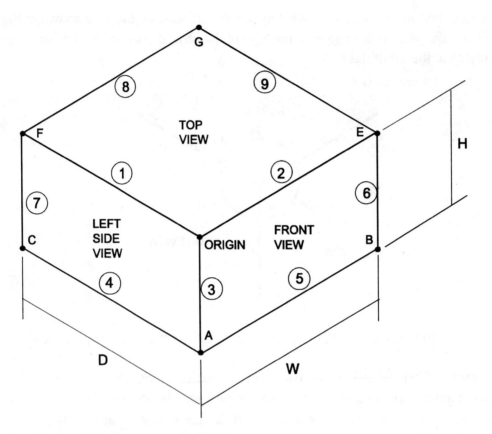

Fig. 6.3 Isometric pictorial of a rectangular block.

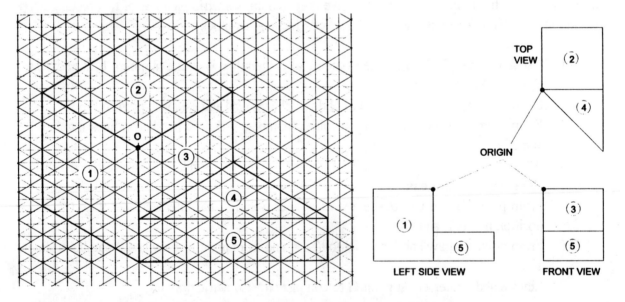

Fig. 6.4 Isometric pictorial of a block with a step and a tapered side.

The example of a rectangular block was too easy. Let's try a more complex geometry as shown in Fig. 6.4. Examine the three-view drawing of an odd shaped block illustrated in Fig. 6.4. It is an unusual three-view drawing since it presents the left-side view instead of the more conventional right-side view. We present the three-view drawing in this manner because the isometric drawing shows the same three-views. For the simple rectangular block, we placed the origin of the isometric axes at the upper left hand corner of the block (front view). For the more complex geometry, we again place the origin at the same location as indicated in Fig. 6.4.

The isometric pictorial, presented in Fig. 6.4, is drawn on isometric paper with only a straight edge to guide the lines. If you have a steady hand, you can sketch the lines without the straight edge. Isometric paper gives many evenly spaced lines parallel to the isometric axes eliminating the need for the 30° triangle and the scale. We begin at point B and draw plane 1 in the region of the left view. We do not draw the entire left-side view, because there is a taper to the block that complicates this view in the isometric representation. We leave the left-side view incomplete, and move to the top view and draw plane 2. Both planes 1 and 2 are clearly defined in the three-view drawing and are easy to construct on the isometric pictorial. Again we do not complete the top view because the step and the taper complicate the geometry. We move next to the front view and add plane 3 which defines the depth of the step. We now return to the top view and draw plane 4 as shown in Fig. 6.4. Now that the top view is complete in the isometric drawing, and the height of the step is established, it is easy to draw plane 5. Note, plane 5 is on the surface formed by the taper; it does not lie in a plane formed by the isometric axes. We locate this non-isometric plane by first drawing the four well defined isometric planes on the pictorial. The location of plane 5 is then clearly established by the position of planes 1 and 4.

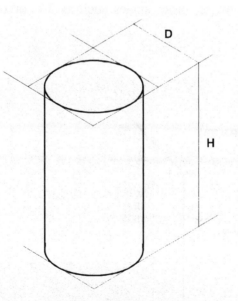

Fig. 6.5 Isometric pictorial of a right circular cylinder of diameter D and height H.

You now understand the procedure for drawing isometric pictorials of simple rectangular blocks and more complex blocks with both a step and a taper. Let's next consider a cylinder of diameter D and height H and illustrate it with an isometric pictorial. To draw the cylinder, lay out two isometric axes located a distance H apart using light construction lines as indicated in Fig. 6.5. On the upper set of axes, draw a square in the top plane with the length of the sides of the square equal to the diameter of the cylinder. Then select an isometric ellipse (35° - 16') from an ellipse template, and draw an isometric ellipse so that it is tangent to each side of the square drawn in the top plane. (If you do not have one of these handy templates, sketch the ellipse in by hand). Note that the ellipse is tangent to the isometric axes at four points as indicated in Fig. 6.5. Move down to the second set of isometric axes which represent the bottom plane. Take the handy template and draw isometric ellipse again. This time draw only the front half of the ellipse, since that portion is all that is visible when we view the cylinder from above. The two ellipses are joined with vertical lines at their outer most points to complete the cylinder. The construction lines in Fig. 6.5 remain to aid in understanding the procedure used in preparing this drawing. To finish, erase the construction lines and shade the cylinder to enhance the visual effect of a three-dimensional object. Shading and shadows will be discussed later in this chapter when we refer to this isometric drawing of a cylinder again.

6.3 OBLIQUE DRAWINGS

Oblique pictorials and isometric pictorials are similar, because both use parallel lines in constructing the three views. The difference between isometric and oblique pictorials is in the definition of the axes. In oblique drawings, we use a x, y, z coordinate system as shown in Fig. 6.6. The three coordinate axes divide the sheet into three regions for drawing the front, top and right-side views. With the axes defined as shown in Fig. 6.6, we are viewing the object from above and observing from right to left. The z axis, which is the receding axis in Fig. 6.6, is drawn with a 45° angle relative to the x axis; however, other angles such as 30° or 60° are often employed to represent the receding axis.

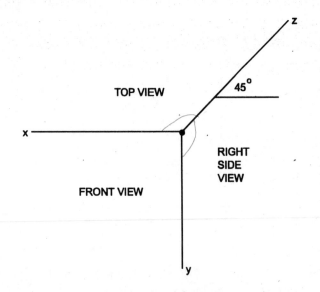

Fig. 6.6 Oblique axes use the x, y, z coordinate system.

Three different types of oblique drawings are frequently used for pictorials as illustrated in Fig. 6.7:

1. Cavalier oblique is drawn with the receding axis at any angle from 0 to 90°, but the measurements along all three axes are the same scale.
2. Cabinet oblique is drawn with the receding axis at any angle from 0 to 90°, but the measurements along this axis are half-scale.
3. General oblique is drawn with the receding axis at any angle from 0 to 90°, but the measurements along all this axis varies from half to full-scale.

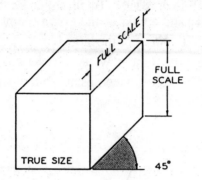

CAVALIER OBLIQUE

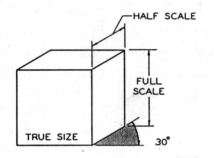

CABINET OBLIQUE

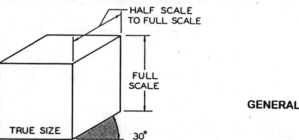

GENERAL OBLIQUE

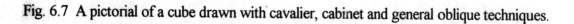

Fig. 6.7 A pictorial of a cube drawn with cavalier, cabinet and general oblique techniques.

If we examine the cube represented by the three types of oblique drawings in Fig. 6.7, it is evident that full-scale (true length) measurements are used in the front view in all three types of oblique pictorials. The difference among them is the scale used along the receding axis. In cavalier oblique, the full-scale measurements are made along the receding axis; however, this scale produces a drawing that is out of proportion. The cube does not look like a cube.

In cabinet oblique, half-scale measurements are made along the receding axis. The resulting drawing is in better proportion than the cavalier oblique, but sometimes it appears that the depth dimension along the receding axis is too short. We prefer the general oblique where the measurement on the receding axis can be varied from half-scale to full-scale. The sale is adjusted between these limits to give what appears to the eye to be the correct proportions.

To illustrate the procedure followed in drawing an oblique pictorial, examine the three-view drawing of a pair of the connected rectangular blocks as shown in Fig. 6.8. To begin an oblique pictorial, draw the x, y, z axes using the lower left hand corner as the origin (point O). First, draw the part of the front view (plane 1) corresponding to the front of the large block. Next, draw part of the top view (plane 2) of the large block. In drawing the front view, use true lengths (on plane 1) to lay-out the width and height of the large block. On the top view, establish the depth of the large block, maintaining proportion by using a scaling factor of ¾.

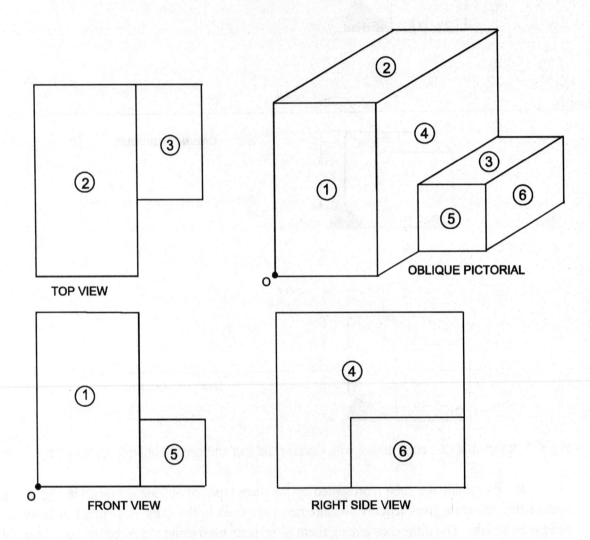

Fig. 6.8 A three-view drawing of a connected pair of rectangular blocks together with an oblique pictorial of the same object.

Note, the right-side view of the large block is obstructed by the smaller block. We handle this obstruction by drawing the small block. Using the information in the three-view drawing, we can place plane 3 on the oblique pictorial. Plane 3 is located by working from the back edge in the top view. The width and height measurements required to locate and size plane 3 in the oblique pictorial are true lengths, but the depth is again scaled by ¾.

After we have drawn plane 3 locating the small block in the pictorial, it is easy to draw plane 4 by referring to the right-side view in the three-view drawing. Complete the drawing of the small block by dropping vertical lines down from the three corners in plane 3 and by closing the sides that form planes 5 and 6.

The resulting oblique pictorial clearly captures the relative proportions and positioning of the two blocks. A comparison of the pictorial with the three-view drawing demonstrates the advantages of the pictorial in visualizing the object—the pictorial is much more effective. The three-view drawing is employed for the precise definition of size and location of all of the features of a component or structure. We dimension the three-view drawing and use it in the shop for manufacturing. Usually the pictorial is not dimensioned and is not used as a substitute for a detailed three-view drawing.

For two final examples of preparing oblique pictorials, consider the drawing of a cylinder or a block with a circular hole as shown in Fig. 6.9. It is easy to draw a cylinder or a circle on an oblique pictorial providing the required circles are placed on either the front plane or any plane parallel to the front plane. Since both the width and height dimensions are true length and the x, y axes are orthogonal on the front view, the circle is not distorted into an ellipse. We can draw the circle with a compass or a circle template. This is a significant advantage.

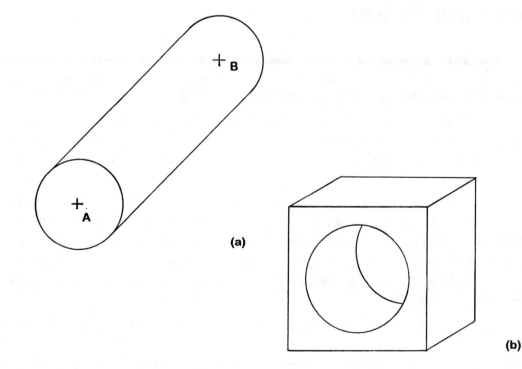

(a)

(b)

Fig. 6.9 Examples of oblique pictorials.
(a) Right circular cylinder. (b) Rectangular block with circular hole.

To represent the cylinder shown in Fig. 6.9a, we draw two circles centered at points A and B. Both points A and B lie on the z (receding) axis, which has been drawn at a 45° angle to the x axis. The circles are drawn in with construction lines with full-scale diameters. The spacing of the two circles along the z axis is ¾ scale relative to the length of the cylinder. Draw two lines parallel to the z axis tangent to the circles for the sides of the cylinder. Draw the front circle with a heavy line, and darken the visible portion of the back circle. The resulting pictorial shows a somewhat distorted cylinder, The distortion is due to the fact that we have drawn the back circle with the same diameter as the front circle. When observing an actual right circular cylinder, the back circle would "appear" to be smaller than the front circle. We will address the apparent difference in the size of features located in different planes in the next section on perspective drawing.

The final example of oblique pictorial, presented in Fig. 6.9b, shows a simple rectangular block with a central hole. We have drawn the block so that the circle of the hole is see in its entirety in the front view. The question is how to draw the circle in the back plane? We first draw the square locating the back plane in the pictorial with light construction lines. We then find the center of this square and draw the second circle with the same diameter as the front circle. Note, the two circles intersect and only a portion of the circle on the back plane is visible. Darken the construction lines on the back circle over that part of the arc that is visible through the hole. The resulting pictorial is an effective three-dimensional drawing showing the appearance of a hole through a rectangular block. Our artist friends would critique the drawing, because it too distorts the visual image of the object. The scale in the back plane and the front plane are the same producing distortion of the image. This distortion is usually acceptable in engineering drawing, but we can improve our pictorials by preparing perspective drawings.

6.4 PERSPECTIVE DRAWINGS

Perspective drawings are pictorials that represent observations made with our eyes or a camera. This method of illustration is critical to the success of an artist making either sketches or paintings. Engineers also use perspective drawings particularly when preparing visuals for those who are not trained to read our more conventional three-view drawings. The tools used are the same as those described previously except for adding a thumb tack and a piece of string to the list.

There is one very significant difference between perspective drawings and isometric or oblique drawings. In both isometric and oblique drawings, the lines defining the edges are parallel to the axes; however, in perspective drawing, the lines defining some of the edges are not parallel. Drawing parallel lines distorts the drawing, because parallel lines appearing to converge as they recede in space. The best illustration of this fact is a pair of railroad tracks, shown in Fig. 6.10. Looking down the tracks, you see that the tracks converge to a point. Also the railroad ties and the trees and poles lining the track appear to become shorter. Everyone knows that the tracks are parallel. What's going on?

To understand what is happening, it is necessary to define four terms used in describing prospective drawing.

1. A picture plane is the surface (i. e. the sheet of paper) of the pictorial. The edges of the paper represent the window through which you "see" a three-dimensional object that is the subject of the perspective drawing.

2. A horizon line divides the sky and the land (or the sea) if you are outdoors. The horizon line is at the level of your eyes and will change with your elevation. In a room where the true horizon cannot be located, because the walls of the block our view, we assume a horizon line at the elevation of our eyes.

3. A viewing point and direction of view depends on the location of eyes relative to the object. You can look directly at the object, from left to right, right to left, downward, upward, etc. What you see changes markedly depending on these parameters. Look at an object from a window, and change where you stand and the direction of your view. Does the view of the object change?

4. A vanishing point is where parallel lines converge to a point as they recede into the distance. You can clearly identify the vanishing point in Fig. 6.10 where the tracks appear to meet.

Fig. 6.10 A photograph of railroad tracks showing the visual effect of converging parallel lines and foreshortening of objects in a distance.

6.4.1 One Point Perspective Drawings

Depending on the view, an object can be represented with a one, two, or three-point perspective. Let's start with the simple one-point perspective and illustrate the approach by drawing a simple rectangular block. In one-point perspective, you place the front view in the picture plane and show its true width and height as illustrated in Fig. 6.11. Then a construction line is drawn to represent the horizon. The location of this line depends on the viewing point and the viewing direction. In Fig. 6.11, you are viewing the block straight on (not from the right or the left); however, you are above the block. Your eyes look downward and observe the top surface of the block. The elevation of your eyes relative to the top of the block is taken into account by raising the horizon line. Next, locate the vanishing point on the horizon line at the center point behind the front view, because we are looking straight on at the block. Draw construction lines from the vanishing point to the top corners of the block in the front view as shown in Fig. 6.11. The back edge on the top view is drawn parallel to the front top edge to establish the depth of the block. Note that the back edge is much shorter in length than the front edge. The shortening of the lines on the recessed planes give an illusion of the third dimension. The edges at the side are darkened to complete the pictorial. Again these edges are converging giving an illusion of depth on the two dimensional sheet of drawing paper.

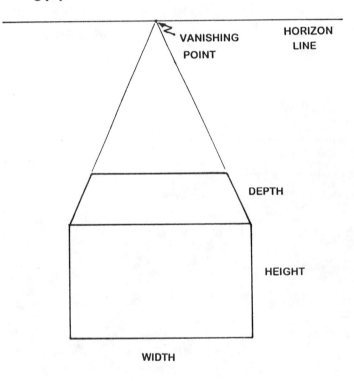
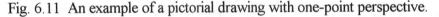

Fig. 6.11 An example of a pictorial drawing with one-point perspective.

Another example of one-point perspective is the drawing of a coffee table presented in Fig. 6.12. Again the front of the table is drawn to scale in the picture plane. The width and height dimensions are true length. Coffee tables are low, so when we stand it is necessary to look downward to view the table. In this example, we are centered relative to the table and looking directly at its top surface. To develop a perspective view with this direction of viewing, draw a horizon line at eye level aligning the vanishing point with the center of the table. Next, draw light construction lines from the top outside corners of the table to the vanishing point. We also draw construction lines from the lower inside corners of the table legs as shown in Fig. 6.12. These construction lines define two triangles. We use the larger of these two triangles to provide the length of the back top edge of the table. Next, we draw the edges of the table top to complete our perspective rendering of the top view. The smaller of the two triangles is used to guide us in drawing the bottom edges of the legs which are visible under the table.

In a normal perspective drawing, we would erase the construction lines, the vanishing point and the horizon line. We have not erased them in Fig. 6.12, because they were used to illustrate the procedure followed in making a one-point perspective drawing.

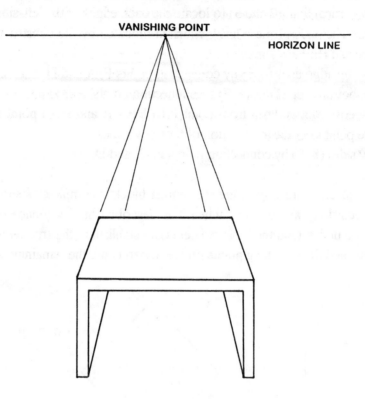

VANISHING POINT

HORIZON LINE

Fig. 6.12 A one-point perspective drawing of a coffee table.

6.4.2 Two-point Perspective Drawings

The one-point perspective drawing is useful when we view an object straight-on with its front view shown in the picture plane. However, if the object is rotated so that neither the front nor side view is in the picture plane, as illustrated in Fig. 6.13, a two-point perspective is required. Again consider a rectangular block to illustrate the approach followed in drawing a two-point perspective pictorial. From Fig. 6.13, it is evident that only one vertical edge of the block lies in the picture plane. We will start with this fact and proceed step by step to illustrate the block with a two-point perspective drawing.

- Draw a vertical line, as indicated in Fig. 6.14, to establish the edge between left-side view and the front view.
- Draw the horizon line to reflect the fact that we are standing above the block looking downward onto the top surface.
- Place two vanishing points on the horizon line. We have spaced the vanishing point located to the right of the vertical line (VP - R) farther from the vertical line than the

vanishing point on the left-side (VP - L), because we are viewing the block at a slight angle from the right toward the left.

- Measure the true length of the vertical line (line 1) and label its ends with the letters A and B.
- Draw four construction lines connecting points A and B with VP - R and VP - L.
- Draw vertical lines (2 and 3) to locate the back edges of the left-side and the front of the block. The ends of these lines, which are located by the construction lines, are labeled (C, D) and (E, F).
- Draw top edge lines (4, 5) by connecting points F, A and D.
- Draw bottom edge lines (6, 7) by connecting points E, B and C.
- Draw construction lines from point F to VP - R and from point D to VP – L. Then locate point G at the intersection of these two lines.
- Draw sides (8, 9) by connecting points F, G, and D.

The two-point perspective drawing of the rectangular block is complete as shown in Fig. 6.14. The procedure may seem long and tedious, when it is described in a sequence of steps, but it is not difficult. When you understand the basic concepts of establishing the true length of the vertical line in the picture plane and the vanishing points on the horizon line, the remaining steps are routine.

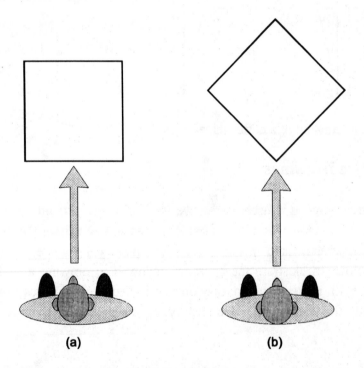

(a)   (b)

Fig. 6.13 Directions of viewing control the number of vanishing points used in perspective drawing.
    a.   Straight-on—one-point perspective.
    b.   Angle view—two-point perspective.

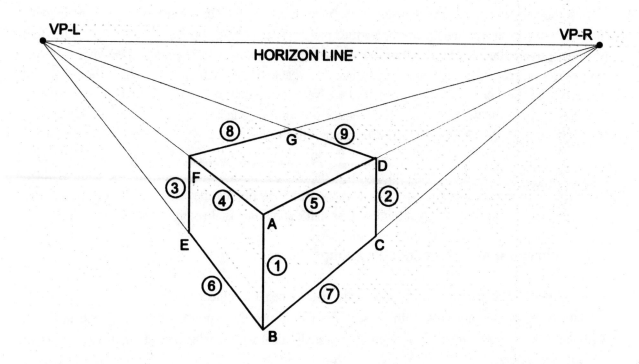

Fig. 6.14 A rectangular block represented with a two-point perspective drawing.

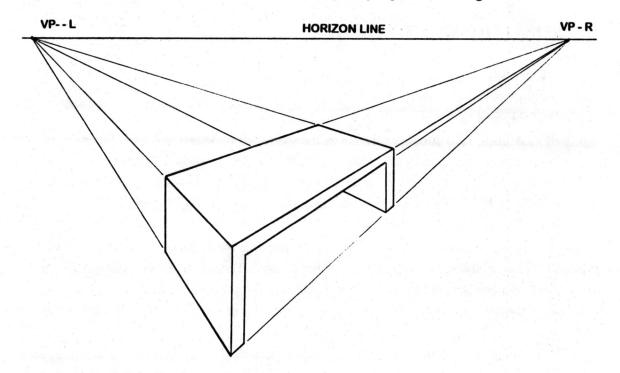

Fig. 6.15 A two-point perspective drawing of a coffee table.

To complete the discussion of the two-point perspective, let's again consider the coffee table. This time, we will use a two-point perspective drawing to illustrate the table. Begin by placing the vertical edge of the table leg in the picture plane as indicated in Fig. 6.15. The edge of the leg is shown in true length on the picture plane. Establish the horizon line and the vanishing points to provide the direction of the view that you are showing. Draw construction lines from the ends of the vertical line to the vanishing points. Vertical lines are then drawn to show the width and the depth of the table. The position of these vertical lines is not established by measurement, because the both the width and depth dimensions are not true length in this two-point perspective. Place the vertical lines in a position that maintains, "to the eye," the correct proportions of the table.

When the vertical lines that bound the left-side view and the front view are drawn, the remainder of the drawing is easy complete. Construction lines are drawn from the ends of these vertical lines to the vanishing points. The construction lines outline the table. It is only necessary to darken those segments of the construction lines that define the visible edges of the table.

6.4.3 Three-point Perspective Drawings

Three-point perspective drawings are used when the object being illustrated is very tall. Architects drawing a city view with skyscrapers would use three-point perspectives and taper the building as it extends into the sky. Engineers usually deal with smaller objects that usually can be represented in pictorials with either one or two-point perspective drawings. For this reason, we will not describe the methods used in preparing three-point perspectives. However, if you are interested in learning much more about perspective drawing, we recommend the excellent book by Powell [2].

6.5 SHADING AND SHADOWS

Add shading and shadows to our pictorial drawings makes them appear more realistic. Let's first distinguish between shading and shadowing. If you light the object, some surfaces are exposed to this light and other surfaces are in the shade. There is a difference in the intensity of light reflected from these surfaces. Those in the shade are darkened slightly. When an opaque object blocks the light, a shadow is produce. The shadow occurring on the floor plane is shown as a dark area in the drawing. We show an example of shading and shadowing of a right circular cylinder in Fig. 6.16. In this example, parallel light rays are illuminating the cylinder from the upper left. (They are included on the drawing only to show the logic for determining the shade and shadow regions). The right half of the cylinder is in the shade and is darkened slightly. A shadow is cast on the plane of the floor upon which the cylinder rests. The depth of the shadow is the same as the depth of the cylinder on the isometric pictorial. The length of the shadow is dependent on the direction of the parallel rays of light. Extending the lines representing the light rays defines the edge of the shadow.

Let's consider another example of shading and shadowing in a two-point perspective drawing of a cube. The drawing, presented in Fig. 6.17, is similar to that shown in Fig. 6.14, so we will assume that you can draw the cube in the two-point perspective pictorial.. In Fig. 6.17, the construction technique is demonstrated for determining the exact size of the shadow when the light is coming from a point source.

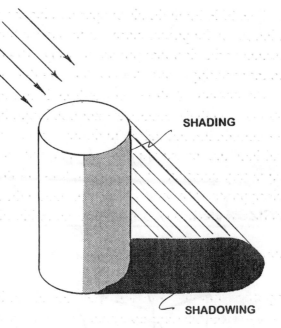

SHADING

SHADOWING

Fig. 6.16 Shading and shadowing of an isometric pictorial of a right circular cylinder.

Suppose you have completed the two-point perspective drawing of the cube and are ready to shade and shadow the drawing. The process is first to select the location of the light source. It is your choice, but you place the light source well above the horizon either to the left or the right of the cube. Second, you must select the vanishing point for the shadow. Again it is your choice as long as it is directly below the light source and in the ground (floor) plane. The reasons for these two constraints are evident. The shadow must vanish when the light source is directly overhead, and the shadow must always lie on the ground (floor) plane. In fig. 6.17, we selected the vanishing point of the shadow on the horizon line; however, it could have been placed anywhere on the ground plane directly under the light source.

Let's consider a step by step procedure for locating the shadow outline shown in Fig. 6.17:

1. Draw the light rays (1, 2, and 3) from the source through three of the top-forward corners of the cube.
2. Draw two construction lines (4 and 5) from the vanishing point of the shadow through the two bottom-side corners.
3. Locate points A and B at the intersections of the lines (1 and 6) and (3 and 7).
4. Draw lines 6 and 7 from the left and right vanishing points through points A and B.
5. The intersection of lines (6 and 7) locates point C.
6. Darken segments of the construction lines (8, 9, 10 and 11) to give the shadow outline.
7. Erase all of the construction lines and darken the shadow region within the outline.

We have completed the shadow, but the front and left-side views of the cube are in the shade. To complete the pictorial, these two sides should be darkened slightly indicating they are in a shadow.

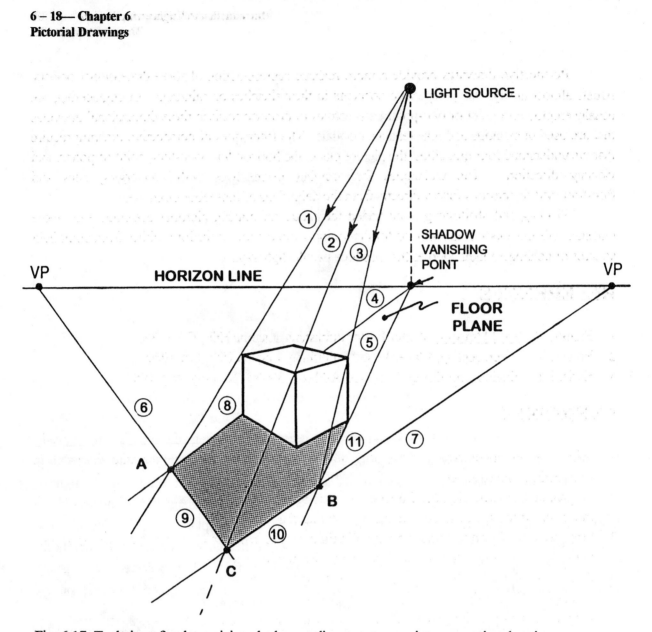

Fig. 6.17 Technique for determining shadow outline on a two-point perspective drawing.

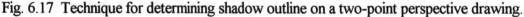

6.6 SUMMARY

Three different pictorial drawings—isometric, oblique, and perspective—have been introduced. Both isometric and oblique pictorials are widely used in engineering to present three-dimension illustrations on a two-dimensional format (a sheet of drawing paper or a computer monitor). The isometric and oblique pictorials are easy to draw because lines are parallel, and most, if not all, of the measurements are to scale. However, exact scaling of dimensions and parallel lines tend to distort these three-dimensional illustrations. The eye or camera "sees" dimensions to different scales, and parallel lines as parallel only if the view depicted is on the picture plane. If the view is on any other plane behind the picture plane, the dimensions are foreshortened. If the parallel lines are on views with receding axes, they tend to converge. We ignore these distortions in engineering drawings when we prepare isometric or oblique pictorials.

Perspective drawings provide a more realistic representation of three-dimensional objects. Artists almost always use perspective concepts in their sketches or paintings. In engineering, we usually employ one-point or two-point perspectives to produce realistic three-dimensional drawings that are used in catalogs and maintenance manuals. The techniques of perspective drawing require that we understand four quantities: the picture plane, the horizon line, vanishing point or points, and viewing direction. The techniques for drawing perspectives with converging lines and foreshortened dimensions follow directly from the definitions of these four quantities.

Shading and shadowing is an added technique for making pictorial drawings even more realistic. We have indicated methods for determining the outline of shadows either from point light sources or collimated light sources that produce parallel light rays.

REFERENCES

1. Franks, G. <u>Pencil Drawing</u>, Walter Foster Publishing, Laguna Hills, CA 1988.
2. Powell, W. F. <u>Perspective</u>, Walter Foster Publishing, Laguna Hills, CA 1989.
3. Earle, J. H., <u>Engineering Design Graphics</u>, Addison Wesley, Reading, MA 1983.

EXERCISES

1. What are the three types of pictorial drawings? Why do we use pictorial drawings in engineering applications?
2. Prepare an isometric pictorial drawing of a rectangular block 10 mm wide by 15 mm deep by 50 mm high. What scale should you use in preparing this drawing? Why?
3. Prepare an isometric pictorial drawing of a right circular cylinder 15 mm in diameter by 35 mm long. What scale should you use in preparing this drawing? Why?
4. Prepare an oblique pictorial drawing of a rectangular block 20 mm wide by 25 mm deep by 70 mm high. What scale should you use in preparing this drawing? Why?
5. Prepare an oblique pictorial drawing of a right circular cylinder 25 mm in diameter by 45 mm long. What scale should you use in preparing this drawing? Why?
6. Prepare an isometric pictorial drawing of rectangular block with a centrally located circular through hole. You are to state the dimensions of the block and the diameter and orientation of the hole prior to preparing the drawing.
7. Prepare an oblique pictorial drawing of rectangular block with a centrally located circular through hole. You are to state the dimensions of the block and the diameter and orientation of the hole prior to preparing the drawing.
8. Prepare an isometric pictorial drawing of the postal scale that your team is developing.
9. Prepare an oblique pictorial drawing of the postal scale that your team is developing.
10. Prepare a prospective pictorial drawing of the J. M. Patterson building. View the building from the intersection of the two frontage streets.
11. Draw a pictorial illustration of a sphere with shading and shadowing.

CHAPTER 7

TABLES AND GRAPHS

7.1 INTRODUCTION

On many occasions we find ourselves with a collection of numerical data that we must present at a design briefing or in an engineering report. At other times, we may have extensive results from a mathematical relationship that must expressed in a meaningful format. We have two choices in presenting numerical data. We can show the numbers in tabular form or on a graph. Both methods of presentation, the table or the graph, have advantages and disadvantages. The method that you select will depend on your audience, the message, and the purpose of the presentation.

In this chapter, we will describe the technique for arranging data in a table using Microsoft's Word 97 as the word processing program. In Chapter 9, we describe data entry into a spreadsheet which is also a convenient method for preparing tables. We then introduce five common methods for representing data in the form of graphs which include:

1. Pie charts
2. Bar charts
3. X - Y graphs.
4. Semi-log graphs
5. Log - log graphs

Each type of graph is used for a different purpose and selection of the correct type is essential in communicating effectively. For instance, the pie chart is used to show distributions—who gets the most and the least. The bar chart is used to compare one set of data with another. X – Y graphs, the most frequently used type of graphing in engineering, show how Y varies with X. When Y varies by very large amounts with small changes in X, we often use semi-log graphs which are capable of covering a very large range of Y values. When both X and Y vary over very large ranges, we display the results graphically by using a log-log representation.

7.2 TABLES

In Chapter 1 we described the responsibilities of Mechanical Engineers in their first position and presented statistical data in a table. Let's reproduce this data in tabular format to show you the ease with which a table may be prepared in Word 97. We first prepare the table heading by selecting the center alignment, identifying the table number, and typing the title. We may choose a bold font for the table heading to attract the reader's attention.

TABLE 7.1 **Responsibilities of Mechanical Engineers in Their First Position**

Okay. We are now ready for the body of the table. With the mouse point to the menu row and click on "Table". From the drop-down menu, we have a choice of whether or not we want to show grid lines in the table. Let's click on the "Gridlines" icon and use grid lines in constructing the table. The gridlines aid us in entering the data and they are also helpful to the reader. If we don't like them they can be removed after the data is entered. Next, click on the "Insert Table" icon. A dialog box that appears on the screen, prompts us to enter information defining the table. It is necessary to understand the arrangement of the table (how many columns and rows). To represent the data for the responsibilities of mechanical engineers, we it is necessary to establish three columns and 15 rows in the arrangement of the table. The initial arrangement is not critical, because we can always add or delete either columns or rows later. Finally, we select auto-format in the dialog box and permit Word control the initial width of the columns. Click the OK button, and an outline of the table appears on the monitor. If the grid lines are light gray when the table first appears on the monitor, they will not print as they are intended only to guide your typing. If you prefer the grid lines to delineate the rows and columns when the table is printed, click on Format in the menu and select Borders and Shading. A dialog box appears on the screen and you click on the "Grid" button; then choose the style and width of the grid lines outlining each cell. The result of our first trial after we have typed the data into the correct cells is shown on the next page.

The result of our first attempt is fair, but it could be improved. The auto-format selection divided the width of the page into thirds for the column width. This partitioning is not the best choice considering the wasted space associated with the numerical entries and the double rows needed to accommodate the assignments. The first column, titled **ASSIGNMENT**, needs more width and the 2nd and 3rd columns are too wide. Let's modify the table by changing the column widths. We accomplish this by clicking on Table in the menu row and selecting Cell Height and Width from the pull down menu. From the dialog box which appears, select the column tab and adjust the width of the first column (3 in.). Next, click the Next Column button and adjust the width of both column 2 and 3 to 1 inch. The appearance of the table has improved and it requires less space on the page. Unfortunately, it is not centered. To center it, click Table from the menu row, click on Cell Height and Width, and select the Rows tab. From the Rows menu select center alignment and click the OK button. The improved table is shown below. As a result of our modifications, it looks very professional.

ASSIGNMENT	% TIME	% TIME
Design Engineering		40
Product Design	24	
Systems Design	9	
Equipment Design	7	
Plant Engineering/ Operations/ Maintenance		13
Quality Control/ Reliability/ Standards		12
Production Engineering		12
Sales Engineering		5
Management		4
Engineering	3	
Corporate	1	
Computer Applications/ Systems		4
Basic Research and Development		3
Other Activities		7

TABLE 7.1
Responsibilities of Mechanical Engineers in Their First Position

ASSIGNMENT	% TIME	% TIME
Design Engineering		40
Product Design	24	
Systems Design	9	
Equipment Design	7	
Plant Engineering/ Operations/ Maintenance		13
Quality Control/ Reliability/ Standards		12
Production Engineering		12
Sales Engineering		5
Management		4
Engineering	3	
Corporate	1	
Computer Applications/ Systems Analysis		4
Basic Research and Development		3
Other Activities		7

Tables are most effective in presenting precise numerical data. We do not need to crudely estimate a number by reading from a curve on an X-Y graph. If we have a reason to show our results with six significant figures, it is very easy to do so. A government agency that probably produces the most widely read tables in the United States is the Internal Revenue Service. They prepare tax tables each year that precisely show our tax obligations. We will present another example later that shows the advantage of a table in conveying numerical data with many significant figures. We will also show the application of spreadsheets to produce both graphs and tables in another Chapter 9.

7.3 GRAPHS

We use graphs to visualize the data in both reports and presentations. The graphs are not intended to give precise results since we use tables for that purpose. The graphs show our audience a trend, a comparison, or a distribution—quickly and effectively. There is no need for the audience to study the data to develop an understanding. Your graph shows the bottom line and eliminates the time and effort required for the reader to analyze the data.

There are many different ways of presenting numerical data in a graphical format. The distribution of goodies is usually shown with pie charts, because we all quickly relate to getting a good size share of the pie. Comparisons among several quantities are made with bar charts with the height of each bar indicative of the magnitude of the quantity being compared. Linear graphs, where we plot a dependent parameter Y as a some function of an independent parameter X, is frequently employed in engineering presentations. The data plotted can be generated with mathematical functions when we know the function Y(X). However, if the mathematical relation is not known, we conduct experiments and measure Y as we systematically change X. In both cases, the X-Y graphs represent the numerical data and show a trend—increasing, neutral or flat, decreasing, oscillating, etc. In some instances, the variations in X, Y or both of these quantities is extremely large; therefore, it is not possible to clearly show the trends over the entire range of either X or Y on a graph with linear scaling. In these cases, we use a non-linear format, namely semi-log or log-log to present the large ranges in the data. Let's consider five types of graphs and learn manual techniques for preparing them. Later in Chapter 9, we will show how a spreadsheet program is used to prepare these same graphs.

7.3.1 Pie Charts

A pie chart is used to show how some quantity is distributed. In Table 7.1, we showed the various assignments for mechanical engineers starting their careers. Let's represent this same data in a pie chart as shown in Fig. 7.1. We begin by drawing a circle to represent the pie, and then we decide what fraction of the pie corresponds to each assignment. This determination is easy if we recall that the whole pie contains an included angle of 360°. The slice of pie representing design engineering is a fraction of the whole pie, i. e. (360°)(0.40) = 144°. The size of the pie slices in terms of degrees for all of the engineering assignments are listed below:

1. Design engineering $\quad\quad\quad\quad\quad\quad\quad\quad$ $(360°)(0.40) = 144°$.
2. Plant engineering, operations and maintenance $\quad$ $(360°)(0.13) = 47°$.
3. Quality Control, reliability and standards $\quad\quad$ $(360)(0.121) = 43°$.
4. Production engineering $\quad\quad\quad\quad\quad\quad\quad$ $(360)(0.121) = 43°$.
5. Sales engineering $\quad\quad\quad\quad\quad\quad\quad\quad\quad$ $(360)(0.050) = 18°$.
6. Management $\quad\quad\quad\quad\quad\quad\quad\quad\quad\quad$ $(360)(0.040) = 14°$.
7. Computer applications and systems analysis $\quad$ $(360)(0.040) = 14°$.
8. Basic research and development $\quad\quad\quad\quad$ $(360)(0.030) = 11°$.
9. Other activities $\quad\quad\quad\quad\quad\quad\quad\quad\quad\quad$ $(360)(0.070) = 25°$.

We construct the pie chart by using a protractor to lay out these angles on the circle. The pie slices are then labeled as shown in Fig. 7.1. When examining the pie chart, we can visualize the importance of design, because the design slice is huge. We can also see the importance of producing products since these activities—plant operations, quality control, and production—combine to produce still another huge slice. The remainder of the pie, which is less that a quarter, is divided between, sales, management, computers and systems, research and development, and other activities. Of course, these activities are important, but they represent opportunities for a much smaller fraction of the engineers beginning their careers. Compare the effectiveness of the data as presented in Table 7.1 and the pie chart of Fig. 7.1. The pie chart, with a visual representation, quickly leads the reader to the conclusion that design and production are where most opportunities exist for entry level positions in mechanical engineering.

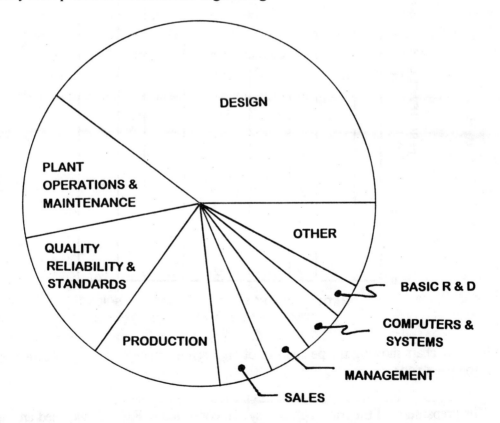

Fig. 7.1 A pie chart showing initial assignments in industry for beginning engineers.

7.3.2 Bar Charts

Bar charts can also be used to show distribution, but they are better suited for illustrating comparisons. As an example of a bar chart, consider the results of a study by the Council of Chief State School Officers on the percentage of high school students graduating with three or more years of secondary mathematics and science in 1982 and 1994. We have prepared a bar chart showing the comparison in Fig. 7.2. The visual effect enhances the comparison. At a glance, we can see that many more students were taking mathematics and science in 1994 than in 1982. The second glance indicates that only about a third of the students were in math and science in 1982 while more than half of them were taking math and science in 1994. Finally, we quickly observe that math was slightly more popular than science in both 1982 and 1994. Three glances and the reader understands the data, because you have presented it in a visual format that effectively carries the message.

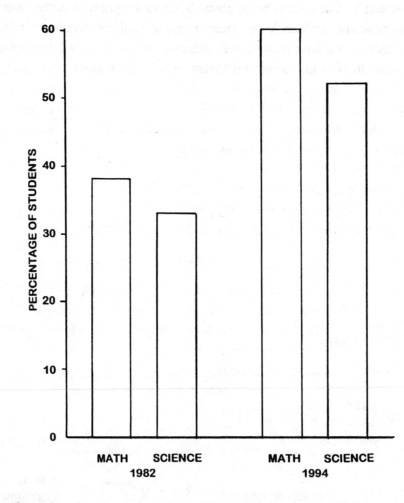

Fig. 7.2 A bar chart showing the percentage of high school students taking math and science in 1982 and 1994.

The preparation of the bar graph is easy. In constructing Fig. 7.2, we used an engineering paper with a barely visible grid that facilitated establishing a suitable scale along the ordinate. Only a straight edge was employed in drawing the bars; the grid lines visible on the engineering paper were used to scale each bar. The time required to prepare the bar chart was about 10 minutes.

7.3.3 Linear X-Y Graphs

X-Y graphs are probably the most frequently employed graph used in engineering because it illustrates trends. The curves are produced by plotting Y, the dependent variable, along the ordinate as a function of X, the independent variable, which is displayed along the abscissa. Connecting the points with a curve or line segments indicates the trend in Y with changes in X. Graphing X-Y can be accomplished using linear scales for both X-Y, a linear scale for X and a log scale for Y, or log scales for both X and Y. We will cover all three of these methods of producing X-Y graphs.

Since numerical results will be necessary for our examples of X-Y graphs, we will generate our numerical data using a very important equation that relates to money, and how you may choose to accumulate or spend it. Let's begin by assuming that you are going to accumulate some money. Sounds good! Suppose your rich uncle purchased a mutual fund for you at your birth, and plans to give you the proceeds of the fund on your 21st birthday. How much will you have accumulated when the big day arrives? The sum S accumulated is given by the following relation:

$$S = P(1 + i)^n \qquad\qquad (7.1)$$

where

S is the sum accumulated.
P is the amount of the initial investment.
i is the interest rate for the compounding period.
n is the number of periods over which the interest accumulates.

Suppose your rich uncle invested P = $5000 in a mutual fund that guaranteed an annual interest rate I = 10 %, and the interest is compounded semi-annually. We will also assume that uncle avoided the tax collectors[1] (Federal, State and Local). Let's calculate exactly how much money you can expect to collect on your 21st birthday. We need to compute S in Eq. (7.1) for P = $5000. The number of compounding periods that will occur from your birth until you are 21 years of age is 42, because the fund compounds the interest earned twice each year. The interest rate I = 10% on an annual basis, but the interest rate i is only 5% per the semi-annual compounding period. Substitute these numbers into Eq. (1) to obtain:

$$S = \$5000(1 + 0.05)^{42} = \$5000(7.761587) = \$38,807.94$$

Wow! This is wonderful. You can buy a very nice car on the good uncle, or defer this option and continue to compound interest accumulating an even larger sum. We have evaluated Eq. 7.1 with a spreadsheet program (Excel) for a total of 80 periods. The results are shown in Table 7.2.

---

[1] Avoidance of taxes is not possible, but the assumption simplifies the analysis.

In examining Table 7.2, we find that the uncle's principal of $5,000 has grown over the years. The table gives accurate values at the end of each compounding period. For example, at the end of the 20th period, your tenth birthday, the principal in the mutual fund was $13,266.49. Note that the table conveys the exact numerical value with as many significant figures as required. If you maintain the fund until you are 40 years old, accumulating interest for 80 periods, you would have a total of $247, 807.21. Quite a growth from the $5,000 seed-money planted by uncle.

Table 7.2					
Accumuated Sum from an Investment of $5000.00					
Interest Rate of 10% Compounded Semi-annually.					
No Taxes Paid on Interest Earned					
Periods	Multiplier	Sum	Periods	Multiplier	Sum
0	1.0000	$ 5,000.00	40	7.0400	$ 35,199.94
1	1.0500	$ 5,250.00	41	7.3920	$ 36,959.94
2	1.1025	$ 5,512.50	42	7.7616	$ 38,807.94
3	1.1576	$ 5,788.13	43	8.1497	$ 40,748.33
4	1.2155	$ 6,077.53	44	8.5572	$ 42,785.75
5	1.2763	$ 6,381.41	45	8.9850	$ 44,925.04
6	1.3401	$ 6,700.48	46	9.4343	$ 47,171.29
7	1.4071	$ 7,035.50	47	9.9060	$ 49,529.86
8	1.4775	$ 7,387.28	48	10.4013	$ 52,006.35
9	1.5513	$ 7,756.64	49	10.9213	$ 54,606.67
10	1.6289	$ 8,144.47	50	11.4674	$ 57,337.00
11	1.7103	$ 8,551.70	51	12.0408	$ 60,203.85
12	1.7959	$ 8,979.28	52	12.6428	$ 63,214.04
13	1.8856	$ 9,428.25	53	13.2749	$ 66,374.74
14	1.9799	$ 9,899.66	54	13.9387	$ 69,693.48
15	2.0789	$ 10,394.64	55	14.6356	$ 73,178.15
16	2.1829	$ 10,914.37	56	15.3674	$ 76,837.06
17	2.2920	$ 11,460.09	57	16.1358	$ 80,678.92
18	2.4066	$ 12,033.10	58	16.9426	$ 84,712.86
19	2.5270	$ 12,634.75	59	17.7897	$ 88,948.50
20	2.6533	$ 13,266.49	60	18.6792	$ 93,395.93
21	2.7860	$ 13,929.81	61	19.6131	$ 98,065.73
22	2.9253	$ 14,626.30	62	20.5938	$ 102,969.01
23	3.0715	$ 15,357.62	63	21.6235	$ 108,117.46
24	3.2251	$ 16,125.50	64	22.7047	$ 113,523.34
25	3.3864	$ 16,931.77	65	23.8399	$ 119,199.50
26	3.5557	$ 17,778.36	66	25.0319	$ 125,159.48
27	3.7335	$ 18,667.28	67	26.2835	$ 131,417.45
28	3.9201	$ 19,600.65	68	27.5977	$ 137,988.32
29	4.1161	$ 20,580.68	69	28.9775	$ 144,887.74
30	4.3219	$ 21,609.71	70	30.4264	$ 152,132.13
31	4.5380	$ 22,690.20	71	31.9477	$ 159,738.73
32	4.7649	$ 23,824.71	72	33.5451	$ 167,725.67
33	5.0032	$ 25,015.94	73	35.2224	$ 176,111.95
34	5.2533	$ 26,266.74	74	36.9835	$ 184,917.55
35	5.5160	$ 27,580.08	75	38.8327	$ 194,163.43
36	5.7918	$ 28,959.08	76	40.7743	$ 203,871.60
37	6.0814	$ 30,407.03	77	42.8130	$ 214,065.18
38	6.3855	$ 31,927.39	78	44.9537	$ 224,768.44
39	6.7048	$ 33,523.76	79	47.2014	$ 236,006.86
40	7.0400	$ 35,199.94	80	49.5614	$ 247,807.21

Let's represent the data shown in Table 7.2, as an X-Y graph. We first decide that the sum S is the dependent variable represented along the ordinate (the Y axis), and that your age is the independent parameter displayed along the abscissa (the X axis). We draw the X and Y axes on our engineering paper (which has a faint grid) and apply a scale to each axis, as shown in Fig. 7.3. Scaling is very important because it determines the size of our graph. If the scale is too small, the graph looks like a postage stamp and you have wasted an opportunity to show the graph to full advantage. However, if the scale is too large, you can not plot all of the data on the paper. To scale the graph correctly, examine the range of the data to be covered for both X and Y. In this case, the range for X varies from 0 to 21 and Y varies from $5,000 to $38,807.94. We have selected a scale of 1 inch equal to 5 years along the X axis, and a scale of 1 inch equal to $5000 along the Y axis. The range displayed for X was from 0 to 25, and for Y from 0 to $40,000. We add numbers to each axis, beginning with zero at the origin and increment in steps of 5 along both axes. Captions are added to both axes to remind the reader of the definitions of X and Y. This scaling results in the graph shown in Fig. 7.3; it is approximately 5 by 8 inches in size, which fits nicely on a single page with sufficient space for the figure caption and the margins.

The data consisting of X and Y coordinates, shown in Table 7.2, locate the points that are plotted on the graph. We have used a drop compass to draw the small points, but the points can be place by free hand if you do not have this instrument. We have only plotted seven points to establish the curve. The number of points plotted depends on the smoothness of the function being graphed. In this case, Y varies monotonically with respect to X. We can draw the curve accurately with only six to eight points because the function is relatively smooth and without peaks or valleys.

Inspecting Fig. 7.3 clearly shows the trend of the accumulation sum S with respect to time (age). The sum increases continuously with time and the rate of the increase (the slope of the curve) is also increasing. The visual effect of the X-Y graph is dramatic. At a glance it is apparent that you are becoming rich. The only question is how rich? If we examine the ordinate and abscissa, we can estimate the sum accumulated for a given age, but the estimate may be in considerable error. The faint grid lines used in plotting the data do not show in the copy presented in Fig. 7.3. We lose accuracy in reading our graph without these grid lines.

If we seek to prepare an X-Y graph that is effective in showing trends, while retaining some accuracy in reading the scales, a good quality graph paper with a fine pitch grid should be employed. An example of this graph, plotted to the same scale on a graph paper with 20 grid lines to an inch, is shown in Fig. 7.4. With these fine pitch grid lines, we can read the scale to:

$$\pm (1/20)(5) = \pm 0.4 \text{ years of age}$$

$$\pm (1/20)(\$5,000) = \pm \$400 \text{ for the sum S.}$$

The resolution of the scale, possible with good quality graph paper, is a big improvement over the graph without visible gridlines. However, the best accuracy in reporting numerical data is obtained with a tabular format.

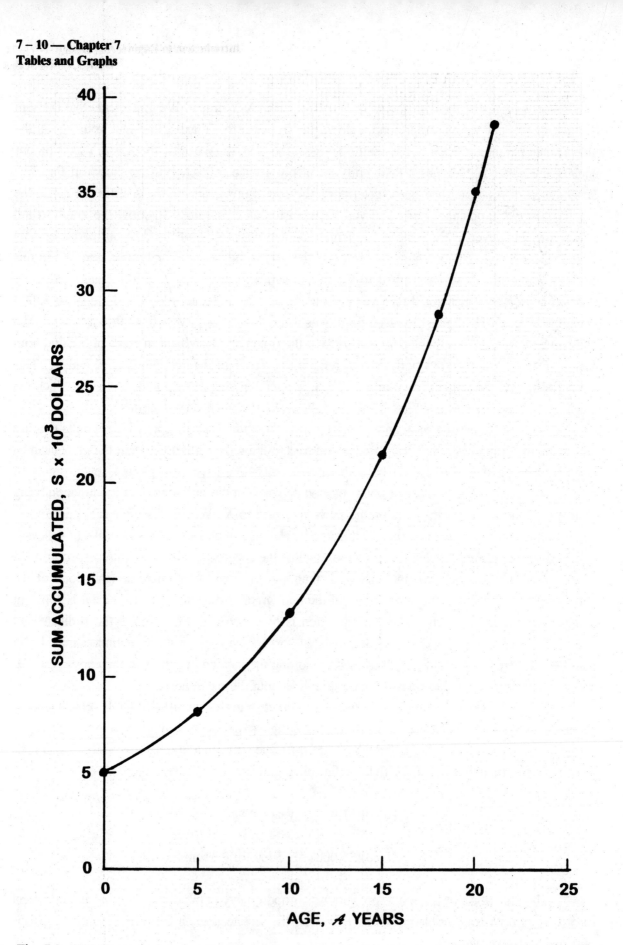

Fig. 7.3 A X-Y graph showing the sum accumulated with time from an initial investment of $5,000.

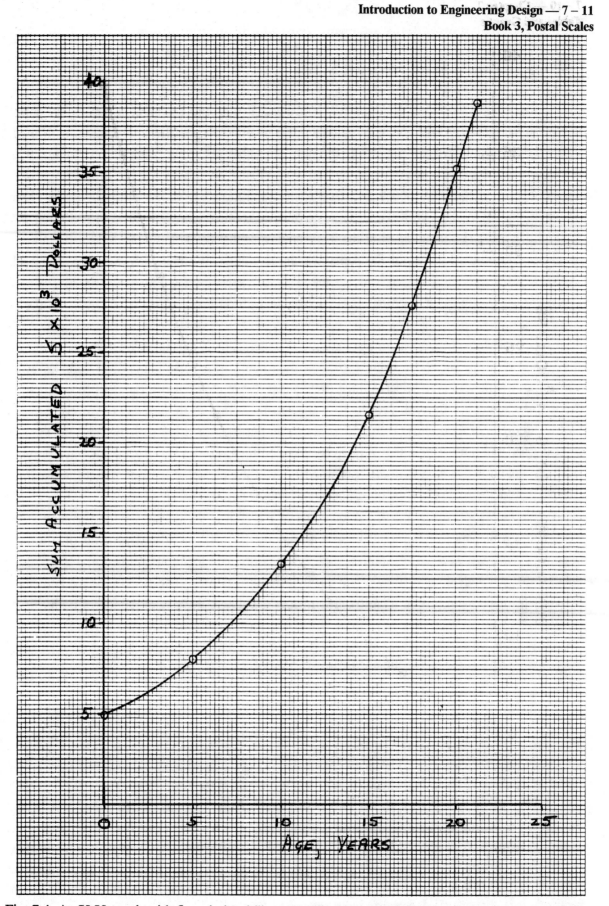

Fig. 7.4 An X-Y graph with fine pitch gridlines used for showing both trends and numerical values.

7.3.4 X-Y Graphs With Semi-log Scales

The results presented in Table 7.2, indicate that you can accumulate some very serious money from a relatively small initial investment if it is invested for a sufficiently long time. If we try to plot the larger numbers associated with the sum S in Table 7.2 on the linear graph of Fig. 7.4, we would go off-scale. We could re-scale, but we would lose resolution along the ordinate where the range in S is very large. A better approach is to use semi-log paper in preparing the graph. Semi-log paper, shown in Fig. 7.5, has a linear scale (in this example along the X axis) and a log scale (along the Y axis). Looking closer at the Y axis indicates that the graph paper is scaled with three log cycles over the 10 inches of the paper which are ruled. We have selected three log cycles because it gives the range needed to cover the range of values for the data presented in Table 7.2. Note semi-log graph paper is commercially available with different numbers of cycles on the log axis.

Let's scale both axes on our semi-log graph paper. The X axis is scaled with one inch equal to 10 years covering the range from 0 to 50 years. The Y axis scale is imposed upon us by the selection of the paper. (We selected three cycle paper so we may cover a range of three decades). The first decade is in thousands (10^3 to 10^4), the second in tens of thousands (10^4 to 10^5), and the third in hundreds of thousands (10^5 to 10^6). Zero can not exist on a log scale because the $\log_{10}(0)$ is not defined. In our graph in Fig. 7.5, we have placed the origin at Y = \$2000 and have provided space for a caption and margin at the bottom of the semi-log paper. It is evident that the maximum value of Y which we can show on our three cycle paper is \$1,000,000 or $\$10^6$.

We have labeled the Y scale at locations of 1 and 5 as designated on the log scale of the preprinted graph paper. The number is multiplied by 10^3, 10^4, or 10^5 depending on the cycle involved. Next, we place points on the graph at the coordinate locations defined in Table 7.2. In this example, we may draw a straight line through the data points. The reason for the linearity of the relation for the sum S on a semi-log graph paper is evident, if we take the log of both sides of Eq. (7.1) to obtain:

$$\log_{10}(S) = \log_{10}(5000)(1 + 0.05)^n$$

$$\log_{10}(S) = \log_{10}(5000) + [\log_{10}(1 + 0.05)]\, n \qquad (7.2)$$

Examination of the right hand side of Eq. (7.2) shows that $\log_{10}(S)$ is linear in the number of periods of accumulation (n). The intercept of the straight line with the Y axis (when n = 0) is \$5000. The slope of the straight line is $\log_{10}(1 + 0.05)$ which is the coefficient of n.

The use of semi-log paper has two advantages. First, it permits us to cover a very large range in the quantity represented along the Y (log) axis. We have used a three cycle paper which covered a range from 2000 to 1,000,000. If the need existed, we could cover even a larger range by selecting a semi-log paper with more log cycles. The second advantage of semi-log representation is that it converts certain non-linear (exponential) functions, such as Eq. (7.1), into linear relations that are much easier to interpret and extrapolate.

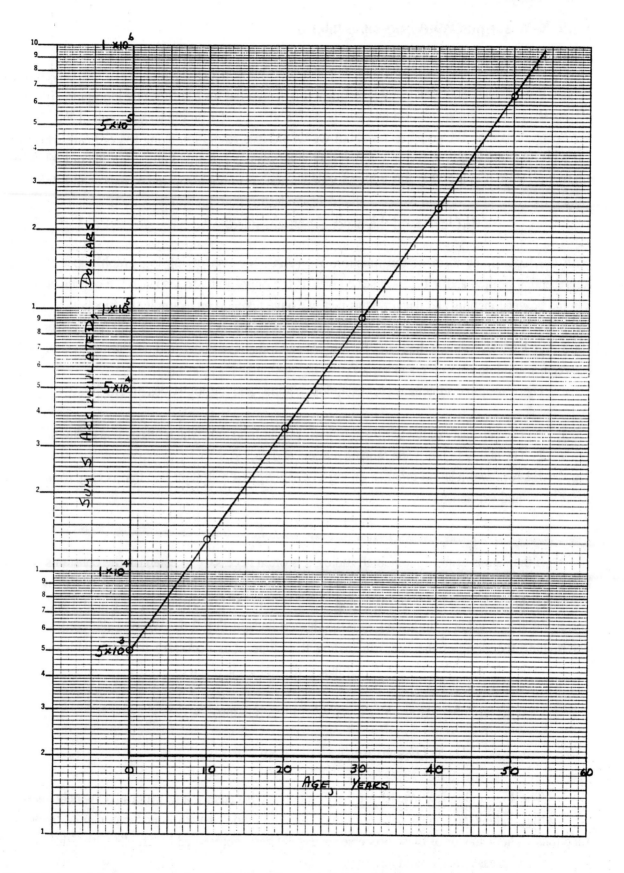

Fig. 7.5 A semi-log representation of the sum S accumulated with time.

7.3.5 X-Y Graphs With Log-log Scales

When you encounter power functions of the form:

$$Y = AX^k \qquad (7.3)$$

it is advantageous to represent both X and Y on log scales in preparing an X-Y graph. Taking the log of both sides of this power function, we obtain:

$$\log_{10}(Y) = \log_{10} A + \log_{10} X^k = \log_{10} A + k \log_{10} X \qquad (7.4)$$

With the logarithmic form of the function $Y = AX^k$, the use of log-log graph paper is ideal. Representing both X and Y on a log scale gives us the opportunity to visualize the constants A and k in the power function. To illustrate this fact, we have taken as an example the function $Y = 2X^{1.4}$ and described it graphically in Fig. 7.6. In this graph, X varies from 0.1 to 100 and Y from 1 to 1000. We have also shown the scales for $\log_{10} Y$ and $\log_{10} X$ in Fig. 7.6 to emphasize the difference between X and $\log_{10} X$ and Y and $\log_{10} Y$.

There are three advantages for using log relations and log-log scales on graphs:

1. The capability of linearizing non-linear power functions which often arise in engineering.
2. The ability to plot wide ranging numerical data.
3. A power relation graphed on log-log paper gives a straight line that is easy to interpret and extrapolate.

If the data is from experimental studies, it is often necessary to establish the constants A and k in a power relation like that shown in Eq. (7.3). The graphical procedure for determining these constants is given in Fig. 7.6. We find the constant A by setting X = 1 and reading the corresponding Y value as indicated with the dashed lines in Fig. 7.6. We find the constant k by determining the slope of the line. In finding this slope, it is necessary to work with the small numbers taken from the log scales. Accordingly, the slope k is given by:

$$k = (\log_{10} Y_2 - \log_{10} Y_1) / (\log_{10} X_2 - \log_{10} X_1)$$

$$k = (2.68 - 0.30)/(1.7 - 0) = 1.40. \qquad (7.5)$$

The log or log-log graphs are very important because many processes, widely used in engineering, exhibit non-linear behavior which may be modeled with either exponential or power functions. The behavior of these processes is usually best represented by graphs with either one or both scales expressed in terms of logarithms.

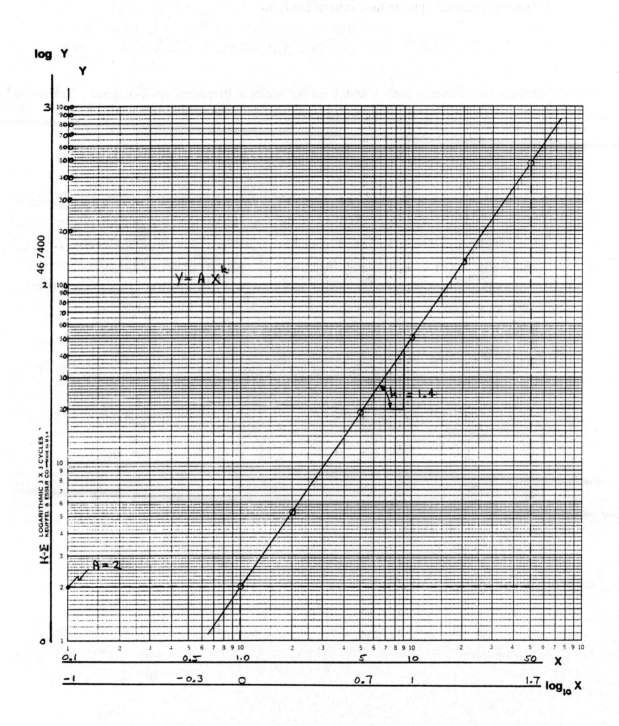

Fig. 7.6 A log-log graph of a power function $Y = AX^k$.

7.3.6 SPECIAL GRAPHS

There are many special graphs that have been developed to help the reader quickly absorb and understand numerical data. This brief chapter on Tables and Graphs does not provide a complete description of the many special and clever graphs that have been devised. However, we will introduce one of these, the geographical graph, which is based on a map of the world, the U. S. or some state. Superimposed on the map are contour lines dividing the entire geographical region into sub-regions. Either the contour lines or sub-regions are labeled to show the geographical distribution of some quantity across the mapped region. An example of a geographical graph, presented in Fig. 7.7, shows the weather prediction by the National Weather Service for the U. S. for Monday, May 25, 1998.

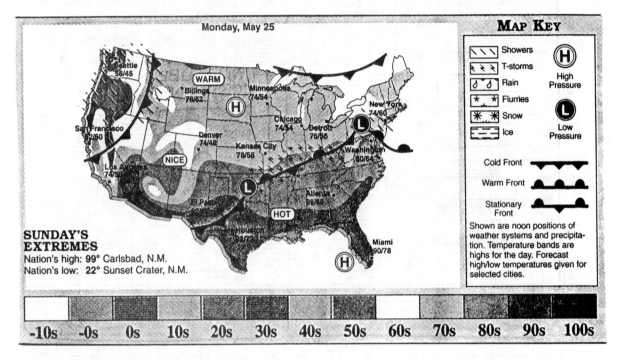

Fig. 7.7 A geographical graph showing the weather prediction for the U. S. for May 25, 1998[2].

7.4 SUMMARY

We have introduced tables as an approach for reporting data. They have the advantage in presenting precise numerical data. If you need six figure accuracy, it is easy to obtain in a tabular representation of the data. We also outlined the procedure for producing very tables presentable tables using Word 97.

Many different types of graphs are used to aid in visualizing data. Graphs are much less precise than tables, but they more effectively show trends, comparisons, and distributions. We first introduced the pie chart which, as the name implies, is a circular area that is divided into pie like

---

[2] The National Weather Service Forecasting Office provides daily weather maps of the U. S. on Web site http://www.nnic.noaa.gov/cgi-bin/page?pg=netcast.

segments to indicate distribution of some quantity. Techniques for determining the size of the slices of pie were described.

We then introduced the bar chart that is used to compare the magnitudes of two or more quantities. The bar chart was illustrated and the methods used in its construction were discussed.

Trends are best illustrated using line graphs including the linear, X-Y linear graphs, X-Y graphs with semi-log scales, and X-Y graphs with log-log scales. Linear X-Y graphs are used to show the trend of one variable with respect to another when the range of both variables is not too great. X-Y graphs are presented on semi-log graph paper when one variable has a very large range while the other variable has a more limited range. Log-log graph paper is employed to represent data fields where both variables change over very large ranges. We have provided examples for these three different types of graphs and have described techniques used in their construction.

Finally, we have discussed special graphs that make use of maps to convey the concept of geography in the presentation of data. The map presented to demonstrate this technique is the weather map. It shows the use of contour lines dividing the U. S. into temperature zones.

EXERCISES

1. Write a memo for a group of young engineers starting their careers with your company that explains the policy of the engineering department relative to reporting data with tables or with graphs.
2. Prepare a table using a word processing program that shows the SAT scores for entering freshman at your College of Engineering for the time period from 1980 to 1997. Report both the math and verbal scores for women and men. Also tabulate the number of women and men in each class. The Office of the Dean of Engineering should be able to provide you with the data that is needed.
3. Use the data from exercise 2 to construct a bar chart comparing the scores of women and men for the years 1987 and 1997.
4. Use the data from exercise 2 to construct two pie charts showing the percentage of men and women enrolled as freshmen in the College of Engineering in 1987 and 1997.
5. Prepare a table like that shown in Table 7.2 showing the sum S with time for the following investment parameters. Use P = $1000, I = 8%, with interest compounded quarterly. Determine the sum after each compounding period for a total of 20 years.
6. Prepare an X-Y graph showing the sum with respect to time using the data generated in exercise 5.
7. Prepare a semi-log graph showing the sum determined in exercise 5 on the log scale and the time on the linear scale.
8. Evaluate the power function $Y = 1.8 X^{1.75}$ and plot the result on a log-log graph. Let X vary from 1 to 100. Try to find log-log paper with a suitable number of cycles for both axes. Confirm that A = 1.8 and k = 1.75 from your graph.
9. Using the Internet print a copy today's weather map for the U. S.

PART III

SOFTWARE APPLICATIONS

CHAPTER 8

Pro/ENGINEER*

8.1 INTRODUCTION

Pro/ENGINEER is a high-end, commercially available, Computer-Aided Design (CAD) program. It is unlike other engineering graphics or drawing programs—like AutoCAD—that are intended to be employed for Computer-Aided Drawing and Drafting (CADD). Pro/ENGINEER is different from these CADD programs because it requires engineers to perform design by creating a computer representation of a solid model. Whereas, typical CADD programs require the user to begin with two-dimensional (2D) drawings, and then create a solid model if necessary. In Pro/E, we start with the solid model and **then** create the 2D drawings by viewing the solid model from different directions as shown in Fig. 8.1.

8.1.1 What is Pro/ENGINEER?

To answer this question, let's continue to differentiate Pro/ENGINEER from typical CADD programs by examining some terms and buzzwords that engineers in the Computer-Aided Engineering (CAE) industry use to describe Pro/E. The five major characteristics of Pro/E are that it is:

1. A Solid Modeler
2. A Feature Based Modeler
3. Parametric

---

* Release 18.0 Pro/ENGINEER, Parametric Technologies Corp., Waltham, MA.

4. Relational
5. Associative

To explain what is meant by these five characteristics, let's consider each individually.

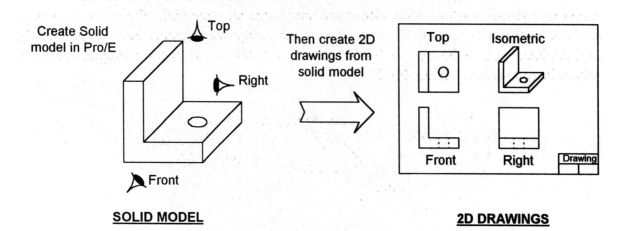

Fig. 8.1   Creation of 2D drawings of a Pro/E solid model.

8.1.1.1 A Solid Modeler

Pro/E does not use 2D projections to represent the part; instead, it requires the user to build the computer representation of the component using solid pieces for its description. This is the reason for identifying Pro/E as a **solid modeler**.

8.1.1.2 A Feature Based Modeler

Those programmers who developed Pro/ENGINEER approached the task from a manufacturing point-of-view. They knew the goal of any computer-aided design program was to ultimately manufacture the component modeled on a computer. In other words, designers do not draw these components for entertainment; their intent is to eventually produce the part being modeled. To capture every manufacturing feature, Pro/E requires the designer to use **features** to capture the **design intent** of the component. We illustrate the concept of **feature based modeling** in Fig. 8.2, where the designer mentally envisions a machine shop and performs the sequence of operations required to manufacture the block illustrated in Fig. 8.2. We recognize that every time a feature is added the value[1] of the component is increased. Features add value to the component and features capture design intent.

We begin the process of manufacturing the block like object of Fig. 8.2 by cutting a piece of metal to the correct size using a band saw. Cutting the block to the correct size is the base feature. The saw cut block has more value than previously when it existed as a plate of

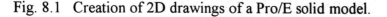

[1] The addition of features in the manufacturing process certainly increases the cost of the component. If the design is sound, these features enhance value to a greater extent than cost.

metal in the stockroom. The next machining operation adding value to the block is the slot feature. We take the rough cut block and go to a milling machine or use an end mill to cut the slot; thereby, adding value to the part. Finally, a drill is used to produce the hole feature adding still more value to the block.

When modeling (notice we **do not** say drawing) in Pro/E, think of having unlimited access to a virtual machine shop with every imaginable machine tool. Every time we use a machine tool, it adds value to the part; i.e., we can charge the customer more money for the final product. Approaching every design in Pro/E with this virtual machine shop in mind, will help us learn to effectively use Pro/E.

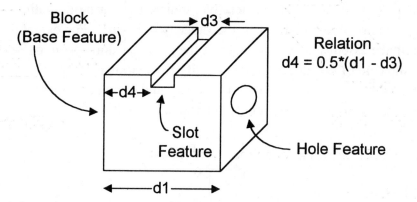

Fig. 8.2 Block base feature with a slot and hole features.

8.1.1.3 A Parametric Modeler

The company that developed and markets Pro/ENGINEER is **called Parametric Technology Corporation**—PTC for short. What does the word **parametric** mean? For Pro/E, it requires us to think of the dimensions placed on the sketch in generic terms; these dimensions are known as **parameters**. These dimension parameters save the designer time during the part's life cycle when the part may be used in a different application with somewhat different dimensions. Since the dimensions are defined **parametrically**, not numerically, the designer can quickly update the drawing by changing the numerical values of the dimensions. For example, consider a family of hex-head bolts. The design intent of the bolt is represented by a single Pro/E model. When a bolt of a different size is required, the designer recalls the model from a file listing and clicks on the dimension parameter requiring modification. By typing a new numerical value (say, change the length of the bolt from one to two inches), and the part is automatically updated. This parametric capability provides significant flexibility and saves time for the designer. It is one of the important advantages of a parametric modeling package such as Pro/E.

8.1.1.4 A Relational Modeler

Extending the power of parametrically defined dimensions is the use of **relations** when we define our part. A **relation** is illustrated on the component presented in Fig. 8.2.

The slot on top of the block is centered regardless of the dimension d1 or the dimension of the slot d3. The designer captures the design intent of the part by using the relation d4 = 0.5*(d1 – d3) to center the slot on top of the block. The d4 dimension "depends" on d1 and d3; hence, d4 is referred to as the "dependent variable" and d1 and d3 are the "independent variables".

8.1.1.5 An Associative System

The final powerful feature of Pro/ENGINEER is its ability to **associate** models regardless of the "mode of operation" in which Pro/E is operating. Pro/E operates in three different modes—parts are defined in the part-mode, assemblies in assembly-mode, and drawings in drawing-mode. Pro/E is **associative**, indicating that modifications of the model in one mode are automatically updated in the other two modes. For example, if we are in operating Pro/E in the assembly-mode and notice the diameter dimension, d1 of the peg is too large to fit into the hole of the block, we can modify the peg diameter in the assembly-mode. The dimension for the peg is corrected in both the part and drawing modes. This capability is known as **bi-directional associativity** and is illustrated in Fig 8.3.

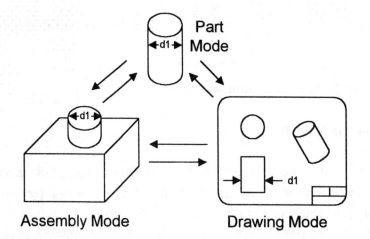

Fig. 8.3 Bi-directional associativity among the three operating modes of Pro/E

8.2 Pro/ENGINEER ENVIRONMENT

8.2.1 Screen Layout

You may start Pro/E from the desktop environment. Once Pro/E is running, and you retrieve a file for the part, a screen layout similar to that presented in Fig. 8.4 appears on the monitor. The screen is divided into four main areas:

1. Main Working Window - Graphics window. The active area is denoted by asterisks
   ****** Pro/ENGINEER Part: BRACKET ******

2. Menu Manager Window - Listing of Pro/E commands available for selection.

3. Message Window   - Messages prompt user for action. Examine the message window for guidance. It is located at the bottom of the screen, and often partially obscured by the main working window.

4. Model Tree Window   - Shows the construction tree of the part model. For the bracket shown, three reference planes (called datum planes 1, 2, and 3) were created, followed by the *base feature* which was the L-shape protrusion. The bracket was finished with an opening by using the slot feature.

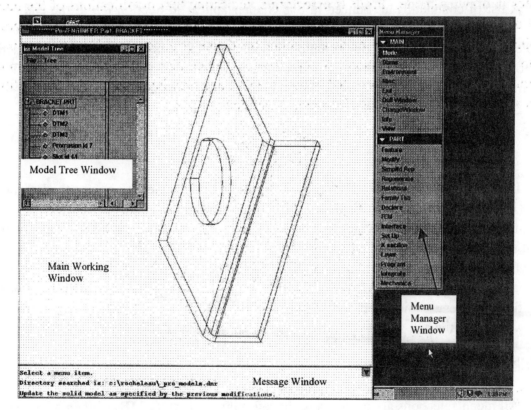

Fig. 8.4 Typical screen layout for Pro/E.

8.2.2 The Message Window

The *Message Window* is the first window examined whenever you encounter difficulty running Pro/ENG. This window is located at the bottom of the screen and is often obscured by the main working window. To observe the contents in the message window, click on its border with the left mouse button (the acronym **LMB** stands for **L**eft **M**ouse **B**utton throughout the chapter). Messages appearing in this window have the following purposes:

• One line help messages explaining the function of a menu item. These messages appear as yellow text in the Message Window.

- One line prompts providing suggestions for your next action. These messages appear as white text above the yellow help line messages.

- Query prompts for keyboard text or numerical input; e.g., numerical value for a dimension. When a keyboard input is required, a prompt will appear in the message window ending with a colon.

8.2.3 Command Line Syntax

A list of commands introduced in this chapter appear in Section 8.8 at the end of the chapter. The command line syntax that is used in this chapter is illustrated in several examples. When Pro/E wants your attention, it sounds a bell; it is waiting for an input action from you. Ringing bells become quite annoying in a lab with 20 students running Pro/E simultaneously! Turn off the bell using the following menu selections:

MAIN	Environment ⇒ ☐ **Bell** (a ✓ in the box turns the bell **on**)	{8.1}

Let's dissect the syntax in Cs.[2] {8.1}. Go to the **MAIN** menu; i.e., use the left mouse button (**LMB**) and click on the title bar of the blue **MAIN** menu window. Repeated selection of the title bar in a menu window toggles between collapsing and expanding the menu choices located in that particular menu window.

Find the **Environment** choice on the **MAIN** menu, and click the **LMB**. A second menu bar entitled **ENVIRONMENT** appears next to the **MAIN** menu. At the top of this menu, find the menu choice **Bell** and left click in the box to the left of the word **Bell**. This action will toggle the bell sound on and off. Go ahead and try.

8.2.4 Selecting Menu Items

All menu selections are made using the left mouse button (**LMB**). When we move our mouse across a certain menu item, it is highlighted in blue. As soon as we select the menu item by clicking, it will turn black. Its sub-menu will pop-up and the default choices on the sub-menus will appear in black. We are free to choose other menu selections, but Pro/E is trying to improve our efficiency by providing guidance with appropriate menu selections hi-lighted in black. Menu choices that are not available will appear dim and grayed out. This appearance indicates that this menu item is not available for this particular menu choice sequence.

To end a sequence of selections within a certain menu, we click on the **Done-Return** or **Quit** options. This returns us to the previous menu or causes us to exit from the current menu. Either command, removes the menu from the screen display.

---

[2] We will use the abbreviation Cs. to indicate command sequence.

8.2.5 File Naming Convention

Pro/E uses many filenames in its operating environment to aid us in keeping track of our work. It uses different filename extensions to denote the operating mode in which we save our file. Recall the three main operating modes introduced in Fig. 8.3:

- Part-mode
- Assembly-mode
- Drawing-mode

In addition to these three modes, we will also be working in a **sketcher-mode**. This mode is a two-dimensional environment in which we sketch profiles of the features we are creating. More information on the sketcher mode will be introduced later.

Pro/E saves files with a particular extension depending on the operating mode environment. It also keeps track of our work with trail files. The files we will see most often in Pro/E are:

*.prt	Part
*.asm	Assembly
*.drw	Drawing
*.sec	Sketcher cross section
*.plt	Plot file for hardcopy output
trail.*	Trail files

8.2.5.1 Creating a Filename for a Part

To illustrate the file naming convention in Pro/E, let's consider an example of creating a part called **block**. We begin our Pro/E session with following command sequence:

MAIN	Mode $\Rightarrow$ Part $\Rightarrow$ Create $\Rightarrow$ Enter part name [PRT0001] : **block**	{8.2}

Again, let's describe the syntax of this command. After the menu selections (**Mode**, **Part**, and **Create**) are made, the message window prompts us for keyboard input for the part name (examine the Message Window). If we simply hit ENTER ↵ on the keyboard, the default name of the part shown in the [ ] brackets will be used. In this instance the part name becomes prt0001.prt. However, if we type the word **block** after the colon prompt as instructed in {8.2}, the part name becomes **block.prt**. Whether we accept the default selection for the filename or assign one of our choice, the extension **.prt** is attached. Since we were operating Pro/E in the Part-mode when creating the file, the **.prt** suffix is automatically attached.

8.2.5.2 Saving Files

Clearly, after working on a file, it is wise to save it. At any time or within any mode of operation, we can save the file to the hard drive by using the following command sequence:

MAIN	**Dbms** ⇒ **Save** ⇒
	Enter object to save [name.xxx]: **<ENTER ↵>** ⇒ **Done-Return** {8.3}

The acronym **Dbms** stands for Database Management System. At the prompt, you may hit ENTER ↵ on the keyboard to accept the default filename in the brackets [name.xxx] or save the object with a new filename by explicitly typing a new filename. While working in the Part-mode, it is recommended that you save your model whenever a new feature is added to the part.

8.2.5.3 Retrieving an Existing Part

Once a part has been created and saved, we may choose to exit Pro/E and come back to work on the same part at some later time. To automatically retrieve an existing part we use the following command sequence:

MAIN	**Mode** ⇒ **Part** ⇒ **Search/Retr** ⇒ **Current Dir** ⇒
	<Use **LMB** to retrieve the part file > {8.4}

This command shows all the *.prt part files and subdirectories of the current directory. You can select a part in the current directory and scroll up and down in the directory by selecting the following symbols in the menu:

 Scroll up in the current directory

 Scroll down in the current directory

 End of listing in the current directory.

We can also move up a directory level by selecting the **UP..** menu selection.

8.2.5.4 Automatic Filename Numbering and Incrementing

Pro/E also adds version numbers for each file; every time you save the file, the version number is incremented by one. The first time we save a part using Cs. {8.3}, Pro/E automatically attaches the first version number to the filename. In our case with the **block**, the filename becomes **block.prt.1**. In fact, every time we execute Cs. {8.3}, the version number of the part is incremented—**block.prt.1** becomes **block.prt.2**, then **block.prt.3**, etc.

This is a nice feature in Pro/E, because none of the previous versions of the file are removed, so at anytime we can go back and retrieve an old version of the file.

8.2.5.5 Purging Old Versions of Files

While the incrementing of file numbers can be a nice feature, it can also be a nuisance. If we are not careful, our hard-drive will rapidly become full. This is especially true if we frequently save the file. To combat overloading our hard-drive with useless versions of a file, Pro/E has the PURGE command. The PURGE command completely removes all old versions of the filename, and retains only the most recent one. We execute the PURGE command using:

MAIN	**Dbms** ⇒ **Purge** ⇒
	Enter object to purge [BLOCK.PRT] : **<ENTER ↵>** ⇒ **Done-Return** {8.5}

Note, by simply using the ENTER ↵ key, we accept the default **block.prt** as the filename to be purged of all old versions.

8.2.6 Environment Summary

This section completes our introduction to the Pro/ENGINEER environment. Hopefully you executed the commands in this section when starting your first Pro/E session. You should now be familiar with the following items:

- Screen layout
- Using the message window for guidance
- Selecting menu items
- Naming files

We will soon begin our tutorial to create our first solid model in Pro/E. However, before we begin the tutorial we need to talk more about **features**. For now, let's exit from Pro/E. Before exiting, save your work with Cs. {8.3}, and execute the following command to exit Pro/E:

MAIN	**Exit** ⇒ Do you really want to exit? [N] : **YES**	{8.6}

You must select the **YES** box in the Confirmation Window to exit Pro/E; note that the default in [ ] brackets is NO, so be certain to click on the **YES** box.

8.3 FEATURE CLASSES

Pro/ENGINEER uses features to create part models. There are two classes of features used most frequently. The first **feature class** assists us in constructing the model with Datum Features. The second **feature class** is used to create solid models with Solid Features. A hierarchy of the commonly selected options from these feature classes is presented in Fig. 8.5. We will now examine the two feature classes.

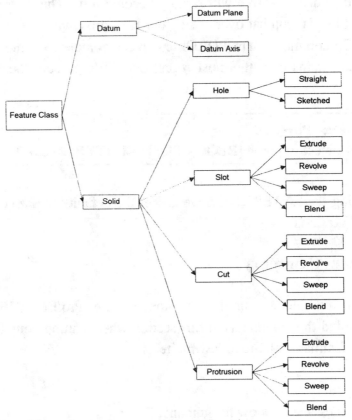

Fig. 8.5 Feature class hierarchy and commonly used options.

8.3.1 Datum Features

A datum feature is the first **feature class** we will encounter. Datum features aid in constructing models of parts. There are two which are used frequently—a datum plane and a datum axis. We will describe the datum plane in this section and defer the description of the datum axis. A datum plane is a two dimensional reference plane. When defining a part, we start by creating datum planes. Though a datum plane appears on the screen as a rectangle, it is an infinite plane extending as far and wide as can be imagined. It has no mass and no volume, and it is infinitesimally thin. We use datum planes to place and locate features in Pro/E. A datum plane has two sides—like a sheet of paper. One side is positive, and when viewed on the screen, it appears with a yellow transparent border. The flip side of the datum plane is negative, and appears with a red transparent border. We usually observe only the positive (yellow) side of the datum plane; however, if we spin our model, the datum planes change from yellow to red.

YELLOW border	≡	+ positive side of Datum Plane
RED border	≡	− negative side of Datum Plane

Datum planes are helpful in establishing the view we are considering when constructing the solid model of some part. Pro/E encourages you to use datum planes to construct and view your model. As an added incentive for you to use them, it has already created three **default** datum planes. The default datum planes intersect each other at 90° angles and are labeled on the screen as DTM1, DTM2, and DTM3. To create the default datum planes, we use the command:

PART	**Feature ⇒ Create ⇒ Datum ⇒ Plane ⇒ Default**	{8.7}

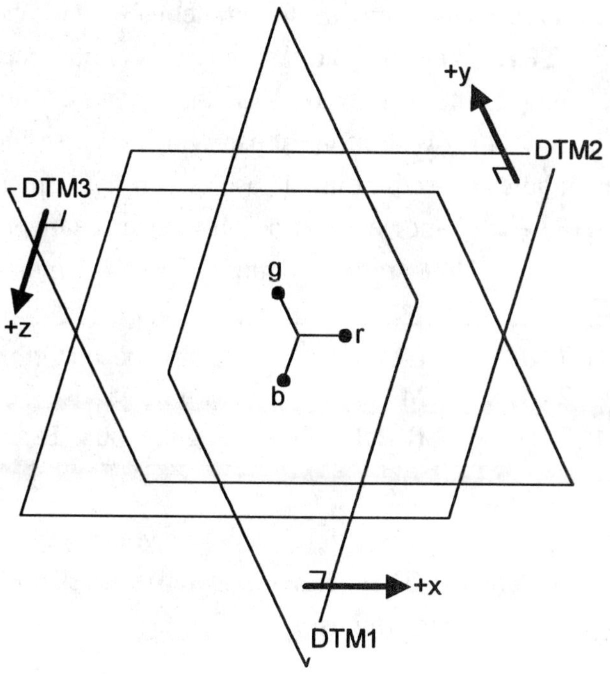

Fig. 8.6 Default datum planes.

The default datum planes appear on the screen as illustrated in Fig. 8.6. All of the datum planes are yellow because we are looking at the "positive" side of each datum plane. Each datum plane is associated with an axis in the xyz Cartesian coordinate system; namely, the outward positive normal. For DTM1 the +x axis is the outer normal. Similarly for DTM2 and DTM3, the outer normals are +y and +z respectively. We do not observe these positive normals on the screen, as shown in Fig. 8.6, but they are implied in the Pro/E environment. The small color triad in the middle of the datum planes is the spin center, and it serves to help orient the part. The red-green-blue color triad corresponds to the +x-axis, +y-axis, and +z-axis, which are the positive sides of datum planes 1, 2, and 3 respectively. A mnemonic rhyme to remember this sequence is:

$$R - G - B \equiv X - Y - Z \equiv 1 - 2 - 3$$

If by chance you execute Cs. {8.7} and the datum planes do not appear as shown in Fig. 8.6, or the spin center does not appear, you have to ☑ (i.e., Turn On) its respective environment variable and set the orientation to trimetric using:

MAIN	Environment ⇒	☑ **Disp DtmPln**	
		☑ **Spin Center**	
		Trimetric ⇒ **Done-Return**	{8.8}

We can check the orientation of the datum planes and the color triad by dynamically changing the display using the keyboard and the mouse simultaneously. For those of you with three-button mice use the <CTRL> key on the keyboard in conjunction with the L≡Left, M≡Middle, and R≡Right mouse buttons to change the display while moving the mouse on the mouse pad. Those of you with a two-button mouse with a center thumb-wheel, the center thumb-wheel acts as the middle mouse button. Those of you with an "antiquated" two-button mouse, hold the <⇑ **SHIFT**> key and left mouse button simultaneously for the equivalent middle mouse button command. Here are the dynamic viewing commands:

ZOOM	<CTRL>	**LMB**	(Left Mouse Button)	
SPIN	<CTRL>	**MMB**	(Middle Mouse Button)	
PAN	<CTRL>	**RMB**	(Right Mouse Button)	{8.9}

When you spin the datum planes far enough, notice that you begin to observe the backside (negative sides) of the datum planes. They show up as red[*]. After you zoom, spin and pan, it is advisable to return to the default view by using:

MAIN	View ⇒ Orientation ⇒ Default ⇒ **Done-Return**	{8.10}

8.3.2 Solid Features

The first **feature class** we considered were datum planes. As is evident in Fig. 8.5, the other **feature class** employed in Pro/E is solids. Under solids feature class, we have four types—the hole, slot, cut, and protrusion.

---

[*] System Administrators, make sure the CONFIG.PRO file has the following line to make sure that datum features remain visible during dynamic spinning of the model: **SPIN_WITH_PART_ENTITIES YES**

HOLE   A hole, as the name implies, **removes** solid material from a part. Usually our holes are "straight". In manufacturing, straight holes are produced by a straight, constant diameter drill-bit. We also have the option to sketch our holes. Sketch holes are similar to holes produced with a tapered reamer in the machine shop.

SLOT   A slot **removes** solid material from a part. A slot is defined using a **closed** section. In manufacturing, we create a slot by cutting into the workpiece with a milling tool to remove the unwanted material.

CUT   A cut **removes** solid material from a part. However, it is defined using an **open** section and can be imagined as removing material from the side of a workpiece using a metal cutting band saw.

PROTRUSION A protrusion **adds** material to the part. It is usually defined using a **closed** section. A protrusion can be imagined as taking metal stock from the shelf and adding to the part by welding, joining, or gluing additional material.

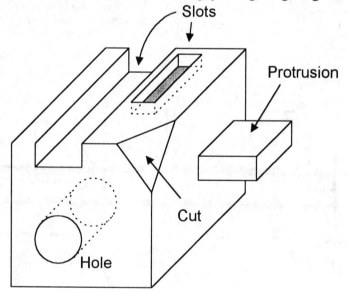

Fig. 8.7 Different types of solid features.

There are many ways to create solid features. The most common options are to extrude, revolve, sweep, or blend two-dimensional profiles to define the solid feature. All of these options are listed under the **SOLID OPTS** menu in Pro/ENGINEER, and they are compared in Fig. 8.8. The definitions of each of these options are as follows:

EXTRUDE   The solid is created by projecting a 2D section perpendicular to our sketching plane. This is the simplest and most frequently used solid option. We have to define the distance (i.e., depth) we wish to project it.

REVOLVE   The solid is created by revolving a 2D section around a centerline. We first have to define the centerline, and then sketch a 2D **closed** section. The closed section **must lie completely on one side of the centerline**. We then complete the definition of the solid by specifying the angle of revolution.

SWEEP   The solid is created by sweeping a 2D section along a trajectory. We first sketch the trajectory path, and then sketch a 2D section that is projected along the trajectory path.

BLEND   The solid is created by blending at least two planar profile sections. We sketch each section, toggling between each as we complete the sketches. There has to be the same number of edges on every section, because edges are blended from one section to the other in a smooth or straight fashion.

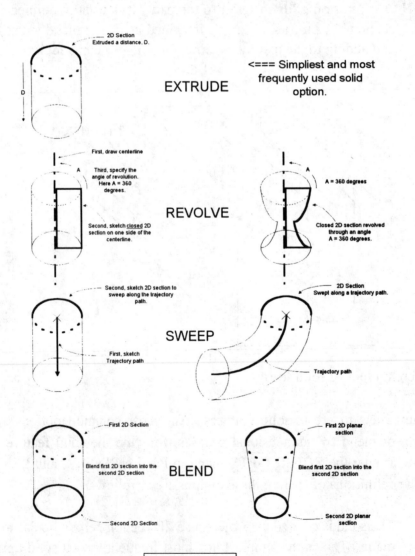

Fig. 8.8 Comparing and contrasting **SOLID OPTS** options for creating solid models.

8.4 ORIENTING MODELS

Pro/ENGINEER allows us to orient our models by selecting **two** planar surfaces that are perpendicular to each other, or by selecting **two** of the datum planes displayed on the screen. In this section, we will orient our models relative to the default datum planes created earlier. We will use the simple part shown in Fig. 8.9 for the model.

Follow along during this exercise to understand the use of datum planes to control orientation. Start Pro/E, and use Cs. {8.2} to create a part **PRT0001**. Next, use Cs. {8.7} to create the default datum planes. You may wish to execute Cs. {8.8} to ascertain that the datum planes are displayed properly. Finally, execute Cs. {8.10} to see the default orientation of the datum planes. At any time during this exercise, we can return to the default orientation shown in Fig. 8.9a by using Cs. {8.10}.

Pro/E orients parts by requiring us to select two mutually perpendicular planes and orienting their **positive** sides toward the FRONT, BACK, TOP, BOTTOM, LEFT, or RIGHT of the screen monitor. Since the datum planes are mutually perpendicular to each other, they are often used when specifying the orientation of a part. Remember, the YELLOW side of a datum plane is the POSITIVE side. As an example, let's consider the procedure for orientating the part as shown in Fig. 8.9b. (Note: in all figures – – – dashed lines denote negative side of datum plane which appears red on the screen).

In Fig. 8.9b we may use any combination of the following three options to obtain the orientation shown in the figure. They are: RIGHT→DTM1; TOP→DTM2; and FRONT→DTM3. What this is saying is that we want the positive (yellow) side of DTM1 facing the right; the positive (yellow) side of DTM2 facing upward; and the positive side of DTM3 facing the front. Any two combinations of these three options will achieve the desired result. We will now look at the explicit command sequence of the first combination (RIGHT→DTM1 and TOP→DTM2).

MAIN View ⇒ Orientation⇒ **Right**	⇒ **Pick** ⇒	\<Click **LMB** on Outline of DTM1>
⇒ **Top**	⇒ **Pick** ⇒	\<Click **LMB** on Outline of DTM2>
⇒ **Done-Return**		{8.11}

If you carefully examine the screen, both the yellow (+ve) and red (−ve) edges of two of the datum planes are visible. In Fig. 8.9d, the red (−ve) side of DTM3 is visible (facing you) because we specified the yellow (+ve) side to face the **back** of the monitor. We suggest that you complete the exercise to determine if you can duplicate the other orientations presented in Figs. 9c and 9d.

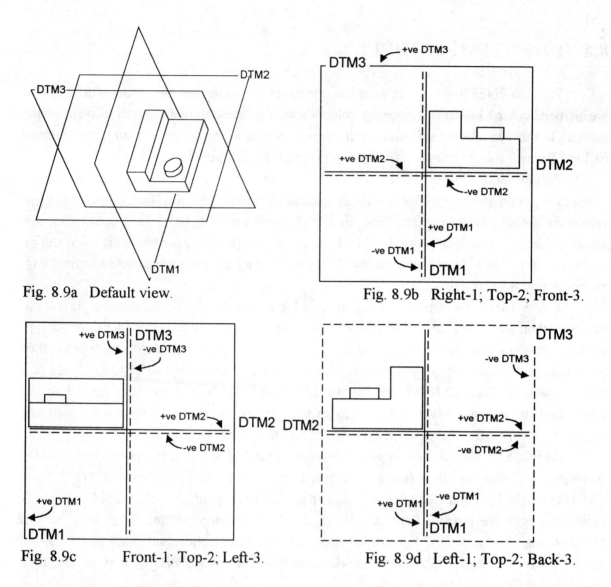

Fig. 8.9a   Default view.

Fig. 8.9b   Right-1; Top-2; Front-3.

Fig. 8.9c   Front-1; Top-2; Left-3.

Fig. 8.9d   Left-1; Top-2; Back-3.

8.5 THE SKETCHER

Whenever we start the creation of a feature, Pro/E automatically puts us in **Sketcher-Mode**. The **Sketcher** is a 2D environment used to create and modify 2D sectional views of our features. Whenever we execute save with Cs. {8.3} from the sketcher mode, Pro/E saves the file with the *.sec extension. The sketcher allows us to sketch our 2D geometry quickly without regard to the exact dimensions. The exact dimensions are placed on the geometry after we **regenerate**, and then **modify** the dimension parameters. The sequence followed is given by the acronym, SADRMR which means:

SKETCH   – Sketch the section geometry.
ALIGN   – Align the section geometry to existing part geometry or datum planes.
DIMENSION   – Place sketcher dimensions, denoted by *sd_* , onto the geometry.

REGENERATE – Regenerate the section. Pro/E uses your alignment and dimension inputs just made, along with its own *assumptions* to arrive at a parametric representation of the feature.

MODIFY – Modify your dimensions by typing in specific numeric values.

REGENERATE – After modifying your dimensions, regenerate the section with the new dimensions.

If we follow this SADRMR sequence, the sketcher is easier to use.

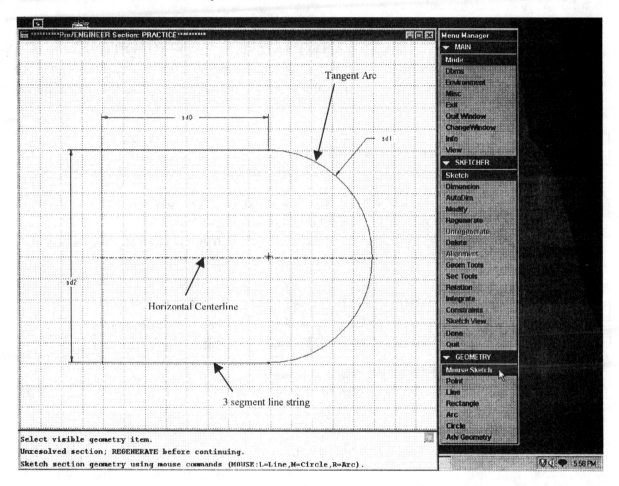

Fig. 8.10 Sketcher-mode.

8.5.1 Sketcher Tools

When entering the sketcher, we employ Sketcher Tools to create the section geometry. To learn to work with these tools, start Pro/E and enter the sketcher to begin a practice session using the following command:

MAIN   **Mode** ⇒ **Sketcher** ⇒ **Create** ⇒ Enter Section name [S2D0001]: **practice** {8.12}

This creates a section file called **practice.sec**. The .sec is automatically appended to the file name to denote that this is a section file. Alternatively, we could have accepted the default section name **S2D0001** by hitting the ENTER ⏎ key when prompted to enter the section name.

After starting the sketcher, a 2D sketcher grid will appear (Fig. 8.10). The sketcher grid is a drawing aid. When the grid is visible, start creating sketcher geometry using the **GEOMETRY** menu located in the bottom right of Fig. 8.10.

8.5.1.1 Mouse Sketch Tool

Upon entering the sketcher, the default selection in the **GEOMETRY** menu is a tool—**Mouse Sketch**. The mouse arrow is parked on this menu pick in Fig. 8.10; note, the display in the message window with this menu selection. We may select this tool any time we are in the sketcher by using:

SKETCHER	Sketch ⇒ **Mouse Sketch**	{8.13}

The mouse sketch tool allows us to use the left, middle, and right mouse buttons to free sketch geometry. (If you only have a two button mouse, the middle mouse button commands are mimicked by holding the <⇑ **SHIFT**> key and clicking the left mouse button simultaneously.) The following mouse sketch commands respond to the respective mouse buttons:

Line Creation -and- Abort Circle Creation	**LMB**	(Left Mouse Button)	
Circle Creation -and- Abort Line Creation	**MMB**	(Middle Mouse Button)	
Tangent **Arc** Creation	**RMB**	(Right Mouse Button)	{8.14}

Practice the creation of the following geometry by using the mouse to sketch line strings, circles and tangent arcs.

Line String: The first LMB click begins line; the second LMB click completes line segment. A MMB click anytime during the line string definition, completes the line string. A 3 segment line string is illustrated in Fig. 8.10.

Circle: The first MMB click defines center of circle; the second MMB click defines the radius of the circle thereby completing the circle. A LMB before the second MMB click aborts the creation of the circle.

Tangent Arc: An existing geometric entity must be used as the start point of tangency; e.g., the end point of a line. RMB click on the end of the entity; move mouse to new location; use a second RMB click to complete arc. A MMB click before the second RMB click, aborts the creation of the tangent arc. This procedure was used to create the tangent arc in Fig. 8.10.

8.5.1.2 Line Tool

The **Line** tool is the third option in the GEOMETRY menu (the second option is the **Point** tool which we will not cover here). It is used to create lines and centerlines. Lines are used to define the boundaries of the feature we are sketching. Centerlines are used to define an axis of rotation and to denote a line of symmetry through a feature. Centerlines appear as a long-dash, short-dash pattern on the screen (note the Horizontal Centerline created in Fig. 8.10). A centerline must be defined when we create a revolve feature.

Under either selection (line or centerline), we have several options. Here are a few of the options available—you are encouraged to explore the other options on your own using the message window as a guide. Read the following options and while in the sketcher, practice sketching these lines.

2 Points: Define a line-string by selecting consecutive endpoints on the screen with the LMB. A MMB click anytime during the line string definition completes the line string.

Parallel: Use the LMB to select an existing line as the one of the two parallel lines. Use the LMB to define the start and end of a new line that is parallel with the selected line. A MMB cancels the line definition.

Perpendicular: Same as parallel, except that the existing line which is to be perpendicular to a new line is the first LMB selection made.

Horizontal: This command starts a line string with a horizontal line segment. Use the LMB to define the starting point of the horizontal line. A second LMB click defines the endpoint of the line, and automatically toggles to the start of a vertical line. The MMB ends the line string definition. This selection was employed to draw the 3 segment line string shown in Fig. 8.10.

Vertical: Same as Horizontal, except the first segment of the line string is vertical.

The last two options are used frequently because most sketcher geometry's starts with either horizontal/vertical line strings. By using these options we can achieve better control of our 2D profile section.

8.5.1.3 Other Sketching Tools

One command you will find very handy when sketching, is the **Delete** command. It allows you to delete one item, several items at once, or all the entities in the **sketch only**. Pro/E even permits you to undelete the last item deleted. The command sequence for deleting a **single** item in the sketcher is:

> **SKETCHER**   Delete ⇒ Delete Item ⇒ Pick ⇒ <LMB click on item to delete>   {8.15}

The other sketching tools we use frequently are **Point**, **Rectangle**, **Arc**, and **Circle**. Now that you know the basics, we suggest that you use these sketching tools along with the delete option to define and erase geometry in the sketcher. If you need assistance or clarification of a menu selection, help is displayed in the message window. Simply park the mouse on the menu selection and read the prompt in the message window to determine the next step required for defining the geometry.

8.5.2 Alignment

Once we have sketched our feature geometry, we have to **align** it to an existing geometry. The purpose of **alignment** is to **locate** the sketch geometry with an existing part geometry. Examples of the sketch and part geometry's are:

SKETCH GEOMETRY – Anything sketched in the sketcher (shows up as light blue on the screen). Includes: Geometry Lines, Centerlines, Endpoints of Geometry Lines, Centers of Circles, Ends of Arcs, etc…

PART GEOMETRY – Existing part geometry appears white on the screen. Includes: Faces, Edges, vertices, datum planes, etc…

When we align sketch geometry with part geometry it is like gluing the sketch geometry to the part. In essence, we assign a zero dimension between the entities. Since the alignment function is **not** a snap function, it will function only if the sketch geometry is **close** to the part geometry. The alignment command is

> **SKETCHER**   Alignment ⇒ Align <LMB click on light blue sketch geometry>
> <LMB click on white part geometry>   {8.16}

If the sketch and part geometry are successfully aligned, the message window will notify us with the message "--ALIGNED --". This option is not required in our practice session because there is no existing part geometry with which to align the sketch. Later, when discussing the tutorial on parts, we will demonstrate the technique for aligning the sketch geometry to the default datum planes.

8.5.3 Dimensioning

Once we have completed our sketch geometry, we can now assign dimensions to it. The command sequence for placing a sketcher dimension (sd_) is:

SKETCHER	Dimension $\Rightarrow$ Normal $\Rightarrow$ Pick
	<**LMB** Select one entity or endpoint on SKETCH>
	OPTIONAL $\rightarrow$ <**LMB** Select one entity or endpoint on PART>
	<**MMB** Place dimension with middle mouse button>{8.17}

The sketcher will automatically assign a symbolic parameter dimension to the sketch starting with sd0, followed by sd1, and so on. In Fig. 8.10, the first dimension defined is on the top horizontal line segment sd0, the second is a radial dimension of the tangent arc sd1, and the third is sd2 for the vertical line segment.

8.5.4 Regenerate

We are now ready to regenerate our sketch. When regenerating Pro/E determines if we have constrained our sketch appropriately. That is, have we provided sufficient dimensions for Pro/E to parametrically represent the part? To assist us in this regeneration process, Pro/E makes some assumptions that include:

- Lines "close" to being **horizontal or vertical** become horizontal or vertical.
- Lines "close" to being **parallel or perpendicular** become parallel or perpendicular.
- Arcs and circle center points "close" to being aligned horizontally and vertically become aligned.
- Arc angle increments "close" to 90° increments become 90°, 180°, 270°, or 360°.
- Lines that are nearly collinear become collinear.
- Line segments of nearly equal length assume the same lengths.
- Arcs and circles of near equal radius assume the same radii.
- Similar entities, with a centerline placed between them, are assumed to be equal distance from the centerline. This is referred to as Centerline Symmetry. (Note, in Fig. 8.10 to observe symmetry about a horizontal axis in the sketched section.

Most of these assumptions may be overridden by placing appropriate dimensions on the part. The command for regeneration is:

SKETCHER	**Regenerate**	{8.18}

If the sketch regenerates, the message window will display—"Section regenerated successfully." Also, yellow numerical values of the dimensions will appear on the feature as well as symbols showing the assumptions that Pro/E made during the regeneration. For example, the symbol // means parallel lines have been assumed. If regeneration is unsuccessful, the message window will display—"Under dimensioned section, Please align to part, or Add dimensions." This simply means that you have not properly **aligned** or **dimensioned** your sketch. It will be necessary to correct these errors.

8.5.5 Modify

After a successful regeneration, Pro/E will automatically place us in the SKETCHER Modify command. Using this command, we change the numerical dimensions displayed on the sketch to the correct values. We dimension by using:

SKETCHER	Modify ⇒ Mod Entity ⇒ Pick ⇒ <LMB Select dimension to modify> ⇒ Enter a new value [00.00] : **nn.nn** <ENTER ↵> {8.19}

Where 00.00 is the old dimension and **nn.nn** is the new dimension.

8.5.6 Regenerate

Before leaving our section and exiting sketcher, we must regenerate the section with our newly modified dimensions specified with Cs. {8.18}. Regeneration should be successful before we leave the section. This completes our practice session on the sketcher. To exit the sketcher, use the following command:

SKETCHER	Quit ⇒ Confirm	{8.20}

This command returns us to the MAIN menu. If desirable, we exit Pro/E with Cs. {8.6}.

8.6 CREATING SOLID MODELS

We are now ready to create our first part using Pro/E. The part is the Stop_Block shown in Fig. 8.11. Start Pro/E and create the part with the filename **Stop_Block** using Cs. {8.2}. Remember, you do not type in the extension .prt since it is automatically appended to the filename when saving the file. After you have executed Cs. {8.2}, the main working window should display ****** Pro/ENGINEER Part: STOP_BLOCK ****** in the title bar. Create the default datum planes by using Cs. {8.7}. We are now ready to create the first solid base feature.

8.6.1 Stop Block Base Feature

Now that the default datum planes have been established, we will create our base feature. Our definition for the base feature is:

The base feature is the first solid feature created in the definition of the part model. It is the foundation for the rest of the model upon which we build.

To work with Pro/E, you should think in manufacturing terms. What piece of metal stock are you going to remove from the supply of stock to use in making your part? Can you add and remove material from the base feature relatively easy? Are there any plans to modify the base feature in the future? If so, how will these changes effect the remainder of the part?

Plan your "manufacturing" operations before you begin; then recall the equivalent Pro/E feature that will add value to your part and to capture the design intent. All subsequent features that are added will depend on the geometry of the base feature; therefore, a moment of planning will save a lot of problems later.

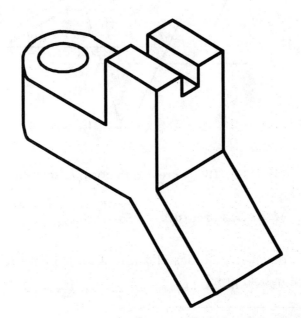

Fig. 8.11 A stop block.

8.6.1.1 Creating Base Feature in Positive-Space

It is good practice to establish your base feature in **positive-space**. This statement implies that we want to create the model on the positive side of every datum plane whenever possible. If the model is on the positive side of the datum planes, the parts come up in the default view when you retrieve parts into a Pro/E session using Cs. {8.4}. In the default view, all the datum planes are yellow indicating the positive-space orientation. A representation of the Stop_Block in positive space is illustrated in Fig. 8.12.

Notice in Fig. 8.12 that the base feature is on the right-side of DTM1 (positive x-space); on most of the top-side of DTM2 (positive y-space); and on the front-side of DTM3 (positive z-space).

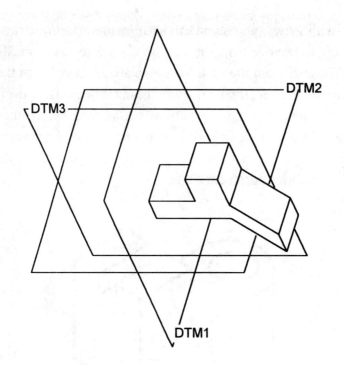

Fig. 8.12   Base feature of the Stop_Block created in positive space

8.6.1.2 Characteristics of the Base Feature

When we create our first solid base feature, it must be a Protrusion.   For emphasis, let's show this statement in bold print.

THE BASE FEATURE <u>MUST BE</u> A SOLID PROTRUSION

Clearly, a protrusion adds material to the model.   We must next decide on the type of protrusion we are going to employ.   There are several options as indicated in Fig. 8.8, but in this example, we will use the **Extrude** option under SOLID OPTS .   The command sequence for the **Extrude** option is:

PART   Feature $\Rightarrow$ Create $\Rightarrow$ Solid $\Rightarrow$ Protrusion $\Rightarrow$ Extrude $\Rightarrow$ Solid $\Rightarrow$ Done{8.21}

Upon completing this command, the display on the screen should appear similar to Fig. 8.13.

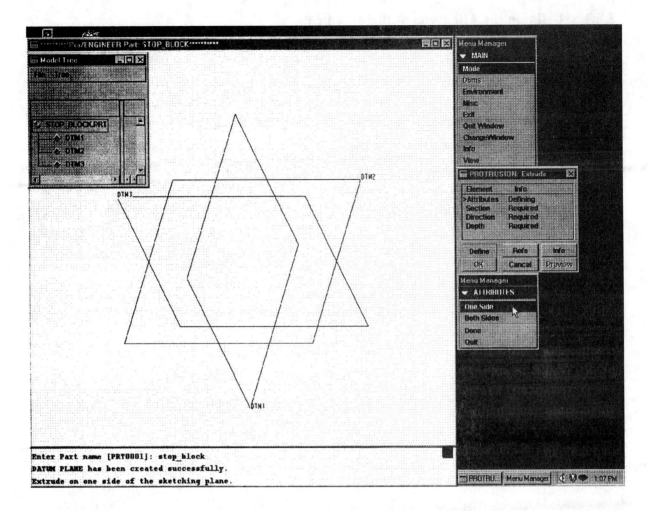

Fig. 8.13 Appearance of screen during protrusion by extruding.

Let's take a moment to describe everything that appears on the screen after executing this command. In the main working window, the title bar displays the active part ****** Pro/ENGINEER Part: STOP_BLOCK ******. In the center of the main working window the datum planes, which we previously created, are displayed in their default orientation. The model tree window, in the top left, displays the current tree structure of our part. Since we have only created the default datum planes, the model tree shows only these three planes. The message window displays the most recently executed commands—the naming of the part STOP_BLOCK, and the definition of the datum planes (default orientation). The message window also displays the help message "Extrude on one side of the sketching plane" because the mouse arrow is parked on the ATTRIBUTES One Side menu item.

The PROTRUSION: Extrude dialog box on the right side reports our status during feature definition. At this moment, we are about to define the attributes of the feature so the ">" symbol is displayed next to the Attributes-Defining. Let's proceed to define the attributes.

8.6.1.3 Defining the Attributes of the Base Feature

With the mouse parked on the **One Side** menu item, it is now time to define the **one side** attribute of the base feature. But first, let's decide the orientation of the sketch of the base feature so it is easy for us to draw in positive space. Pro/E defaults by adding material to the part by GROWING THE FEATURE <u>TOWARD</u> YOU (out of the screen). When we remove or cut material, Pro/E defaults by orienting the part so material is removed away from you and into the screen. Since we are adding material whenever we create the base feature, the feature will grow outward toward us. Because we wish to grow the feature from **one side** of DTM3, select the **one side** attribute. To extrude the feature in the positive direction and to achieve the orientation of the base feature shown in Fig. 8.12, we perform the following actions:

- Sketch on DTM3 (faces **Front**)
- Select the positive (yellow) side of DTM1 to face **Right** *-or-* Select the positive (yellow) side of DTM2 to face the **Top**.

(If you do not understand the last sentence, you are encouraged to review section 8.4 on Orienting Models.) We will now return to the command sequence in Fig. 8.13 and define the **one side** attribute of the feature with the following menu selection:

ATTRIBUTES	**One Side** ⇒ **Done** ⇒ **Setup New** ⇒ **Plane** ⇒ **Pick**
	Select border of DTM3 to grow the feature toward you on <u>one side</u> (the positive side) of DTM3.
	< **LMB** click on DTM3 Border> {8.22}

A red arrow appears denoting the direction of the extrusion. Since the arrow direction is out of DTM3, we are satisfied and okay the direction with:

DIRECTION	**Okay**	{8.23}

Upon accepting the direction, Pro/E prompts us to orient the sketch. Here we choose:

SKET VIEW	**Right** ⇒ **Plane** ⇒ **Pick**
	Select border of DTM1 to face its positive (yellow) side toward the <u>Right</u> of the screen.
	< **LMB** click on DTM1 Border> {8.24}

Alternatively, we may orient the sketch with the following sequence:

SKET VIEW	**Top** ⇒ **Plane** ⇒ **Pick**
	Select border of DTM2 to face its positive (yellow) side toward the <u>top</u> of the screen.
	< **LMB** click on DTM2 Border> {8.25}

Either command sequence will produce the sketch orientation shown in Figs. 8.10 and 8.14.

8.6.1.4 Creating the Base Feature Section

Use the **Sketcher** to create the 2D **section** view of the feature. The PROTRUSION: Extrude dialog box indicates that we have defined the attribute of the feature to be <u>one side</u>. This means that the protrusion will grow out of the positive side of DTM3. The dialog box is now informing us that the next step of feature creation is to define the <u>section</u>. This is denoted by the ">" symbol next to "Section Defining" in this dialog box.

The default background color of the main working window is royal blue[3]. Your screen should look like Fig. 8.14 with a white grid displayed on the royal blue background along with:

DTM1 – Cutting vertically through the window. Close inspection shows the left side of DTM1 is red (–ve); and the right side of DTM1 is yellow (+ve).

DTM2 – Cutting horizontally through the window. Close inspection shows the top side of DTM2 is yellow (+ve); and the bottom side of DTM2 is red (–ve).

DTM3 – Square border of the window. DTM3 is displayed in yellow, denoting that we are viewing the positive side of this datum plane.

We are prepared to sketch the geometry of the base feature. Following the acronym SADRMR, use the sketching tools discussed in 8.5.1 to **S=sketch** the 2D section presented in Fig. 8.14. Use Cs. {8.15} to delete any unwanted sketch geometry. **A=Align** the left vertical edge to DTM1 and the bottom horizontal edge to DTM2. The align option is a **glue** option, <u>not</u> a **snap** option. The sketch geometry to be aligned with the datum planes should be drawn very close to these datums for the align command to work. The message window will display "--ALIGNED--" when you have successfully aligned the sketch boundaries with the appropriate datum planes.

The next operation is to **D=dimension** the sketch. There are several ways to dimension a 2D section that are acceptable to Pro/E. One dimensioning scheme that produces a successful regeneration is illustrated in Fig. 8.14. All the dimensions in Fig. 8.14 were placed by using Cs. {8.17}. The middle of each line segment was picked with a <u>single</u> LMB click, and the dimension was placed using the MMB. The angular dimension, sd4, was defined by first selecting the bottom horizontal line segment with the LMB, followed by picking the lowest slanted line segment with the LMB. The placement of the angular dimension was made with the MMB.

---

[3] The screen-captures in this chapter were performed with a white background color for printing reasons.

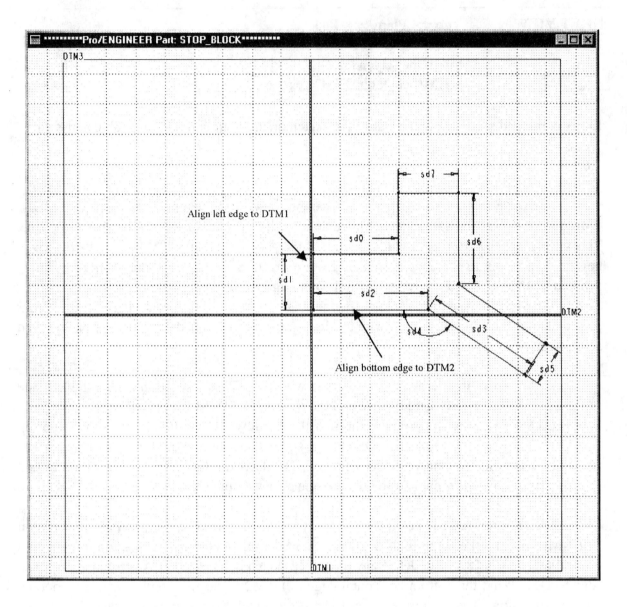

Fig. 8.14   Sketcher dimensions for the base feature.

After all the dimensions have been defined, we **R=Regenerate** the section using Cs. {8.18}.   After regeneration a successful regeneration, we see the message "Section regenerated successfully" displayed in the message window.   We also see the assumptions the sketcher made during the regeneration of the sketch.   Several of these assumptions; i.e., V≡Vertical line; H≡Horizontal line; //≡Parallel lines; and ⊥≡Perpendicular lines are illustrated in Fig. 8.15.

After a successful regeneration, the sketcher automatically places us into the **M=Modify** command.   We use Cs. {8.19} to modify the dimensions to correspond to the numerical values shown in Fig. 8.15.   We **R=Regenerate** one more time for Pro/E to update the sketch with the new dimension values.   Upon our last successful regeneration, we are ready to define the depth of the protrusion (i.e., how much the 2D section profile grows out of the screen toward us). We do this by looking for the **Done** menu item at the bottom of the **SKETCHER** menu and choose it.

SKETCHER	**Done**	{8.26}

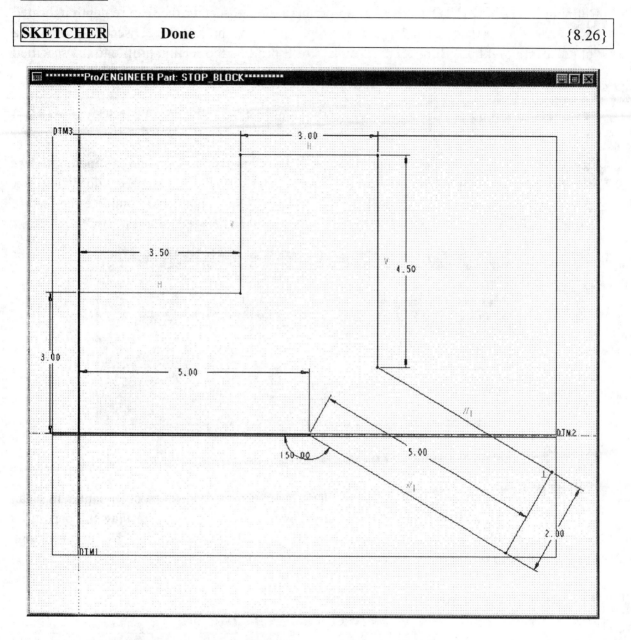

Fig. 8.15   Regenerated and modified dimensions of the base feature.

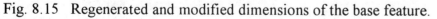

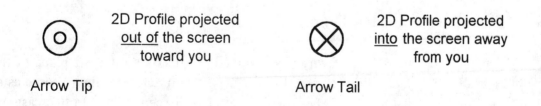

2D Profile projected <u>out of</u> the screen toward you

Arrow Tip

2D Profile projected <u>into</u> the screen away from you

Arrow Tail

Fig. 8.16   Arrow indicates the direction of projection for the feature.

Following the execution of Cs. {8.26}, you are prompted to input the depth of the protrusion. Examine the PROTRUSION: Extrude dialog box and note the ">" symbol is next to "Depth Defining." The SPEC TO menu pops up prompting us to specify the feature depth indicated by the direction of the red arrow tip or tail (Fig. 8.16). We project the 2D section profile to a "blind" depth of 4.00. "Blind" indicates that Pro/E projects the section profile to the specified depth. Enter the following command to specify the feature depth:

| SPEC TO | Blind $\Rightarrow$ Done $\Rightarrow$ Enter depth [nn.nn] : **4.0** $\Rightarrow$ **<ENTER ↵>** | {8.27} |

Pro/E now displays the following message: **"All elements have been defined. Select element(s) or action(s) from the dialog box."** The PROTRUSION: Extrude dialog box appears as shown in Fig. 8.17. LMB click on the **OK** button to complete the feature definition. Pro/E now displays **"PROTRUSION has been created successfully."**

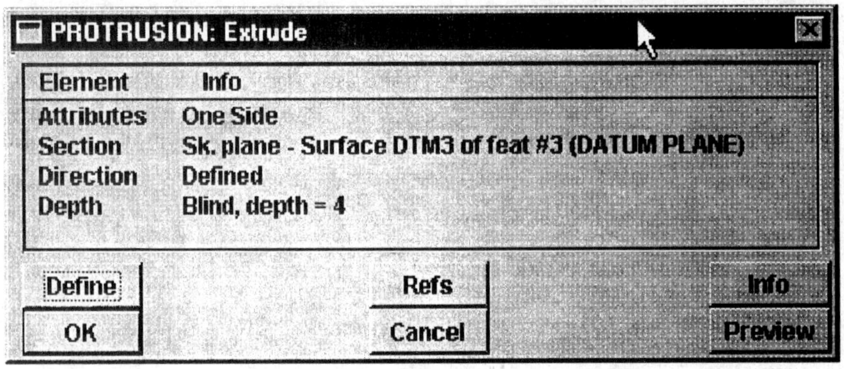

Fig. 8.17 Complete dialog box for the base feature.

Before we create our next feature, let's change some environment variables to clean up our current view so that the model appears as in Fig. 8.18. First, display the part in its isometric default view by using Cs. {8.10}. Next, turn off the datum planes by "un-checking" the Disp DtmPln environment variable and display the model using hidden lines.

| MAIN | Environment $\Rightarrow$ ☐ **Disp DtmPln** | |
| | **Hidden line** $\Rightarrow$ **Done-Return** | {8.28} |

We are now ready to create our next feature.

8.6.2 Revolve Protrusion Feature

The next feature we will create is a revolved protrusion on the left plane of the base feature (see Fig. 8.18). Features created by revolving require a centerline to serve as an axis of rotation, and a **closed** section that is revolved around the centerline. In this case, we will revolve the feature 180° about the centerline to form the semi-circular surface evident in the solid model presented in Fig. 8.11. To create the revolved protrusion, we execute a command similar to Cs. {8.21} with the exception that a protrusion is to be **revolved**. The command to accomplish this action is:

PART Feature $\Rightarrow$ Create $\Rightarrow$ Solid $\Rightarrow$ Protrusion $\Rightarrow$ Revolve $\Rightarrow$ Solid $\Rightarrow$ Done{8.29}

We will add the revolve feature by sketching on the left plane of the base feature. To select this plane, we use the **query select** command.

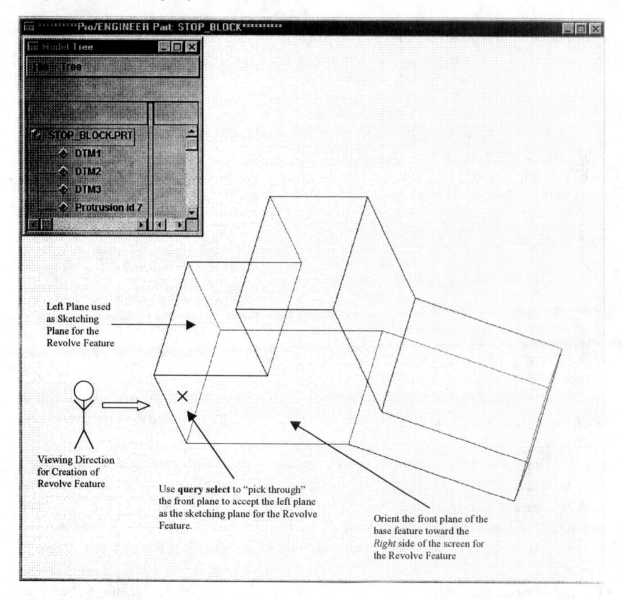

Fig. 8.18 Base feature result (datum planes off, hidden line displayed).

First, we need to assign the **One Side** attribute, and then use **query select** to "pick through" the front plane of the base feature to get at the left plane of the base feature.

ATTRIBUTES	One Side ⇒ Done ⇒ Setup New ⇒ Plane ⇒ Query Sel
	Select <u>through</u> the front plane of the base feature to get to the left plane of the base feature (Fig. 8.18).
	< LMB click on Left face of base feature> {8.30}

The front face will highlight in red. To move to the left face, which is behind the front face, toggle through the selections using the **query select next**, **previous**, and **accept** options.

CONFIRM	Next ⇒ Previous ⇒ Next … ⇒ Accept
	Upon seeing the face we want highlighted in red, use the **Accept** option to choose the left face. {8.31}

A red arrow appears denoting the direction of feature creation. Since the arrow direction is out of the left face, we are satisfied with this direction and **Okay** the direction using Cs. {8.23}. Upon accepting the direction, Pro/E prompts us to orient the sketch. Choose the front plane so that it faces the right-hand side of the screen (Fig. 8.18).

SKET VIEW	Right ⇒ Plane ⇒ Pick
	Select front plane of the base feature for it to face toward the <u>Right</u> side of the screen.
	< LMB click on Front Plane of base feature> {8.32}

This command sequence produces the sketcher grid orientation shown in Fig. 8.19.

Follow the SADRMR rule to complete the section. **S**ketch the vertical centerline using the sketching techniques described in Section 8.5.1.2. (Note, you only have to click the LMB once, to create the vertical centerline). Use the **Sketch ⇒ Rectangle** command to create a rectangle on the right-hand side of the centerline. **A**lign the top, bottom, and the right-hand side of the rectangle to the existing part geometry (Fig. 8.19). Remember, you have to sketch the rectangle near the existing part geometry for the alignment tool to be effective. Because we have aligned three of the four sides of the rectangle to existing part geometry, it is necessary to specify only one **D**imension before we **R**egenerate the sketch. The rectangle's horizontal dimension is 2.0 which is one-half of the base feature depth of 4.0. To demonstrate the use of arithmetic statements when entering numerical values in Pro/E, we **M**odify the dimension using the dimension 4.0 divided by 2 (**4.0/2**).

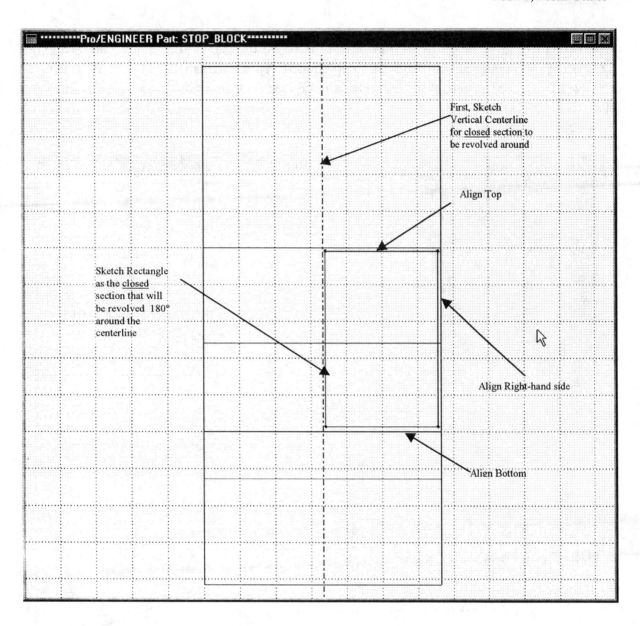

Fig. 8.19 Sketcher view of the revolve feature.

SKETCHER   **Modify** ⇒   <**LMB** Select dimension to modify> ⇒
Enter a new value [1.90] : **4.0/2** = 2.00 <**ENTER** ↵>{8.33}

Pro/E automatically displays "= 2.0" after <ENTER ↵> indicating the numerical value calculated from the arithmetic statement. **R**egenerate the section and then revolve the section 180° with the following command:

REV TO   180 ⇒ **Done**   {8.34}

Upon completing the revolved protrusion feature, orient the part in the default view with Cs. {8.10}. The drawing at this stage of completion is represented in Fig. 8.20. Whenever a revolved feature is created, Pro/E automatically creates a datum axis for future referencing— the axis is labeled A_1 in Fig. 8.20. This is convenient because we will use this datum axis to create the hole feature.

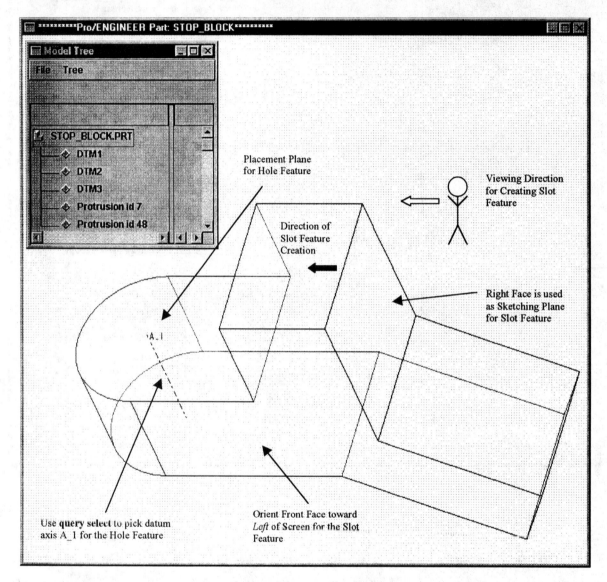

Fig. 8.20 Results for Stop_Block after adding the revolve feature
(Datum Planes <u>Off</u>, Datum Axis <u>On</u>, Hidden Line Display, and Default View.)

8.6.3 Hole Feature

A hole feature is very easy to create when a datum axis exists at the location of the centerline of the hole. To create the hole feature, it is not necessary to enter the sketcher. To start the definition of the hole, use:

PART	Feature $\Rightarrow$ Create $\Rightarrow$ Solid $\Rightarrow$ Hole	{8.35}

This command sequence will bring up the **HOLE OPTS** menu; then select the "straight" option using:

HOLE OPTS	Straight $\Rightarrow$ Done	{8.36}

which pops up the HOLE: Straight dialog box on the right-hand side of the screen along with the **PLACEMENT** menu. On the **PLACEMENT** menu execute the following command to place the hole coaxial with datum axis A_1:

PLACEMENT	Coaxial $\Rightarrow$ Done	{8.37}

You may use query select to pick datum axis A_1 (Fig. 8.20).

GET SELECT	Query Sel $\Rightarrow$ <**LMB** click on datum axis A_1>	
	CONFIRM $\Rightarrow$ Accept	{8.38}

We select the placement plane for the hole as the top surface of the revolved protrusion (Fig. 8.20).

GET SELECT	Query Sel $\Rightarrow$ < **LMB** click on top surface of revolve protrusion >	
	CONFIRM $\Rightarrow$ Accept	{8.39}

We next specify a 2.0 diameter hole to start on one side of the base feature and direct it to go through all the surfaces it encounters.

SIDES	One Side $\Rightarrow$ Done	
	SPEC TO Thru All $\Rightarrow$ Done $\Rightarrow$	
	Enter DIAMETER [0.nn00] : **2.0** $\Rightarrow$ <ENTER ↵>	{8.40}

Pro/E will display the following message: **"All elements have been defined. Select element(s) or action(s) from the dialog box."** LMB click on the **OK** box on the HOLE: Straight dialog box to complete the definition of the hole feature.

8.6.4 Slot Feature

Our last step in producing a solid model of the STOP_BLOCK is to incorporate the slot feature. The slot **removes** material from the top surface. Recall, the default orientation of the sketch section when we are **removing** material from a part is to orient the sketching plane **facing away** from the viewer. Therefore, we orient the part such that the sketch is on the right vertical face of the part. We require the front face of the part to be toward the left-hand side of the screen (See Fig. 8.20 to locate this plane). The command sequence to begin the slot feature is:

| **PART** | Feature ⇒ Create ⇒ Solid ⇒ Slot ⇒ Extrude ⇒ Solid ⇒ Done | {8.41} |

We cut (extrude) the slot into the screen on **One Side.**

ATTRIBUTES	One Side ⇒ Done ⇒ Setup New ⇒ Plane ⇒ Pick
	Select the right plane as the sketching plane.
	< LMB click on Right face of base feature> {8.42}

The red arrow appears indicating the direction of feature creation. Note, Pro/E removes material inward toward the solid. **Okay** this as the viewing direction with Cs.{8.23}. Orient the front face toward the left side of the screen (Fig. 8.20) with the following command sequence:

SKET VIEW	Left ⇒ Plane ⇒ Pick
	Select the front plane of the base feature to face toward
	the Left side of the screen.
	< LMB click on Front Plane of base feature> {8.43}

The orientation of the sketch plane is shown in Fig. 8.21.

Sketch a vertical centerline directly on top of axis A_1. The centerline is used to denote that the rectangle is symmetrical about the vertical axis of the centerline. Align the centerline to the axis A_1. Next, use the rectangle tool to **S**ketch a **closed**. Note: to denote symmetry, the rectangle needs to be drawn such that the **distance from the left side of the** rectangle to the centerline is the same as the distance from the right side of the rectangle to the centerline. **A**lign the rectangle to the top face. Assign **D**imensions, **R**egenerate, **M**odify, and **R**egenerate again before leaving the sketcher.

To specify the depth of the slot, several options may be used. Feasible options on the SPEC TO menu are—**Thru Next** and **Thru All**. Let's use the first option—**Thru Next**—to create a slot feature that goes through to the intersection of the next surface.

| **SPEC TO** | Thru Next ⇒ Done | {8.44} |

Pro/E provides a message that the slot feature has been created successfully. Select the **OK** button on the SLOT: Extrude dialog box to complete the definition of the slot.

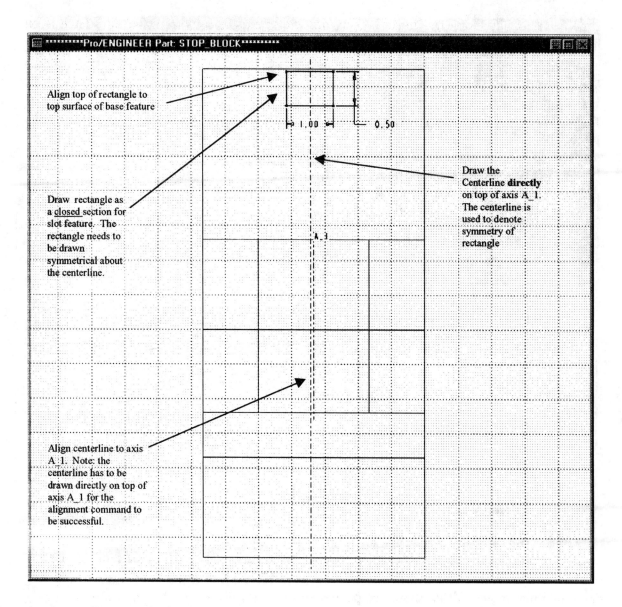

Fig. 8.21   Sketcher view of slot feature.

8.6.5 Completed Part

We have now completed our part and are ready to tackle the Drawing Tutorial. Make sure you save your part with Cs. {8.3} before you proceed. Turn off the datum planes and datum axis environment variables, and change the display to a hidden line view with Cs. {8.28}. The part in the default viewing orientation (Cs. {8.10}), should look like Fig. 8.22. Note how the *Model Tree* window keeps track of all the features we created and assigns them a unique identification number. Your numbers will most likely be different.

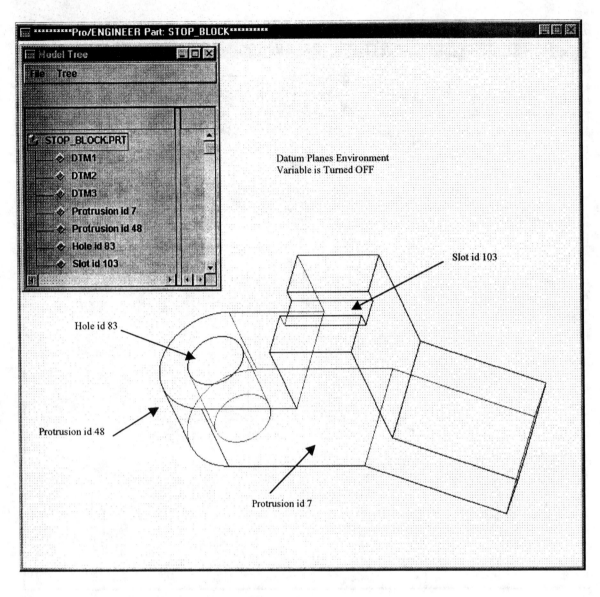

Fig. 8.22 Solid model of the completed part.
Datum planes off, datum axis off, hidden lines displayed, and default view.

8.7 MULTI-VIEW DRAWINGS

Now that we have completed the solid model of the STOP_BLOCK, it is necessary to prepare engineering drawings. These engineering drawings are provided to manufacturing personnel so they can produce the part. Recall, when saving the drawing files, Pro/E automatically appends the .drw suffix to the filename (see Section 8.2.5 on file naming convention).

Drawings have bi-directional associativity with their respective parts as discussed in 8.1.1.5 and illustrated in Fig. 8.3. This means that whenever we change the dimensions or features on the part in part-mode, the drawing file is automatically updated. The multiple views in the drawing file are also associative—whenever we change the dimension in one mode, all the other modes are modified accordingly. This is the power behind Pro/E.

8.7.1 Creating the Drawing

Let's start a new Pro/E session and create a multi-view drawing. When starting Pro/E, it is necessary to go into *drawing-mode* to develop a multi-view drawing.

MAIN Mode $\Rightarrow$ **Drawing** $\Rightarrow$ **Create** $\Rightarrow$ Enter Drawing name [DRW0001] : **s_block**{8.45}

Let's name the drawing **s_block**, and store the file as s_block.drw. Note, the drawing name does <u>not</u> have to be the same as the solid model stop_block.prt; however, in many cases, it is wise to retain the same name for tracking purposes. Pro/E then prompts us to, **"Select drawing size."** The recommended drawing size format is a landscape view of an 8-1/2 by 11 inch sheet.

GET FORMAT   **Set Dwg Size**
                     DWG SIZE TYPE   Landscape
                                         DWG SIZE   A   {8.46}

8.7.2 Creating Drawing Views

Once the drawing type and size has been defined, we add different views. Pro/E offers several choices of different types of drawing-views. Of these many types of views offered, we will only describe the general and projected views. These two views are sufficient to produce three-view drawings commonly employed in representing components to be manufactured.

General View –   A view that is placed **independent** of any other view. The view is oriented by the user and is usually the first view created. After placing this view, it is oriented and then used to generate projected views.

Projected View –   A view that is an orthographic projection of an existing view that is observed from its front, back, side, top, or bottom. The projected view is dependent on the general view that was defined first.

The command to add our first view is:

DRAWING Views $\Rightarrow$ Enter Model name [ ] ? : **stop_block**   {8.47}

In Cs. {8.47}, we enter the part name **stop_block** for the first view to be placed on the multi-view drawing. We also have the option of typing a question mark followed by the Enter-Key (i.e., **?** followed by <**ENTER** ⏎>) to bring up a listing of parts in the current directory. We scroll through the parts until we find **stop_block.prt** and then LMB click on the file name to select it. After selecting the part, Pro/E displays the sheet information at the bottom of the main working window as illustrated in Fig. 8.23.

The type of view is specified next. In this example, let's add a general view, orient it, and then generate projected views from the general view. To add a general view we use:

$\boxed{\text{VIEW TYPE}}$ General $\Rightarrow$ Full View $\Rightarrow$ No Xsec $\Rightarrow$ No Scale $\Rightarrow$ Done	{8.48}

Pro/E prompts us to, **"Select CENTER POINT for drawing view."** Pick the lower left quadrant of the sheet to place the view. This will add the general view to our sheet as shown in Fig. 8.23. Pro/E will automatically adjust the scale of the sheet to an appropriate size for the drawing. In Fig. 8.23, we note that Pro/E employed a scale factor of 0.167.

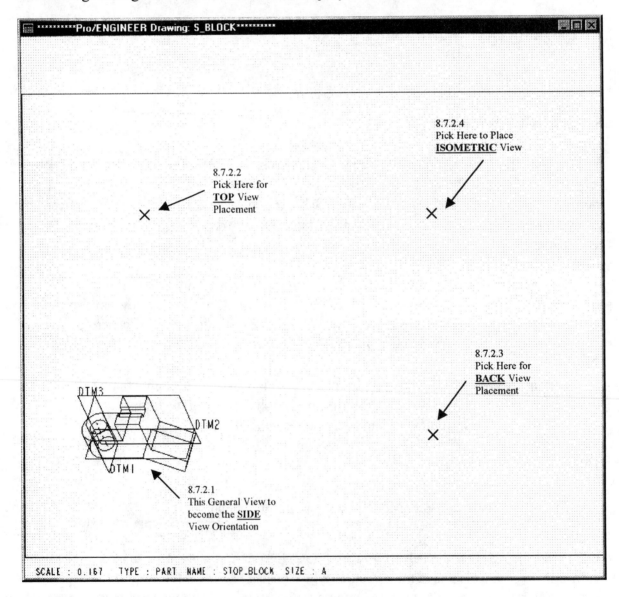

Fig. 8.23 Arrangement of the drawing sheet in Pro/E.

8.7.2.1 Side View

We are not satisfied with the default orientation that Pro/E has provided for the general view. We prefer to have a different view of the part; more specifically, a view of he front side. We change the general view by orienting it so the positive side of DTM3 faces toward the front, and the positive side of DTM1 faces toward the right. The positive side of DTM2 facing toward the top will produce the same result.

ORIENTATION	**Front** $\Rightarrow$ **Query Sel**
	< **LMB** select and then Accept DTM3 to face Front >
	Right $\Rightarrow$ **Query Sel**
	< **LMB** select and then Accept DTM1 to face Right > {8.49}

8.7.2.2 Top View

After orienting the general view to provide a SIDE (front) view, we now add a **projected** TOP view and **projected** BACK view of the stop block. Both of these views are dependent on the general view (i.e., the SIDE view). To add the projected TOP view use:

VIEWS	**Add View** $\Rightarrow$ **Projection** $\Rightarrow$ **Full View** $\Rightarrow$ **No Xsec** $\Rightarrow$ **No Scale** $\Rightarrow$ **Done** {8.50}

At the prompt for selecting the center point for this view, select a location directly above the SIDE (front) view in the upper left quadrant as indicated in Fig. 8.23.

8.7.2.3 Back View

The BACK view is created in a similar fashion. Use Cs. {8.50} again, but when prompted for the center point for the drawing view, select the location to the right of the SIDE view in the lower right quadrant of the drawing sheet as shown in Fig. 8.23. The BACK view is automatically aligned horizontally so that it is in line with the SIDE view.

8.7.2.4 Isometric View

The ISOMETRIC View is created as another general view; i.e., it is not dependent on any other view. Add the ISOMETRIC View in the top right quadrant of the sheet using the following commands.

VIEWS	**Add View** $\Rightarrow$ **General** $\Rightarrow$ **Full View** $\Rightarrow$ **No Xsec** $\Rightarrow$ **No Scale** $\Rightarrow$ **Done** {8.51}

Followed by:

DRAW VIEW	**Done-Return** (To accept the **Default** Orientation) {8.52}

8.7.3 Modifying the Drawing Views

Four different views of he stop block have been created. We will now enhance these views to make them easier to interpret by machine shop personnel. First, turn off the datum planes environment variable with Cs. {8.8}. Next, display the 3 projected views (SIDE, TOP, and BACK) with hidden lines displayed in gray. The command to accomplish this is:

VIEWS **Disp Mode** ⇒ **View Disp** ⇒ **Pick** <**LMB** each of the 3 Projected Views> ⇒ **Done Sel** ⇒ **Hidden Line** ⇒ **Done**   {8.53}

Execute Cs. {8.53} on the ISOMETRIC view; however, for any isometric view hidden lines are not shown. We must display the isometric with invisible hidden lines. We manage this by choosing the **No Hidden** option in the **VIEW DISP** menu.

If you are not satisfied with the location of your views, they can be moved. However, the movements of the projected views are dependent on each other. The ISOMETRIC view is a general view so we can move it anywhere we please. The command for moving any view is:

VIEWS **Move View** ⇒ **Pick** <**LMB** view to move> ⇒ <**LMB** to set views>   {8.54}

Finally, we may want to change the sale of the drawing that is displayed in the lower left corner of the drawing. (In Fig. 8.23, the scale is 0.167). To change the scale factor, we use:

DRAWING **Modify** ⇒ <**LMB** click on "SCALE : 0.167" in the lower left sheet corner> The Scale will highlight in red and the message window will prompt you: Enter value for scale [0.167]: **0.30** <**ENTER** ↵>   {8.55}

After using Cs. {8.55} to change the scale factor to 0.30, it may be necessary to move the views to better suit the new scale factor.

8.7.4 Dimensions

Next, we add dimensions to the views. Dimensions are referred to as **details** in Pro/E vernacular. We use the Show/Erase Dialog Box (Fig. 8.24) that pops up on the right-hand side of the screen to show or erase dimensions.

DRAWING **Detail** ⇒ **Show/Erase**   {8.56}

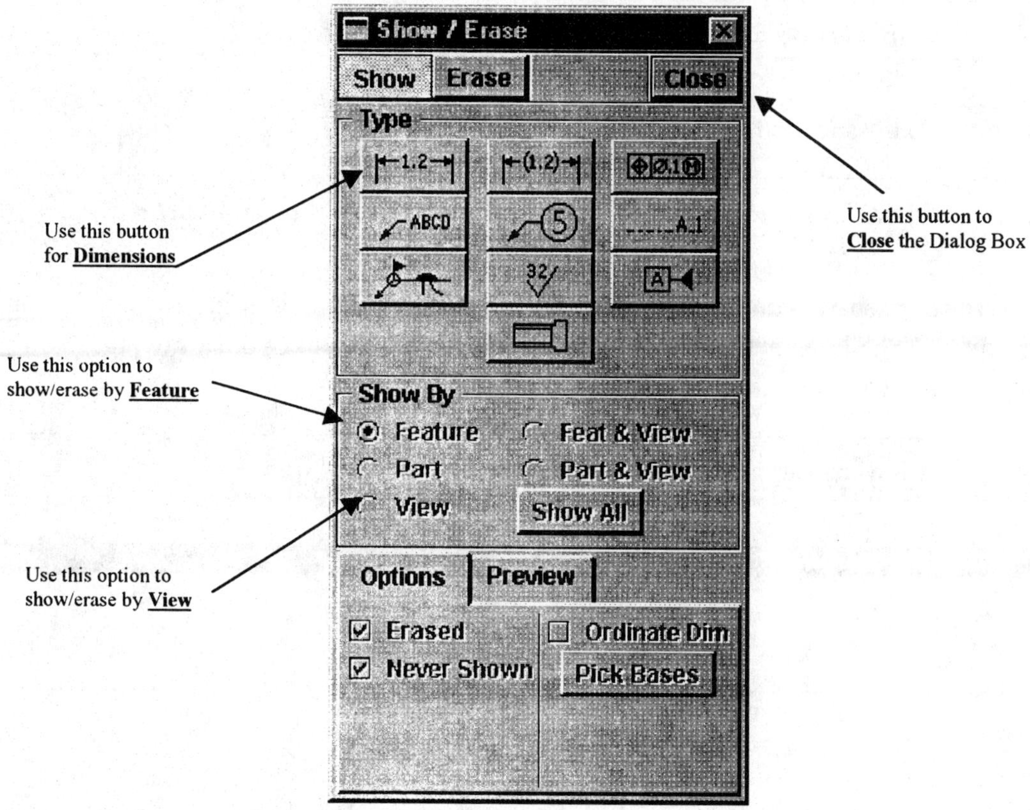

Fig. 8.24 The Show/Erase dialog box used for dimensioning.

Use the **Show** or the **Erase** button together with the dimension button 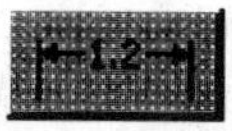 to display or erase dimensions. Two of the options are to show/erase dimensions by **View** or by **Feature**. When you have made your selections in the Show/Erase dialog box, use the **LMB** to select views or features to dimension. Having completed the selection, close the dialog box to exit the Show/Erase Dimensions command.

To modify the dimensions, we can **move** them to new locations **within** views. We can also use the same command to trim/extend the witness lines for a dimension.

DETAIL	**Move** ⇒ **Pick** <**LMB** dimension to move or witness line end to trim/extend>
	⇒ <**LMB** to set dimension or set end of witness line> {8.57}

We can **switch dimensions** that are displayed in one view and display them in another.

DETAIL	**Switch View** ⇒ **Pick** <**LMB** the dimensions to be switched to another view>
	⇒ **Done Sel** <To end selecting dimensions>
	⇒ <**LMB** select view to switch the dimensions to> {8.58}

We can **flip arrows** on the dimensions so the endpoints of the arrows point toward or away from each other. Try this.

| **DETAIL** | Flip Arrows ⇒ Pick <LMB dimension to flip arrows> | {8.59} |

We can **move text** associated with a dimension.

| **DETAIL** | Move Text ⇒ Pick <LMB dimension text> ⇒ <LMB to set> | {8.60} |

As an exercise, prepare your drawing to resemble that shown in Fig. 8.25. During this exercise some of the lines may disappear when you are modifying your drawing. To repaint the main working window, use the repaint command.

| **MAIN** | View ⇒ Repaint ⇒ Done-Return | {8.61} |

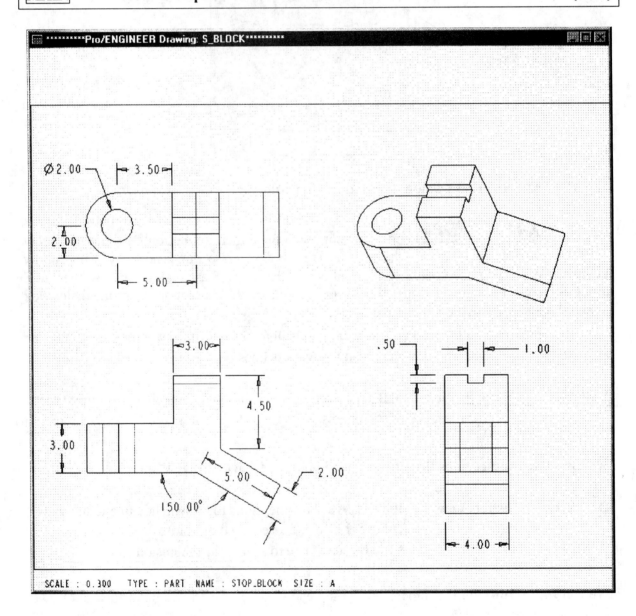

Fig. 8.25 Completed multi-view drawing

8.7.5 Producing Hardcopy

To produce a hardcopy of your drawing you may have to consult with the individual responsible for printing in the computer laboratory. The following generic approach will work in most situations. Do not get frustrated if it is not effective for you—seek help.

Pro/E produces a hardcopy output as a *.plt plot file. It will capture everything on your drawing—including datum planes and datum axes. If you do not want these datums to appear in your hardcopy output (probably you do not) turn off their respective environment variables by using Cs. {8.8}. After the drawing looks like the print that you would like to make, initiate the following command to bring up the Plot Dialog Box shown in Fig. 8.26.

DRAWING Interface ⇒ Export ⇒ Plotter	{8.62}

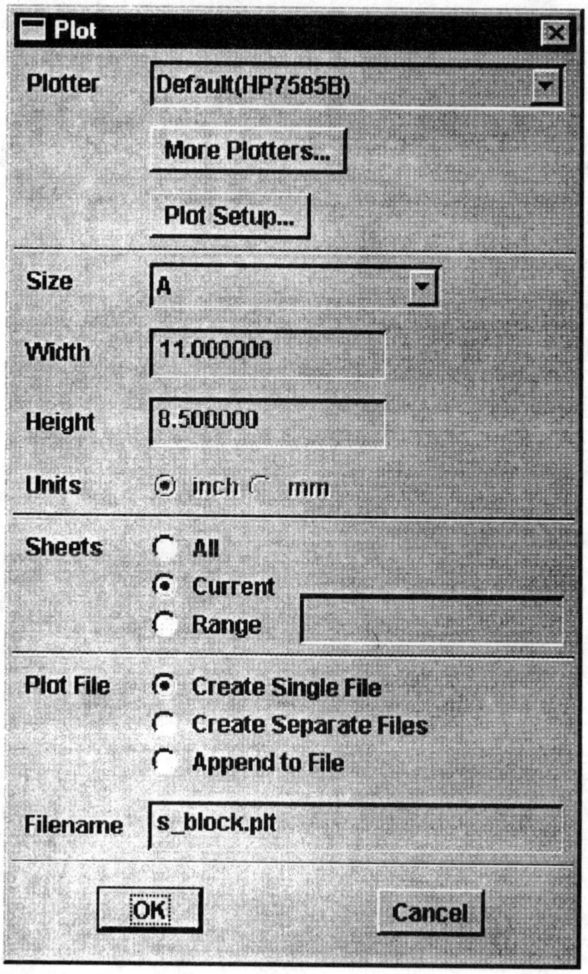

Fig. 8.26 The plot dialog box

The default plotter in the dialog box in Fig. 8.26 is the HP7585B. This may not be the plotter in use in your system, and it is necessary to change it by selecting the **More Plotters...** button. Scroll through the list of compatible plotters until you find one corresponding to the plotter

installed in the computer system you are employing. After selecting the plotter, choose the
Plot Setup... button and set the scale to less than 1.00 (e.g. 0.90). Lastly, close the plotter
dialog box by selecting OK . This will generate a *.plt output that, depending on your
system configuration, will be sent directly to the plotter or will be written to a file on your
hard-drive. If the file is written to your hard-drive, you will have to plot out the file by
executing the print command directly from the DOS prompt, Windows NT, or Unix
environment.

8.8 FREQUENTLY REFERENCED COMMAND SEQUENCES

Throughout this chapter, several command sequences are referenced. To minimize the
time required to find these frequently employed commands, the following list is presented.
We suggest placing a tab on this page (or photocopy it) for quick reference.

MAIN   Mode ⇒ Part ⇒ Create ⇒ Enter Part name [PRT0001] : **block**   {8.2}

MAIN   Dbms ⇒ Save ⇒ Enter object to save [name.xxx]: **<ENTER ↵>**   {8.3}

MAIN   Mode ⇒ Part ⇒ Search/Retr ⇒ Current Dir ⇒
<Use **LMB** to retrieve the part file >   {8.4}

MAIN   Exit ⇒ Do you really want to exit? [N] : **YES**   {8.6}

PART   Feature ⇒ Create ⇒ Datum ⇒ Plane ⇒ Default   {8.7}

MAIN   Environment ⇒ ☑ Disp DtmPln ⇒ Done-Return   {8.8}

MAIN   View ⇒ Orientation ⇒ Default ⇒ Done-Return   {8.10}

SKETCHER   Delete ⇒ Delete Item ⇒ Pick ⇒ <LMB click on item to delete>   {8.15}

SKETCHER   Dimension ⇒ Normal ⇒ Pick
<LMB Select one entity or endpoint on SKETCH>
OPTIONAL → <LMB Select one entity or endpoint on PART>
<MMB Place dimension with middle mouse button>{8.17}

SKETCHER   Regenerate   {8.18}

| SKETCHER | Modify $\Rightarrow$ Mod Entity $\Rightarrow$ Pick $\Rightarrow$ <LMB dimension to modify> $\Rightarrow$ Enter a new value [oo.oo] : **nn.nn** <ENTER ↵> | {8.19} |

| PART | Feature $\Rightarrow$ Create $\Rightarrow$ Solid $\Rightarrow$ Protrusion $\Rightarrow$ Extrude $\Rightarrow$ Solid $\Rightarrow$ Done {8.21} |

| DIRECTION | Okay | {8.23} |

| MAIN | Environment $\Rightarrow$ ☐ Disp DtmPln $\Rightarrow$ Hidden line $\Rightarrow$ Done-Return | {8.28} |

| VIEWS | Add View $\Rightarrow$ Projection $\Rightarrow$ Full View $\Rightarrow$ No Xsec $\Rightarrow$ No Scale $\Rightarrow$ Done {8.50} |

| VIEWS | Disp Mode $\Rightarrow$ View Disp $\Rightarrow$ Pick <LMB each of the 3 Projected Views> $\Rightarrow$ Done Sel $\Rightarrow$ Hidden Line $\Rightarrow$ Done | {8.53} |

| DRAWING | Modify $\Rightarrow$ <LMB click on "SCALE : 0.167" in the lower left sheet corner> The Scale will highlight in red and the message window will prompt you: Enter value for scale [0.167]: **0.30** <ENTER ↵> | {8.55} |

8.9 SUMMARY

This description of Pro/E is intended to introduce you to this powerful computer-aided design tool. The treatment is brief and is aimed at developing entry level skills. You should be able to duplicate the solid model of the stop block and its multi-view drawing in from 4 to 5 hours. For those readers who wish to extend their ability to employ Pro/E, we suggest the much more detailed coverage in Reference [1].

The characteristics of Pro/E were described to show the advantages designing with a feature based solid modeler. The additional characteristics such as parametric, relational and associative provided with Pro/E are important time savers for engineers.

The Pro/E environment was described in detail to acquaint you with the various windows and the menus. File naming, file saving and the extensions for the various files maintained in Pro/E were also discussed.

Since Pro/E employs features in the construction of solid models, these features were categorized as datum or solid types. Datum planes were described in detail since they are essential in creating the solid model. Then the solid model features such as protrusion, cut, slot and hole were discussed. Command sequences for developing a solid model with these solid features were defined and demonstrated.

The very important sketcher mode of operation is discussed in detail with descriptions of the sketcher tools. The acronym SADRMR is defined to aid in preparing a sketch of the section of a solid feature. The acronym means:

- Sketch
- Align
- Dimension
- Regenerate
- Modify
- Regenerate

An example of a stop block is introduced to demonstrate the techniques and command sequences employed to develop a complete solid model of a part in the Pro/E environment. The example is detailed and intermediate screens are displayed to shown expected results during the construction of the model from appropriate solid features. The command sequences frequently used are bordered and numbered to facilitate in subsequent applications of Pro/E.

Multi-view drawings are easy to create in pro/E after the solid model of the part is complete. All of the information for creating multi-view drawings is in the files. As operators of the CAD program, we extract this information and position the various views at appropriate locations on the sheet of drawing paper. The command sequences for recalling files and locating the various optional views are defined. Dimensioning the traditional three-view drawing is described, and the stop block is used as an example in a demonstration of these techniques.

Finally, a summary of the frequently referenced command sequences are given in a convenient listing.

REFERENCES

1. Zhang, G. Engineering Design and Pro/ENGINEER, College House Enterprises, Knoxville, TN, 1998.

EXERCISES

1. Describe the meaning of the characteristics of Pro/E such as:

- Solid modeler
- Feature based
- Parametric
- Relational
- Associative

2. Start Pro/E and create a filename for a part called block1.
3. Prepare a solid model of a cube 3 inch on each side.
4. Cut a slot in the top of the solid model of the cube prepared in exercise 3 similar to that shown in Fig. 8.2. The slot is to extend across the entire top of the cube with a width of one inch and a depth of ½ inch.

5. Drill a hole in the solid model of the cube of exercise 3 and 4 similar to that shown in Fig. 8.2. The hole should be centered on the right side of the cube; it should extend completely through the cube, and be ¾ inch in diameter.

6. Prepare a three-view drawing of the cube of exercise 5 showing the front, top and right side view. Incorporate dimensions on these views.

7. Prepare an isometric view of the cube of exercise 6.

8. Create a solid model of the drawing presented in Fig. 5.12. Print a hard copy of it.

9. Prepare a multi-view drawing of the solid model of exercise 8 showing four views including; front, top, right side, and isometric.

10. Prepare a dimensioned three-view drawing of the solid model of exercise 8.

11. Prepare a solid model of the television set in your home.

12. Prepare a multi-view drawing of the solid model of exercise 11.

CHAPTER 9

MICROSOFT EXCEL–97

9.1 INTRODUCTION

Microsoft® EXCEL–97 is a spreadsheet program that is an extremely important tool in engineering, because it is useful in so many different applications. We will cover techniques for three of these applications including, preparation of a parts list, performing calculations, and drawing graphs. There are several different spreadsheet programs on the market, namely EXCEL® by Microsoft, QUATTRO PRO® by Borland, and LOTUS 1 • 2 • 3® by Lotus. All of these programs have a similar format and they all have essentially the same capabilities. If you learn to use one program, it is relatively easy to change to a different program with a modest investment in time. We have selected EXCEL–97 because it is up to date, has been adopted by many universities, and you will probably have access to it on the computer net during your tenure in college.

The purpose of this chapter is to show you how to navigate on the EXCEL spreadsheet, to make tables, perform calculations, plot several types of graphs, print the output, and save a file. You should recognize that these objectives are limited. We are concentrating on developing your entry level skills. Hopefully, you will find the spreadsheet tool important enough to develop a higher skill level with independent study. More complete and detailed treatments are given in references [1, 2].

9.2 THE SCREEN

You load EXCEL from WINDOWS by clicking on the EXCEL icon and the program opens with the spreadsheet display shown in Fig. 9.1. Let's explore this display. At the top you will note five rows, then below in the center is a big table that is the spreadsheet proper, and finally at the bottom two more rows.

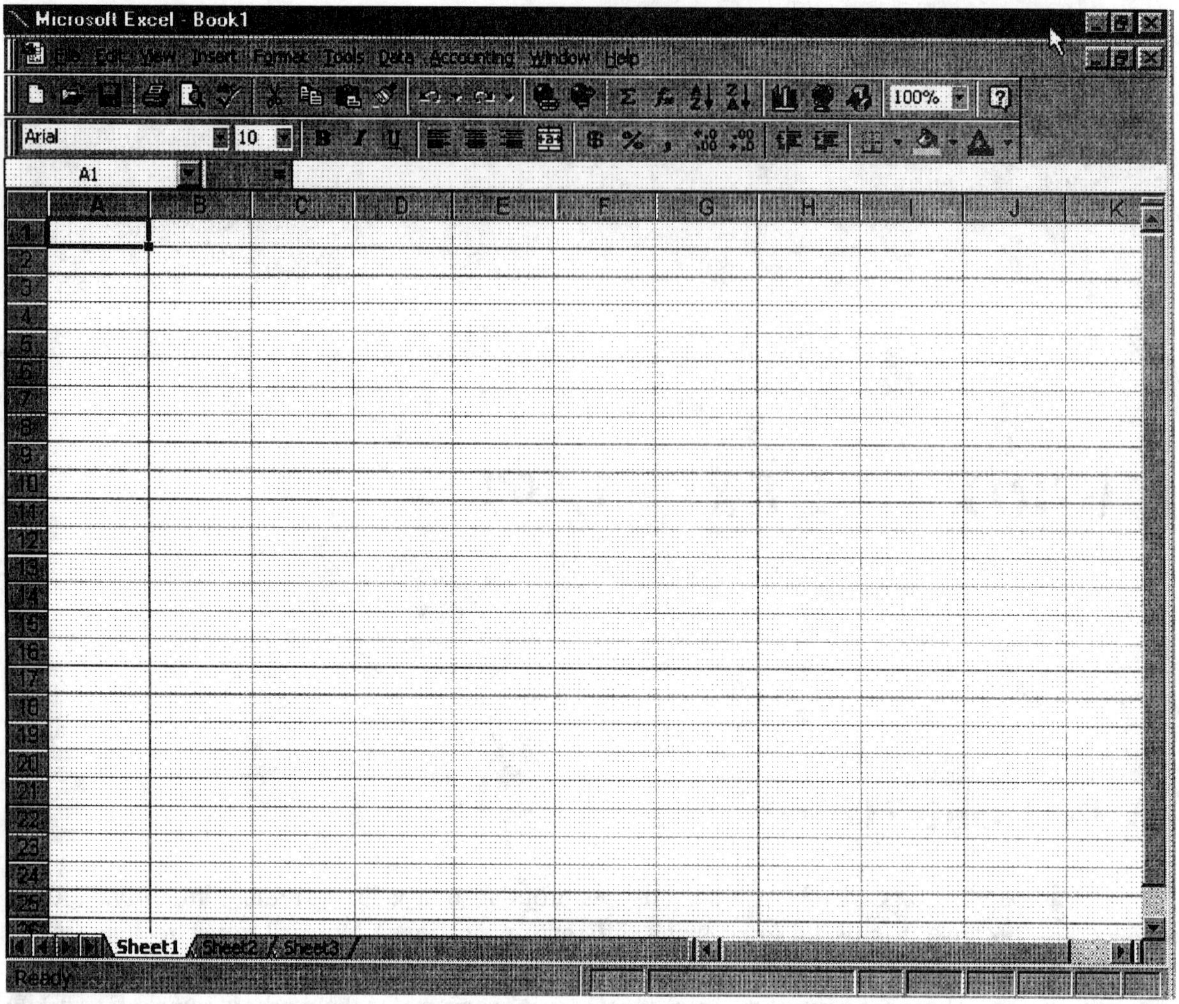

![Microsoft Excel screen display showing a blank Book1 spreadsheet with menu bar, toolbars, and cells labeled A through K and rows 1 through 25]

Fig. 9.1 The screen display in Microsoft Excel

9.2.1 The Five Top Rows

The top most row, called the title bar, shows that you are working in EXCEL and the name of the open file is Book 1 (until you save and rename the file). The three buttons to the far right side of the title row are common to all Microsoft® programs.

The left hand button with the dash symbol places the program aside so that you can work in a different program. The middle button toggles to permit you either to expand your view of the screen or to shrink it. The right button with the X symbol closes both EXCEL and the file upon which you are working. Do not use this button if you intend to use EXCEL again during your current session. Instead, use the dash button and set the active program aside. Click on each of these three buttons until you understand their functions.

The second row from the top is the called the menu bar; it lists ten different features included in the program beginning with File and ending with Help. The menu bar is similar to that found in other Microsoft® programs. Using your mouse, select one of the menu items and click. A pull down menu is displayed. We will not list the items on each of these pull down menus because you will learn faster by clicking than by reading. The purpose of each of the menu entries is listed below:

1. File is to open or close, save, or print a file or to exit EXCEL.
2. Edit is to change entries that you have already made.
3. View is to change the appearance of the screen or to add a header or footer.
4. Insert is to add cells, rows and columns to the spreadsheet.
5. Format is to change the height of rows and the width of columns.
6. Tools are to check spelling, audit calculations and for developing macros.
7. Data is to provide a means for sorting, filtering and forming data entries.
8. Accounting is to access import, report and what if wizards.
9. Window is to enable you to open a new window.
10. Help is to provide answers to your questions as you attempt to learn the particular program that is loaded.

The third row from the top exhibits a toolbar with about twenty buttons. Again, this is a standard toolbar (unless someone has customized it) with many icons common to other Microsoft® applications. The icon figures give a good clue as to the purpose of each button; however, but if you are in doubt, use the mouse and point to a particular button. Do not click. Wait for a moment and a small label appears to tell you the function of each button. Again, point and click until you understand.

The fourth row down is the formatting tool bar—again many of its icons are common to other Microsoft® applications. Icon labels on each button give us a hint of the purpose of each button, but if we are in doubt, point to the button and a label will appear defining the function. Try changing the font type and size.

The fifth and last row down from the top bar provides information about the spreadsheet. The block to the left defines the cell location on the spreadsheet that is active (A1 shows when you begin a new book or file). The longer block on the right, called the formula bar, is activated when you make an entry into a cell. The numbers or letters that you have entered into the cell appear in this block. If you wish to edit the entry without completely retyping it, activate the cell, point with the mouse to place the cursor in the formula block, and change the entry.

Okay! We have briefly defined the top five rows, but it is necessary for you to develop an understanding of the menu items and the icons by pointing and clicking. Don't be afraid to make the computer beep—that is part of the learning process.

9.2.2 The Spreadsheet

Most of the screen is occupied by the spreadsheet, which is simply a big table with many columns and rows. The columns are labeled across the top with letters A, B, C, to K and beyond, and the rows are labeled down the left side with numbers 1, 2, 3, to 26 and many more. Each little block, called a cell, is identified with its coordinates (i.e. In Fig. 9.1, the cell A1 is outlined with a border indicating it is active). Note that we put the letter (column location) first and the number (row location) second. The cells are locations in a very large table (spreadsheet) where we enter numbers, text or formulas.

We can move about on the spreadsheet using either the mouse or the key board. When the pointer is on the spreadsheet, the location is identified with a large plus sign. We may use the mouse to point to any cell on the screen; however, if the cell of interest is not visible on the screen, the scroll buttons located to left or below the spreadsheet are used to bring this cell into the field of view. When the big plus sign is pointed at the correct cell—click; the cell becomes active and is ready to receive the data we enter. We can also move the active cell by using the arrow keys, the tab key, the shift-tab keys, and the control-arrow keys. Try them and note how the active cell moves about the spreadsheet.

The spreadsheet is much larger than it looks on the screen, which shows only columns A through K and rows 1 through 26. (The number of columns and rows visible on the screen will depend on the size of the monitor. A 17 inch monitor was used to record the image shown in Fig. 9.1). Explore down on the spreadsheet by holding the control + the arrow-down key, and note that the number of the last row is 65,536. Depressing the control and the arrow-right keys moves the active cell to the far right of the spreadsheet, and indicates a column heading of IV which is the 256th column. We have a total of $(256)(65,536) = 16,777,216$ cells on the spreadsheet. Our spreadsheet is huge with many cells—let's hope we never have to use all of them!

9.2.3 The Bottom Rows

In the next to last row at the bottom of the spreadsheet, we have three tabs that indicate sheet numbers 1, 2 and 3. When opening EXCEL and starting a file, we begin with a workbook for a given project. The program automatically establishes many sheets (pages) in our workbook[1], although only three are visible to us. If our project requires more than three worksheets, we add more by clicking on Insert in the menu and then on Worksheet in the drop-down menu. We may enter data and perform calculations on several different worksheets in this file by clicking on the tabs to move from one sheet to another. If you right click on the buttons to the left of these tabs, a drop-down menu allows you to select an active sheet number.

The row at the very bottom of the spreadsheet is called the status bar. It provides a brief statement about the function of any button that is being activated and other information about the status of the program. In Fig. 9.1, the status message is "Ready" indicating that the EXCEL program is prepared to accept your data entry in cell A1.

---

[1] I added 35 worksheets without establishing the maximum number available.

9.3 PREPARING A PARTS LIST

Let's begin to learn techniques for using EXCEL by preparing a parts list for the postal scale. Load the program from **WINDOWS** by clicking the Start button, select Programs and then Microsoft Excel. The screen display should resemble that shown in Fig. 9.1, indicating that we are ready to start. We recommend that you title the file before beginning to develop the parts list. Click on File and select Save As. When the Save As dialog box page is displayed, enter a suitable file name for the project (PARTS LIST) and click the save button. The file name is now displayed on the top row of the spreadsheet.

We start the parts list by typing its title in cell A1. Since the title should be obvious and easy to read, we use bold Arial font and increase its size to 16 point. Skip a row and position the cursor on cell A3 to begin organizing the section headings as shown below:

PARTS LIST FOR A POSTAL SCALE				
NUMBER	NAME OF ITEM	DESCRIPTION OR DRAWING NO.	QUANTITY	PRICE
1	BASE SUPPORT	WOODEN SHEET---- DWG. NO. 100-01	1	$ 0.32
2	HOUSING	SHEET METAL -------DWG. NO. 100-02	1	$ 0.83
3	FACE PLATE	PLASTIC SHEET WITH LABEL FOR SCALE	1	$ 1.15
4	PLATFORM	SHEET METAL -------DWG. NO. 100-03	1	$ 0.35
5	TOP LEVER	SHEET METAL -------DWG. NO. 100-04	1	$ 0.43
6	LOWER LEVER	SHEET METAL -------DWG. NO. 100-05	1	$ 0.55
7	U --- SUPPORT	AL BAR 1/16 X 3/8 ---DWG. NO. 100-06	1	$ 0.14
8	HELICAL SPRING	1/2 OD X 2 IN. LONG WITH END LOOPS	1	$ 0.27
9	CURVED BAR	AL BAR 1/16 X 3/8 ---DWG. NO. 100-07	1	$ 0.05
10	TARE SCREW	8-32 BRASS THUMB SCREW 1/2 IN. LONG	1	$ 0.14
11	POINTER	1/32 SHEET METAL --- DWG. 100-08	1	$ 0.08
12	RIVETS	0.40 IN. DIA X 1/8 IN. LONG SOFT STEEL	10	$ 0.02

We have five section headings (columns) in the parts list:

    A. A number which refers to the part number; every unique part has its own part number.
    B. The name of the item.
    C. A description of the item or a drawing number defining the part.
    D. The quantity of parts that will be needed to build a single prototype.
    E. The unit price for the item.

When we type these column headings in cells A3, B3, C3, D3, and E3, it becomes clear that the columns are not the correct width to accommodate the text. The column widths are adjusted by pointing with the mouse to the line between A and B on the column headings. When the pointer is positioned correctly, you will observe a short vertical line with arrows to the left and right. Press the left mouse button and drag the column boundary either to the left or the right to adjust the width of column A. Move to the position of the cursor to a position on the column headings between column B and C; repeat the process to size the width of column B. Continue to adjust all of the column widths until they all are the proper width to best show the data included in the parts list.

In our example parts list, we have shown 12 part numbers. You may have more part numbers, because the number of parts depends on the complexity of the design. We type the necessary information for the parts list into the table that is provided by the spreadsheet. Our only concern in this process is that we have activated the correct cell before making an entry. If one of our entries is too long, we can change the column width at any time to accommodate short descriptions. It is important to keep the descriptions short because the parts list should fit the page.

When you have completed the entries, work on the spreadsheet to improve its appearance. We have centered the part numbers, the quantities, and the prices within the respective columns. We center by marking all of the entries in a given column (point and drag with the mouse and watch a black background color fill the column). When the black covers the column entries you wish to mark, move the pointer to the formatting bar and click on the centering button. On the column for price, we initially made the entries with the default format; however, prices are in terms of dollars. We changed the format of the column by marking it, and then clicked on the $ button located on the formatting toolbar to show the prices in terms of a currency.

We are ready to print the parts list, but we want to check to see if it fits the page, has the correct margins and is readable. We click on File then Print Preview, to examine the appearance of the spreadsheet when it is printed. To change its appearance, click on Setup button that is located on the bar at the top of the page preview screen; then choose the sheet tab. From this dialog box, point and click on gridlines, so that all of the cell boundaries are visible when the spreadsheet is printed. Printing the cell gridlines makes the parts list much easier to read. Next, click on the page tab, and select either the portrait or landscape mode. The default setting is the portrait mode because it is more common, but if the spreadsheet is much wider than it is long, the landscape mode is preferred. You can also adjust the size and the quality (dots per inch) of the printing on this page. Finally, click on the margin tab and change the margins so that the spreadsheet is positioned properly on the page.

When you have made all of the necessary modifications on the print preview screen, the spreadsheet should look professional (that means very good). Next, click on the print button, select the number of copies, and print the spreadsheet. Your parts list should compare favorably to the one shown above.

9.4 PERFORMING CALCULATIONS

One of the most important advantages of any spreadsheet program is its capability to perform calculations. EXCEL performs not just one calculation, but a whole sheet full of them. Moreover, EXCEL performs the calculations almost instantaneously and without error. The results can be displayed in two different ways—as an accurate table or a suitable graph. In this section, we will show you two simple examples illustrating the computing power of EXCEL and illustrate the results in tabular format.

9.4.1 The Rich Uncle Example

For the first example, consider your rich uncle that we introduced in Chapter 7 on Tables and Graphs. Recall that we discussed the compound interest which you earned from his initial investment of $5000. We will now describe the results of this table, presented on the next page, which is an analysis of the growth of the investment. Let's recall the accumulation relation and the input data from the previous chapter where:

$$S = P(1 + i)^n = PM \tag{9.1}$$

Uncle contributed P = $5,000.00, and the interest rate (i) for the six month compounding period was 5%. (The annual interest rate was 10%). The sum S accumulated for a total of 80 compounding periods (40years) is to be tracked. Recall, n is the number of compounding periods and M is a multiplier dependent on n.

Load EXCEL and Save As—Compound Interest—to establish the file name. In the results shown on the next page, we titled the spreadsheet, and entered text describing the parameters that control the results in the first five rows (A1 to A5). We skip a row and define the headings for Periods, Multiplier and Sum in row 7. Initially, we define only three columns (A, B and C). We place the active cell at location A8 and begin by entering the number of the period starting with zero. We could type all 40 entries into column A, but it is easier to fill in the entries automatically. To let EXCEL do the work, position the active cell on the zero entry in cell A8, click on Edit, select Fill, and then Series. A dialog box is displayed; you select Column to indicate that you are filling entries into column A. You also select Linear because the series that you will use to fill column A is a linear series (i. e. 1, 2, 3, etc.). At the bottom of the dialog box, note two small boxes—one for Step and the other for Stop. We enter 1 for step and 40 for stop and then click on the OK button. The program fills all of the cells from A8 to A48 with numbers starting with 0 and increasing in steps of one until it stops at the number 40. The Fill command listed in the Edit menu item saves a lot of time.

Let's go next to column B and activate cell B8. We want to calculate the multiplier M in this column. The multiplier M is a number that increases with the number of the compounding periods. It is given by:

$$M = (1 + i)^n = (1.05)^n \tag{9.2}$$

With cell B8 active, we begin by notifying EXCEL that we intend to feed it a formula by typing =; then the formula. In this case, our entry would be =(1.05)^A8. We use the cell address A8 because it contains the value of n for the column of calculations being considered. We use the ^ symbol to indicate that we are raising (1.05) to a power. Just to the left of the formula bar, note the buttons labeled X, ✔ and f_x. After you have entered and carefully checked the formula, click the ✔ button and examine cell B8. It should read one, because your entry =(1.05)^A8 = $(1.05)^n$ = $(1.05)^0$ = 1.

DETERMINING THE RUNNING SUM						
FOR AN INITIAL INVESTMENT OF $5000.00						
ANNUAL INTEREST RATE OF 10%						
COMPOUND INTEREST SEMI-ANNUALLY						
NO TAXES PAID						
Periods	Multiplier		Sum	Periods	Multiplier	Sum
0	1.0000	$	5,000.00	40	7.0400	$ 35,199.94
1	1.0500	$	5,250.00	41	7.3920	$ 36,959.94
2	1.1025	$	5,512.50	42	7.7616	$ 38,807.94
3	1.1576	$	5,788.13	43	8.1497	$ 40,748.33
4	1.2155	$	6,077.53	44	8.5572	$ 42,785.75
5	1.2763	$	6,381.41	45	8.9850	$ 44,925.04
6	1.3401	$	6,700.48	46	9.4343	$ 47,171.29
7	1.4071	$	7,035.50	47	9.9060	$ 49,529.86
8	1.4775	$	7,387.28	48	10.4013	$ 52,006.35
9	1.5513	$	7,756.64	49	10.9213	$ 54,606.67
10	1.6289	$	8,144.47	50	11.4674	$ 57,337.00
11	1.7103	$	8,551.70	51	12.0408	$ 60,203.85
12	1.7959	$	8,979.28	52	12.6428	$ 63,214.04
13	1.8856	$	9,428.25	53	13.2749	$ 66,374.74
14	1.9799	$	9,899.66	54	13.9387	$ 69,693.48
15	2.0789	$	10,394.64	55	14.6356	$ 73,178.15
16	2.1829	$	10,914.37	56	15.3674	$ 76,837.06
17	2.2920	$	11,460.09	57	16.1358	$ 80,678.92
18	2.4066	$	12,033.10	58	16.9426	$ 84,712.86
19	2.5270	$	12,634.75	59	17.7897	$ 88,948.50
20	2.6533	$	13,266.49	60	18.6792	$ 93,395.93
21	2.7860	$	13,929.81	61	19.6131	$ 98,065.73
22	2.9253	$	14,626.30	62	20.5938	$ 102,969.01
23	3.0715	$	15,357.62	63	21.6235	$ 108,117.46
24	3.2251	$	16,125.50	64	22.7047	$ 113,523.34
25	3.3864	$	16,931.77	65	23.8399	$ 119,199.50
26	3.5557	$	17,778.36	66	25.0319	$ 125,159.48
27	3.7335	$	18,667.28	67	26.2835	$ 131,417.45
28	3.9201	$	19,600.65	68	27.5977	$ 137,988.32
29	4.1161	$	20,580.68	69	28.9775	$ 144,887.74
30	4.3219	$	21,609.71	70	30.4264	$ 152,132.13
31	4.5380	$	22,690.20	71	31.9477	$ 159,738.73
32	4.7649	$	23,824.71	72	33.5451	$ 167,725.67
33	5.0032	$	25,015.94	73	35.2224	$ 176,111.95
34	5.2533	$	26,266.74	74	36.9835	$ 184,917.55
35	5.5160	$	27,580.08	75	38.8327	$ 194,163.43
36	5.7918	$	28,959.08	76	40.7743	$ 203,871.60
37	6.0814	$	30,407.03	77	42.8130	$ 214,065.18
38	6.3855	$	31,927.39	78	44.9537	$ 224,768.44
39	6.7048	$	33,523.76	79	47.2014	$ 236,006.86
40	7.0400	$	35,199.94	80	49.5614	$ 247,807.21

Okay! You have successfully employed EXCEL to compute the first multiplier M; let's determine the other 40 multipliers by copying the first one. Activate B8, and then click the Copy button on the toolbar. You have just stored the formula =(1.05)^A8 in temporary memory; it will be stored there until it is replaced the next time you use the copy command. Next, mark (select) the B column starting with B9 and continuing to B48 by dragging the pointer down the column. Check that the region is black where you want to copy the multiplier term and then click on the Paste button. The cells B8 to B48 show the multiplier M superimposed on the black background color of the column. The black background is eliminated by clicking on any cell outside this region. In performing the calculations down the B column from the Copy—Paste commands, EXCEL modified the formula =(1.05)^A8. Activate cell B9—note the formula has changed to =(1.05)^A9. EXCEL modifies the formula to accommodate the changes in n as we move down the column. In this instance, it was necessary for EXCEL to make this modification. However, in the next example, this type of modification leads to errors. To avoid errors in some extended calculations, we will show the technique necessary to control the way EXCEL modifies the formulas in the Copy—Paste command sequence.

Inspecting the results for M in column B, we note that the number of decimals is not consistent from cell to cell. Select any cell in column B, click on either one of the two decimal buttons on the formatting toolbar and observe the number of decimals change. Click on one or the other of these buttons until you display the multiplier with four digits after the decimal point.

To determine the sum S that is accumulated with time, it is necessary to multiply the multiplier M by the initial investment of P = 5000. We select cell C8 and type =5000*B8. The symbol * is used to indicate multiplication. When we click on the check button on the formula bar, the entry in cell C8 changes from =5000*B8 to 5000. EXCEL noted that cell B8 contained the number 1, performed the calculation (5000)(1), and displayed the results in cell C8. When we wish to incorporate a cell address into a formula, it can be typed or we can point to the cell with the correct address. After pointing at the cell, click the left mouse button to enter its contents into the formula. In the point and click approach, one of the math symbols (+, −, *, /, ^) must precede the pointing.

Look again at cell C8, and note that the result 5000 is not in the correct format. It is necessary to show the results in as a currency indicating the sum accumulated from Uncle's investment is in terms of dollars. To change the format:

1. Mark the entire column from C8 to C48 by dragging the column with the cursor.
2. Check that the black background covers only this region.
3. Point and click on the $ button located on the formatting bar.
4. Clear the black background by clicking on any cell not in the black region.

When we converted the C column into a currency format another problem occurred. The results for several cells changed into #######. This symbol indicates that the column is not wide enough to display the result. To widen column C, we point to the column heading row at a location between C and D. When the line with the double arrows replaces the pointer, press the left mouse button and drag to the right to widen column C. With a wider column the result in cell C8 reads $5,000.00.

We have a result for the sum S in cell C8. Let's arrange for EXCEL to complete the calculations. Activate cell C8 because it has the required formula, and click on the copy button on the toolbar. Next mark the cells C9 through C48 where we want the formula to apply. The black background verifies that the region is marked correctly. Next, click on the paste button and EXCEL applies the formula to all of the marked cells in the C column calculating the results in each cell. Scroll down the column and check to check if all of the results are displayed. If you encounter #######, it is necessary to widen the column. The result in cell C48 should read $35,199.94.

We have computed the accumulated sum S for 40 periods, but we originally stated that S was to be determined for 80 periods. Why did we stop at n = 40? We divided the calculation into two parts so that the output, when printed, would fit on one page. It is easy to extend the calculation to n = 80 by using the copy command.

To complete our work, mark the block with corners at A7 and C8 and click the copy button. We now have our headings and the initial formula in temporary storage (memory). Move the cursor and activate cell E7. We skip column D to provide space in the table between the two lists of results. After activating cell E7, click on the Paste button and the results from the A7 to C8 block are copied (with modifications for the shift in columns) into the E7 to G8 block. Look at cell E8 and note that the period shown is n = 0. Let's change that value to 40 by editing the numbers displayed on the formula bar. When cell E8 is changed to 40, the results in E9 and G9 check with previous results for the period n = 40. This fact implies that the program in EXCEL has adjusted the formulas to account for the fact that we have changed the columns in which the calculations are made. Activate cells F8 and G8 and compare the formulas in them with those in cells B8 and C8. Have you noted the changes automatically made by EXCEL as we copied formulas from one column to another?

To complete the table, activate cell E8 and fill in the entries for n = 40 to 80. We again use the automatic number generator in EXCEL. Click on Edit, select Fill, and then Series. On the dialog box, click on Column, Linear, and indicate a step of 1 with a stop at 80. Click on OK and then scroll down the spreadsheet to check that you have filled in the correct entries in column E. Next, mark cells F8 and G8 and copy them. Mark the F8 – G48 block and paste the formulas from F8 and G8 into this block. Clear the black background and scroll down the G column. You will probably note the symbol ####### for many of the results. Widen column G until all of the results are shown in the currency format. As a check, you should have the result $ 247,807.21 in cell G48.

We are now ready to print the table showing all of the results. Click on File, then on Print Review. The screen shows the spreadsheet as it will appear when printed. We note that it could be improved with gridlines and a larger left hand margin. To show the gridlines, click the Set-up button. From the dialog box which appears on the screen, select the sheet tab. Click on the gridlines square and note that a check appears indicating the gridlines will be printed. Next, click on the OK button to return to the preview of the spreadsheet. To adjust the margins, simply point, click and drag a margin line until the table is centered in the sheet shown in the print review screen. Finally, click on the Print button to print the spreadsheet.

We will return to this spreadsheet later when preparing a graph of the results.

9.4.2 Strain on a Simply Supported Beam

In Chapter 3 on Mechanical Scales, we introduced an equation for determining the strain ε in a simply supported beam when it is subjected to a central load of W_u. The relationship for the strain is given below:

$$\varepsilon = 3\ W_u\ S/(2bh^2\ E) \tag{9.3}$$

where: W_u is the load (weight) in pounds or Newton.
S is the span of the beam in inch or meter.
b is the depth of the section of the beam in inch or meter.
h is the height of the section of the beam in inch or meter.
E is the elastic modulus in psi or Pascal.

Let's program this relation in EXCEL to determine the strain induced in the beam at a location under the load point. In examining Eq. (9.3), note that the strain ε depends on five variables—S, b, h, E and W_u. For this example, let's fix three of these variable at reasonable values such as those shown below:

$$S = 4 \text{ in.},\ b = \tfrac{1}{4} \text{ in. and } E = 10.6 \times 10^6 \text{ psi. (An aluminum alloy)}.$$

We will treat the weight W_u and the height h as variables, and explore solution space for the strain ε. Let's consider the load W_u increasing from zero to a maximum of 4 lb in steps of 0.25 lb. Also, consider the height h of the cross-section of the beam varying from 0.01 in. to 0.07 in. in steps of 0.01 in. To determine the strain, we display the variable W_u in column A, and the other variable, height, along the 8th row. This arrangement of the spreadsheet is shown below:

DETERMINE STRAIN							
FOR A SIMPLY SUPPORTED BEAM							
SPAN S =4 in., DEPTH b = 1/4 in., ELASTIC MODULUS E = 10.6 x 10^6							
WEIGHT Wu AND HEIGHT h ARE VARIABLES							
STRAIN = 3WuS/(2bh^2E)							
HEIGHT	0.01	0.02	0.03	0.04	0.05	0.06	0.07
Wu (lb.)	h = 0.01 in	h = 0.02 in	h = 0.03 in	h = 0.04 in	h = 0.05 in	h = 0.06 in	h = 0.07 in
0.00	0.000000	0.000000	0.000000	0.000000	0.000000	0.000000	0.000000
0.25	0.005660	0.001415	0.000629	0.000354	0.000226	0.000157	0.000116
0.50	0.011321	0.002830	0.001258	0.000708	0.000453	0.000314	0.000231
0.75	0.016981	0.004245	0.001887	0.001061	0.000679	0.000472	0.000347
1.00	0.022642	0.005660	0.002516	0.001415	0.000906	0.000629	0.000462
1.25	0.028302	0.007075	0.003145	0.001769	0.001132	0.000786	0.000578
1.50	0.033962	0.008491	0.003774	0.002123	0.001358	0.000943	0.000693
1.75	0.039623	0.009906	0.004403	0.002476	0.001585	0.001101	0.000809
2.00	0.045283	0.011321	0.005031	0.002830	0.001811	0.001258	0.000924
2.25	0.050943	0.012736	0.005660	0.003184	0.002038	0.001415	0.001040
2.50	0.056604	0.014151	0.006289	0.003538	0.002264	0.001572	0.001155
2.75	0.062264	0.015566	0.006918	0.003892	0.002491	0.001730	0.001271
3.00	0.067925	0.016981	0.007547	0.004245	0.002717	0.001887	0.001386
3.25	0.073585	0.018396	0.008176	0.004599	0.002943	0.002044	0.001502
3.50	0.079245	0.019811	0.008805	0.004953	0.003170	0.002201	0.001617
3.75	0.084906	0.021226	0.009434	0.005307	0.003396	0.002358	0.001733
4.00	0.090566	0.022642	0.010063	0.005660	0.003623	0.002516	0.001848

We have selected the ranges for the load and the height to cover the region of interest in solution space. The weight in the specifications described in Chapter 2 for a particular type of postal scale goes to a maximum of 4 lb. Also, the calibration will be conducted using small increments of weight to verify the accuracy of the scale. The height of the beam is critical to the strain developed. For small heights (0.01 to 0.05 in.) the resulting strains, even at small loads, are excessive. However, when the height is increased to values of 0.06 to 0.07 in., the strains are much lower.

In the spreadsheet illustrated above, we have entered the data for the height of the beam in row 8, columns B to H. In rows 1 to 4, we provided the title of the spreadsheet and information related to the analysis of the strain in the beam. On row 6, the equation used in determining the strain is given. Both rows 5 and 7 are blank to provide space above and below the equation. In row 9, we have entered the heading for W_u and its units and repeated the data for the height of the beam together with its units. The spreadsheet is now organized with an arrangement that displays the two variables and their units of measure.

To begin the programming of the equation for the strain, let's fill column A with information about the weight W_u. Type zero in cell A10 and keep this cell active. Then click Edit on the menu bar, select Fill and Series. On the series dialog box, select Column, Linear, and enter step = 0.25 and stop = 4. Finally, click the OK button and check to determine if the numbers filling column A are correct.

Next, activate the B10 cell and type the formula for the strain as:

$$=3*A10*4/(1*(1/4)*10.6*10^6*(B8)^2)$$

This entry is correct for cell B10, but it will give us problems later when we try to use the copy and paste commands to extend the calculation of strain to other cells in the spreadsheet. Let's proceed by copying the formula in cell B10 into cells B11 to B26. To copy the equation into other cells in column B, position the cursor on cell B10 and click on Copy. Then mark the column from B11 to B26. Next, click on Paste. When we examine our results, they are clearly in error. To trouble shoot the problem, activate cell B11 and examine the formula displayed in the formula bar. You observe

$$=3*A11*4/(1*(1/4)*10.6*10^6*(B9)^2)$$

Unfortunately this relation is not correct. In the copy-paste operation, EXCEL modified our formula indexing the entries for row locations in both columns A and B by 1. This indexing was correct for the entries in column A, but the indexing produces an error in the entries in column B. Instead of indexing the entries from column B, we want to use the data in cell B8 for all the calculations in column B. To prevent unwanted indexing, we modify the formula in cell B10 to read:

$$=3*A10*4/(1*(1/4)*10.6*10^6*(B\$8)^2)$$

The \$ symbol before the number 8 locks the value contained in the cell B8 into the equation. When we perform the paste operation and EXCEL indexes down the B column in modifying the formula, the value of the entry in cell B8 is used in all of the calculations. We now copy cell B10 and paste down the B column from B11 to B26 to generate the correct results for the strain corresponding to a beam with a height of 0.01 in.

The copy and paste operation worked well on column B after we fixed the B$8 entry. Let's try to copy the modified formula in cell B10 into the row of cells C10 to H10. If you Copy, Paste, and examine the formulas in these cells, you will note that EXCEL has not provided the correct relations. Again you must modify the entry in cell B10 to properly perform the Copy and Paste operation along row 10. The equation in cell B10 reads:

$$=3*A10*4/(1*(1/4)*10.6*10^{\wedge}6*(B\$8)^{\wedge}2)$$

When we copy along a row to the right, the cell address A10 changes to B10, C10, D10 etc. Because it is necessary to always multiply by the weight which is listed in column A, it is imperative that the A column designation not be changed in the Copy-Paste operation. To modify our entry in cell B10, we type:

$$=3*\$A10*4/(1*(1/4)*10.6*10^{\wedge}6*(B\$8)^{\wedge}2)$$

The $ symbol before the A locks column A into the formula; it remains fixed as we change columns in the Copy-Paste operation.

Now that we have corrected our entry into cell B10, we Copy and Paste the formula to cells C10 to H10 along row 10. Then Copy this row (C10 to H10) and Paste to the block C11 to H26. The program executes these calculations and displays the results as shown in the previous spreadsheet. For the maximum load of 4.00 lb, the strain varies from 0.090566 for a beam 0.01 in. high to 0.001848 for a beam 0.07 in. high. You should always use your calculator to manually check a few values of the strain to insure that errors were not made in programming the equation into the EXCEL spreadsheet. Finally, the spreadsheet does not indicate the units for the strain. In this case, the absence of units does not present a problem because strain is a dimensionless quantity.

You may also want to change the number of decimals used in reporting the results. We usually use six decimal places to represent the strain. This is a very large number of decimals places to carry, but strains are usually small and often two or three of these places are expended carrying zeros. You can check the spreadsheet results for the strain to verify this fact. You may change the number of decimals by marking the block containing the results (B10 to H26). Then click on one or the other of the increasing or decreasing decimal keys on the formatting toolbar until the number of decimals shown is six.

Try these examples and understand and appreciate the power of EXCEL in performing not one but an entire sheet full of calculations. Use EXCEL to solve the homework problems assigned in Physics and Math courses. You can determine the single answer usually sought in an assigned problem and then easily explore solution space in EXCEL. Try exploring solution space; understand the significance of the determining all reasonable solutions in a single analysis.

9.5 GRAPHS WITH EXCEL

EXCEL can be employed to produce several types of graphs easily and quickly after you have learned the procedure. To show you the techniques used for producing graphs, several examples will are used to demonstrate the capabilities of EXCEL in preparing pie, bar, and X-Y graphs. Let's begin by discussing the pie chart.

9.5.1 Pie Charts

In Chapter 7, Tables and Graphs, we demonstrated a manual technique for preparing a pie chart. If you remember, this chart is used primarily to show the distribution of some quantity. The larger the piece of the pie the larger the share of the distribution. To illustrate how to make a pie chart in EXCEL, we open a new book and title it—PIE CHART—using the As Save selection under the File menu item. We use the top rows to title the spreadsheet entering the tabular data that describes the distribution of time in various assignments for Mechanical Engineers in their initial positions in industry in cells B7 to B15. These entries are shown in the spread sheet presented in Fig. 9.2.

PIE CHART

RESPONSIBILITIES OF MECHANICAL ENGINEERS
FIRST ASSIGNMENT IN PERCENT TIME

ASSIGNMENT	PERCENT
1. DESIGN ENGINEERING	40
2. PLANT ENGINEERING, OPERATIONS, MAINTENANCE	13
3. QUALITY CONTROL, RELIABILITY, STANDARDS	12
4. PRODUCTION ENGINEERING	12
5. SALES ENGINEERING	5
6. MANAGEMENT	4
7. COMPUTER APPLICATIONS, SYSTEMS ANALYSIS	4
8. BASIC RESEARCH AND DEVELOPMENT	3
9. OTHER ACTIVITIES	7
TOTAL	100

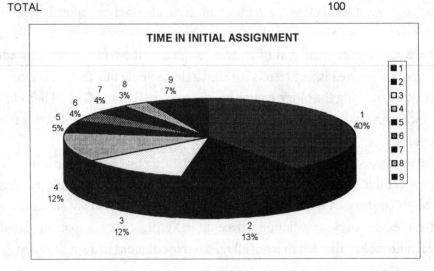

Fig. 9.2 Spreadsheet and a pie chart.

This pie chart looks very good. How did we generate it? After the numerical data on the distribution has been entered in cells B7 to B15, we mark them. Then click on the chart wizard button on the toolbar. The chart wizard presents a series of four dialog boxes that lead us through the steps necessary to convert the data displayed on the spreadsheet into a graph of one type or another. The first dialog box, presented in Fig. 9.3 guides us in the selection of the type of chart or graph that we choose to represent the data listed on the spreadsheet. We find that EXCEL offers many options from which to select. We select the pie chart and note that they can be constructed with six different options. Let's select the three dimension pie chart by clicking on its symbol and then click on the Next > button which brings the second dialog box to the screen.

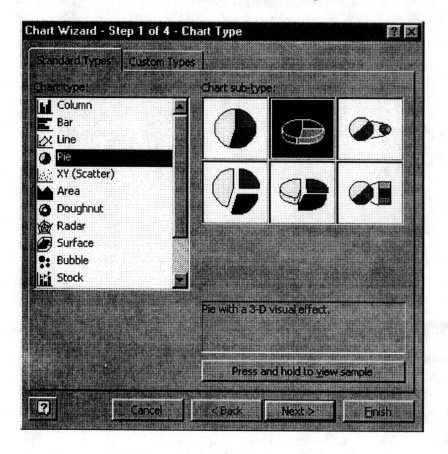

Fig. 9.3 The first of four dialog boxes presented by the chart wizard.

The second dialog box, titled source data, is presented in Fig. 9.4. Its purpose is to identify the range of data to be employed in constructing the pie chart. In our example, we have already marked the data in column B—cells B7 to B15. We confirm that these cells are correct when clicking on the Next > button.

The dialog box for step 3, illustrated in Fig. 9.5, previews our pie chart. Since it looks perfect, we do not alter the default settings for the range of data. Also, this dialog box assists us in titling the graph and providing legends to identify the different slices of the pie. We click on the title tab and add the title. The legend tab yields a dialog box that provides several choices for the placement of the legend.

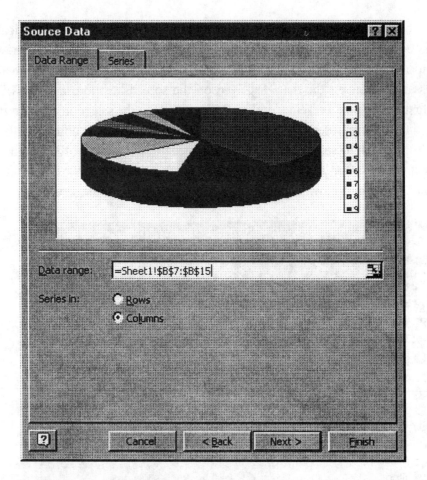

Fig. 9.4 The second dialog box in the chart wizard identifies the range of data to be employed.

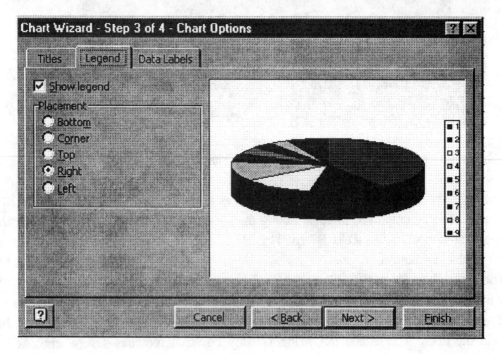

Fig. 9.5 Step three—placement of the legend, title, and data labels and a preview of the chart.

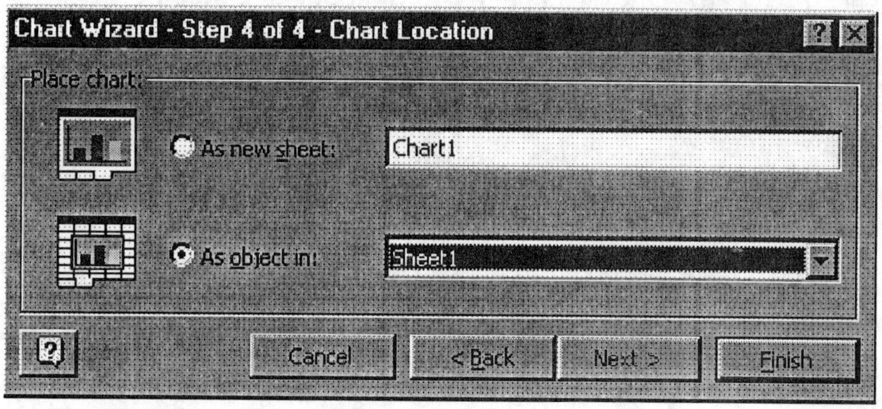

Fig. 9.6 The final step offered by the chart wizard is to place the chart.

Except for sizing the chart and locating it on the spreadsheet, we have completed the pie chart. We click the Next > button on the Step 3 dialog box and the final page from the chart wizard, presented in Fig. 9.6, provides a choice for the placement of the chart. We have decided to show the chart as an object on Sheet 1 of the spreadsheet. Click the finish button and examine the results on the monitor. If the chart is too small or too large, click and drag on a corner of the chart to adjust the size.

When you are satisfied with the appearance and location of the chart, click on File and select Print Review. You may add gridlines and adjust margins on the Print Review display. Print the spreadsheet and the pie chart and observe its professional appearance (see Fig. 9.2).

9.5.2 Bar Charts

Bar charts are usually employed for showing comparisons. In the Chapter 7, Tables and Graphs, we introduced an example of a bar chart that compared the percentage of high school students graduating with three or more years of secondary mathematics and science in 1982 and 1994. It is a very simple comparison, but it indicates the power of visual graphics in carrying a message.

We begin by loading EXCEL which opens to a new book. Again, we title this book by clicking on File, Save As and then name the file—BAR CHART. The title of the new spreadsheet is entered in the first five rows, and then the data for the graduating high school students is entered in the block A7 to C9, as illustrated in Fig. 9.7. In the spreadsheet, the column headings are in row 7, and the years used in the comparison are given in column A. We begin to create our bar chart by marking the region on the spreadsheet where the numerical data exist—A8 to C9. We then click on the Chart Wizard button; the first dialog box shows illustrations for many different chart types. One of the options is for bar charts and another is for column charts They are nearly the same. The bar charts display comparison categories along the Y axis and numerical values along the X axis. The column charts display comparison categories along the X axis and numerical values along the Y axis. We consider both the column and the bar chart representations in EXCEL to be suitable for preparing bar charts. Take your pick; we have selected the three dimensional bar chart in this example. The first dialog box, presented in Fig. 9.8, shows these selections. Near the bottom of this dialog box is a long key—when pressed a preview the bar chart is presented. It advisable to use this feature to determine if the data range has been marked correctly. If the chart is not correct, cancel and revise your selection of the data to be included in the bar chart.

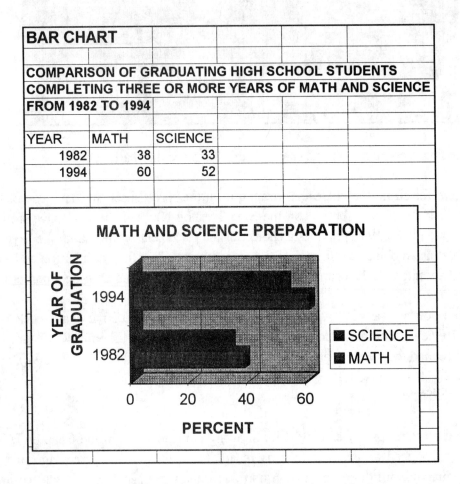

Fig. 9.7 Spreadsheet and bar graph illustrating both tabular and graphical presentation of data.

The next dialog box (Step 2), provides two tabs (two pages) for data entry. The first is for the data range which we marked prior to initiating the chart wizard. However, the series is in a column format and this choice should be made as indicated in Fig. 9.8. Click on the series tab for the Source Data dialog box and identify the series as math and science by pointing with the mouse to the headings of the respective columns. The "category (x) axis labels" is confusing. In EXCEL the (x) axis is the ordinate on a bar chart. We identify the (x) axis labels by giving the cell addresses for the years 1982 and 1994. The series page of the source data dialog box is presented in Fig. 9.8.

The dialog box associated with step three provides several tabs (pages) for entering information on the bar chart concerning the title, defining the axes, providing gridlines, showing legends, labeling data and showing the data table. A preview screen in the dialog box is helpful in selecting axes and in deciding on gridlines. We show the page associated with the titles tab in Fig. 9.8. Click on the various tabs and explore the various techniques for modifying the bar chart.

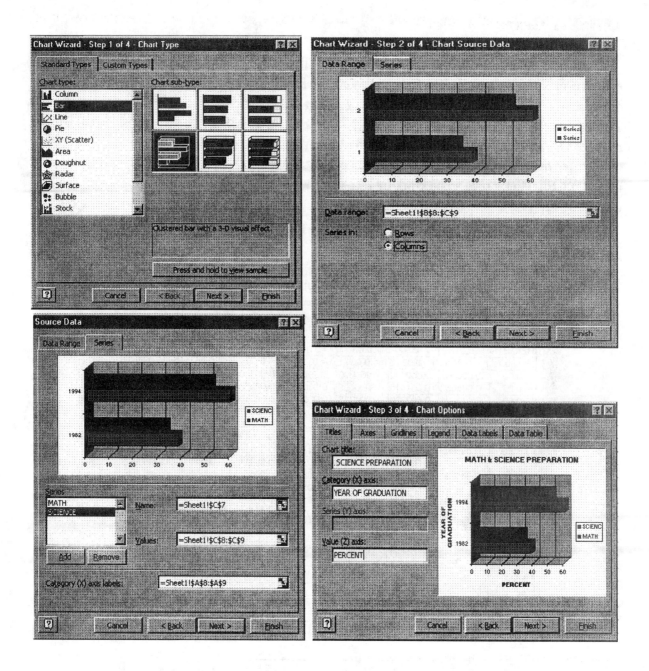

Fig. 9.8 Chart wizard dialog boxes showing data entries used to create the bar chart of Fig. 9.7.

The final dialog box, Step 4, is identical to that shown in Fig. 9.6; it provides the opportunity for you to selection of the location of the bar chart. We clicked on the circle which places the bar chart as an object in Sheet 1. Inspection of fig. 9.7 shows the chart together with the spreadsheet. When you are satisfied with the appearance of the spreadsheet, click on File and Print Preview. Add gridlines, adjust the margins as you wish and then print the output. Try to duplicate the spreadsheet and chart shown above.

9.5.3 X-Y Graphs

The X – Y graph is the type of visual most frequently employed by engineers. It is very effective in visually conveying trends indicated by numerical data. Is the trend increasing or decreasing—is it level or oscillating? The X-Y graph quickly and dramatically shows these trends. To demonstrate the method for producing X-Y graphs in EXCEL, we will consider the strain produced by a centrally loaded simply supported beam as predicted by Eq. (9.3).

We have previously used Eq. (9.3) to demonstrate the technique for performing calculations using EXCEL. Let's save some work by recalling the spreadsheet showing the results obtained in evaluating Eq. (9.3). These calculations provide the numerical data necessary for preparing our X-Y graph. Using the chart wizard, we develop an X-Y graph from the results as shown in Fig. 9.9.

DETERMINE STRAIN							
FOR A SIMPLY SUPPORTED BEAM							
SPAN S =4 in., DEPTH b = 1/4 in., ELASTIC MODULUS E = 10.6 x 10^6							
WEIGHT Wu AND HEIGHT h ARE VARIABLES							
STRAIN = 3WuS/(2bh^2E)							
HEIGHT	0.01	0.02	0.03	0.04	0.05	0.06	0.07
Wu (lb.)	h = 0.01 in.	h = 0.02 in.	h = 0.03 in.	h = 0.04 in.	h = 0.05 in.	h = 0.06 in.	h = 0.07 in.
0.00	0.000000	0.000000	0.000000	0.000000	0.000000	0.000000	0.000000
0.25	0.005660	0.001415	0.000629	0.000354	0.000226	0.000157	0.000116
0.50	0.011321	0.002830	0.001258	0.000708	0.000453	0.000314	0.000231
0.75	0.016981	0.004245	0.001887	0.001061	0.000679	0.000472	0.000347
1.00	0.022642	0.005660	0.002516	0.001415	0.000906	0.000629	0.000462
1.25	0.028302	0.007075	0.003145	0.001769	0.001132	0.000786	0.000578
1.50	0.033962	0.008491	0.003774	0.002123	0.001358	0.000943	0.000693
1.75	0.039623	0.009906	0.004403	0.002476	0.001585	0.001101	0.000809
2.00	0.045283	0.011321	0.005031	0.002830	0.001811	0.001258	0.000924
2.25	0.050943	0.012736	0.005660	0.003184	0.002038	0.001415	0.001040
2.50	0.056604	0.014151	0.006289	0.003538	0.002264	0.001572	0.001155
2.75	0.062264	0.015566	0.006918	0.003892	0.002491	0.001730	0.001271
3.00	0.067925	0.016981	0.007547	0.004245	0.002717	0.001887	0.001386
3.25	0.073585	0.018396	0.008176	0.004599	0.002943	0.002044	0.001502
3.50	0.079245	0.019811	0.008805	0.004953	0.003170	0.002201	0.001617
3.75	0.084906	0.021226	0.009434	0.005307	0.003396	0.002358	0.001733
4.00	0.090566	0.022642	0.010063	0.005660	0.003623	0.002516	0.001848

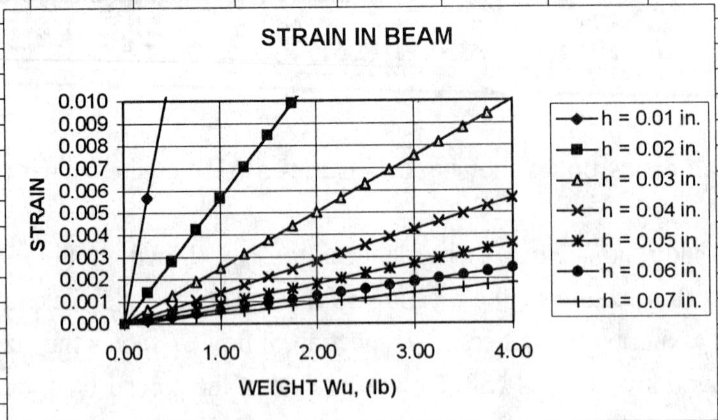

Fig. 9.9 Spreadsheet and X-Y graph associated with the calculation of strain in a beam.

Let's develop the step by step procedure for converting the numerical data in the spreadsheet to an X-Y graph. Start by marking the data block—A9 to H26. Note, we have included the headings for the seven columns of Y data that are given columns B to H of the data block. Remember that the data for the caption along the X axis is listed in column A. The results for the strain will be plotted automatically by EXCEL along the Y axis. After the data block is marked (A9 — H26), click the chart wizard button and select the type of chart from the options listed in the first dialog box. Do not select the line graph from the available options because it often distorts the data along the X axis. Instead, select the X-Y scatter graph type as shown in Fig. 9.10. Press and hold the long button on this dialog box to review the X-Y graph.

Click on the Next > button, and the dialog box associated with Step 2 appears as presented in Fig. 9.10. The tab for the data range is active, and we check that the data range is correct. We also indicate that the series to be plotted is in columns. Also note that the legend is shown in the preview of the graph with the beam height h assigned. Click on the Series tab, and verify the location of the series of numerical results to be plotted. In this example, a series of numerical results is available for each of the seven beam heights. The cell locations of the values plotted for X and for Y are given in a box for data entry. If these cell locations do not identify the correct parameter they can be changed by clicking on the button on the right side of this box. The correct series designations are indicated in Fig. 9.10.

The next dialog box, Step 3 in Fig. 9.10, provides an opportunity to add the title and caption the X and Y axes. We added the title **STRAIN IN BEAM**, and captioned the X and Y axes as **WEIGHT, W$_u$ (LB)** and **STRAIN,** respectively. We may also add or remove grid lines and move the position of the legend. The draft of the graph is essentially completed when we click the Next > button.

In the fourth and finally dialog box, we choose to show the graph on the same worksheet as the numerical data (see Fig. 9.9). We click on the graph to activate it, and then drag the corner and/or side markers of the graph to enlarge it. When the graph is the correct size it is positioned with the pointer to the most suitable location on the spreadsheet. Before printing the graph, click on File then Print Preview, and check to determine if you have arranged the display so that it conveys the numerical data in tabular form on the spreadsheet as well as the X-Y graph.

The combination of the spreadsheet and the graph is an effective technique for presenting the results of theoretical or experimental data. However, in some instances we want to show the graph without including the spreadsheet. We can show the graph alone without the spreadsheet by displaying it on another worksheet. Click on the graph to activate it and then click on the Copy icon. Next, click on the tab for Sheet 2 to bring a new spreadsheet onto the screen. Finally, click on the Paste button. The graph from Sheet 1 is copied onto Sheet 2. Click on the graph to activate it, then drag its corners or sides to position and size it on the sheet. Examine the appearance of the graph with Print Review prior to printing. The result is shown in Fig. 9.11.

A close examination of Fig. 9.11 shows that the graph is not exactly the same as that presented in Fig. 9.9. The scales on both the abscissa and the ordinate are different and the gridlines for the x axis are missing. The choice of scale for the X and Y axes is very important. The very high values of strain shown in Fig. 9.11 tend to distort the graph. These strains are excessive and indicate that the beams with small heights would fail by plastic bending long before achieving a load of four lb.

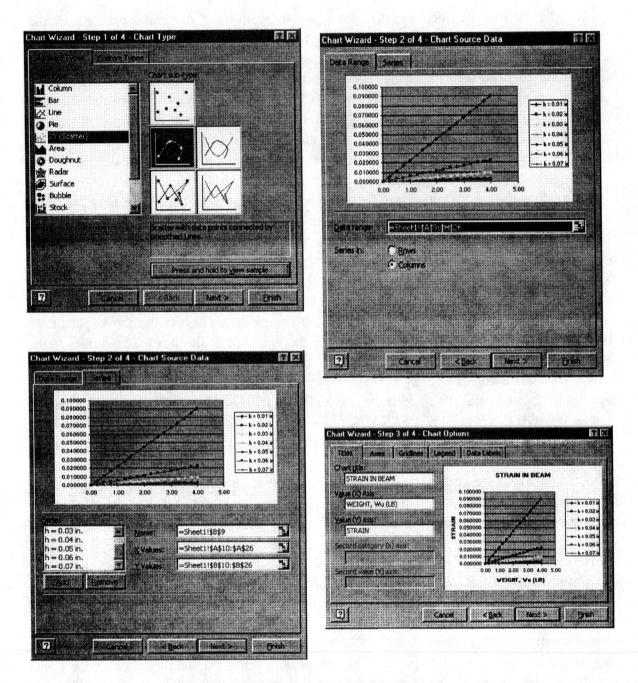

Fig. 9.10 Dialog boxes associated with plotting the X-Y graph.

To change the scale on the graph in Fig. 9.11, point to the 5.00 on the caption for the X axis and double click the left mouse button. A dialog box appears with five different tabs. Select the tab for scale and enter the number 4.00 for the maximum value for the X axis. To change the scale on the Y axis, point to the number 0.10 in the caption for the ordinate and repeat the process. Again, select the scale tab and enter the value of 0.01 for the maximum strain to be plotted along the Y axis. When the new selections are made, the graph is rescaled so that it displays more meaningful values of strain as shown in Fig. 9.12.

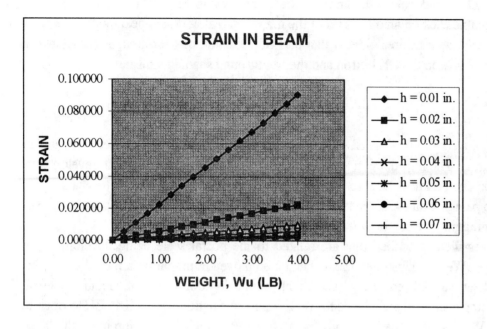

Fig. 9.11 Appearance of the X-Y graph before editing.

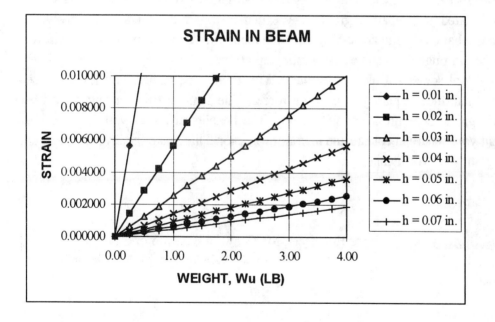

Fig. 9.12 Example of the edited version of the X-Y graph.

A close examination of graph in Fig. 9.11 indicates that the gridlines for the X axis are missing. To correct this oversight, click on the graph to activate it; then click on the chart wizard icon. Proceed quickly through the dialog boxes which are complete because the graph is active. On the third dialog box click on the gridlines tab and select gridlines for the major markings along the X axis. Click on the finish button to add the grid lines.

Finally, the background color (shading) of the chart area in Fig. 9.11 is too dark. To remove the shading double click on an open area of the graph. A dialog box appears with selections for the color of the border and the area. Under that part of the dialog box for area, select white for the background color. Click on the OK button and the background shading vanishes.

9.6 SUMMARY

We have introduced you to EXCEL–97 one of the most popular spreadsheet programs in use in the United States. The introduction was limited because our purpose was to develop only entry level skills. You should know how to interpret the screen, the menu, and the toolbars. You should also be able to navigate on the spreadsheet with the keyboard or the mouse and to mark cells or blocks of cells by dragging the cursor with the mouse.

Three examples were described that are related to the postal scale being developed in a project for this course. We first illustrated the technique employed in preparing a parts list. In this instance, the spreadsheet was utilized to organize a table containing both descriptive and numerical data. The second example showed techniques for performing calculations using EXCEL to operate on simple formulas. We first illustrated methods for performing a single calculation in a cell. Then we copied this calculation and performed a line of calculations by varying one parameter. Finally, we performed a spreadsheet full of calculations when we permitted two parameters to vary. The third example was related to preparing graphs. We demonstrated the techniques involved in preparing a pie chart, a bar chart, and an X-Y graph. We made extensive use of the chart wizard in EXCEL that leads you through the technique in a four step process.

EXCEL is a tool somewhat like a hammer. With time and practice, you learn to use a hammer; you hit the nail more squarely with more force. The same is true for any spreadsheet program. You will grow more proficient with time. The benefits provided with a spreadsheet program throughout your tenure in college and in your professional life make the investment of time in learning the program well worthwhile.

REFERENCES

1. Gottfried, B. S., Spreadsheet Tools for Engineers, Mc Graw Hill, New York, NY, 1996.
2. O'Leary, T. J. and L. I. O'Leary, Microsoft Excel 7.0a for Windows® 95, Mc Graw Hill, New York, NY, 1996.

EXERCISES

1. Explore the Excel screen and learn the purpose of the icon buttons.
2. Explore the Excel menu and learn the options available to you by employing this spreadsheet program.
3. Prepare a parts list for the postal scale that your team is developing.
4. Suppose that you have a rich grandmother, who has provided an inheritance for you. Unfortunately the money is in a trust, and you cannot place your hands on the loot until you are a mature 50 years of age. If the amount of the inheritance was $ 7,000.00, and the trust yields

7% per annum compounded semi-annually, determine the value of the trust each year until your fiftieth birthday. Also prepare an X-Y graph showing the value of the trust with time. Neglect the effect of taxes on your analysis.

5. Duplicate the work involved in producing the spreadsheet on determining strain shown on page 9 – 11.

6. Prepare a bar chart comparing the percentage of men and women in your high school graduating class with the percentage in the freshman engineering class in 1996. Your high school year book will provide information on the gender of your graduating class, and the Office of the Dean of Engineering will provide data on the freshman class.

7. Suppose that you decide that the chart in Fig. 9.12 covers too large of a range of height of the beam. Prepare a new chart covering the range in height from 0.04 in. to 0.07 in.

8. Suppose that you decide that the chart in Fig. 9.12 covers too small of a range of weight to be placed on the postal scale. Prepare a new chart covering the range in weight from zero to 10 pounds.

9. Suppose that you decide that the chart in Fig. 9.12 covers too small of a range of height of the beam. Prepare a new chart covering the range in height from 0.01 in. to 0.12 in.

10. Suppose that you decide that the chart in Fig. 9.12 covers too large of a range of weight to be placed on the postal scale. Prepare a new chart covering the range in weight from zero to one pound.

11. Evaluate the relation $Y = a + bX$ when $a = 5$ and $b = 3$. Let X vary from zero to 20. Prepare an $X - Y$ graph showing the results.

12. Evaluate the relation $Y = a + bX^2$ when $a = 3$ and $b = 2$. Let X vary from zero to 10. Prepare an $X - Y$ graph showing the results.

13. Evaluate the relation $Y = a + bX^2 + c X^3$ when $a = 0$, $b = - 4$ and $c = 2$. Let X vary from zero to 8. Prepare an $X - Y$ graph showing the results.

14. Evaluate the relation $Y = ae^X$ when $a = 4$. Let X vary from zero to 6. Prepare an $X - Y$ graph showing the results.

15. Evaluate the relation $Z = a + b Y + c X$ when $a = 4$, $b = 3$ and $c = 5$. Let $X = 1$, 2 and 3 and $Y = 1$, 2, and 3. Prepare a graph showing A as a function of X and Y.

CHAPTER 10

MICROSOFT PowerPoint

10.1 INTRODUCTION

During this course, your team will be responsible for two design reviews; you will be required to make a presentation before the class and your instructor. In the design briefing you will describe the product specification, the important features that your team is incorporating in the design, and illustrate your concepts with engineering drawings. If you want to make a very good presentation, it is important that the audience closely follow your descriptions of the team activities. To help in this regard, you should use graphics which enable transmission of information through two senses—audio and visual. This chapter provides instruction for a graphics presentation program, (GPP) that facilitates the preparation of the visual aids needed for your presentations.

Microsoft PowerPoint is a graphics presentation program, that markedly reduces the time needed for you to prepare professional quality visual aids. It has a wide range of capabilities from simple overhead transparencies to sophisticated on-screen electronic displays. GPPs assist you in producing either black and white or colored transparencies, 35 mm slides, full screen projected electronic slides, and valuable support materials such as hard copy printouts, notes and outlines.

Before beginning to learn about PowerPoint, it is necessary to understand that preparing a presentation is a five step process.

1. Plan the presentation to provide the information necessary for a design review. In planning, determine the amount of time that you have to speak, the size and layout of the room, and the type of equipment available for visual aids.
2. Formulate the presentation by preparing an outline that identifies all of the topics, that will be described.

3. Prepare the slides needed in your presentation by using a GPP such as PowerPoint to guide you. As you compose the slides, edit them and make certain that there are no misspellings. After you have completed the slides, review them and make modifications for their enhancement. Can you add graphics or bullets, to better capture the attention of the audience? Is the font size correct? Have used the bold, underline or shadow options to the best advantage? If you are using electronic displays, significant latitude exists for enhancing the visuals and controlling the pace of the presentation.

4. Finally, you must rehearse the presentation. It is imperative that you completely understand the material on your slides. Nothing is more deadly than a speaker reading his or her material from the slide to the audience. Rehearse until the presentation is smooth and polished. You should feel confident about the message to be conveyed and your ability to handle the slides and the projector.

10.2 THE PowerPoint WINDOW

Let's begin by loading PowerPoint. If the Tip of the Day dialog box shows on the screen, read and close it to reveal the PowerPoint Window as shown in Fig. 10.1. There are four important rows or bars located above the body of the window. The top row, called the title bar, displays the file name that you assign to the presentation when you first save your file. At the right side of the title bar, you will find three buttons: the left one sets the program aside, the middle button changes the size of the display, and the button on the right closes the file. The second row contains Microsoft's standard menu items. Click on one of these items and a pull down menu is displayed. Click and explore the menu items until you are comfortable with each item and have a good understanding what can be accomplished with each of them.

The third row is the standard toolbar that displays 21 buttons and a zoom control arrow to change the size of the display. Point to any button and note that its purpose is defined in a small box attached to the pointer. The icon buttons on the left side of this toolbar are common to all of the Microsoft applications; however, those on the right side of the bar are unique to PowerPoint. We will introduce them later in an example presentation for a design review. The fourth row down from the top is the formatting bar. Again, most of the icon buttons are common with other Microsoft applications. With the mouse, point to some of the icon buttons on the right side of the formatting bar and the purpose of each icon is displayed.

The center area of the window is the workspace where you will prepare the slides. We have shown a slide in the workspace in Fig. 10.1. A vertical drawing toolbar is displayed on the left side of this workspace. We can prepare drawings in PowerPoint to enhance the visual impact of the slides. A vertical scroll bar is located on the right side of the window. The two buttons at the bottom of the scroll bar with the double ▲ permit us to move from one slide to the next—either forward or backward.

At the bottom of the window, we have two more rows. The top one is for the horizontal scroll controls. To the left of the scroll bar, you will find five buttons that control the presentation views. These buttons are very helpful as we work to develop a multi-slide presentation because they:

1. Present the slides individually.
2. Present the outline of all the slides, showing all of the titles and text.
3. Rearrange the slides with a slide sorter.
4. Add notes to the speaker's copy with a note page.
5. Present a slide show enabling you to review the complete presentation.

These buttons will be described in more detail later when we edit a presentation prepared for a design review. The bottom row contains the status bar where helpful messages are displayed. To the right of the status bar are shortcut buttons used to introduce a new slide into the presentation and to select a specific slide layout.

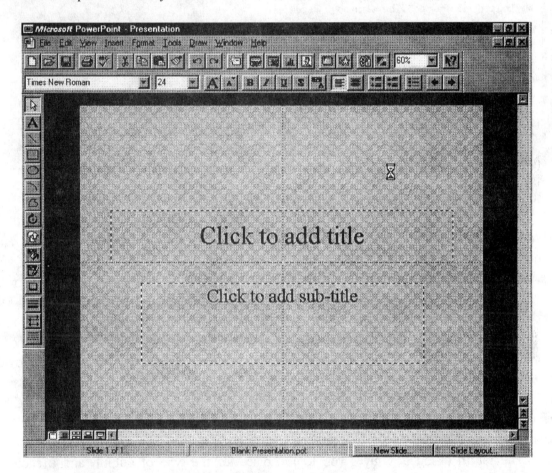

Fig. 10.1 A PowerPoint display showing toolbars and working window.

10.3 PLANNING AND ORGANIZING WITH PowerPoint

The PowerPoint program has built-in features to assist you in planning, organizing and timing your presentation. To show these features, let's load PowerPoint, click on File, and then New. A folder is displayed with three tabs named General, Presentation Design, and Presentations as shown in Fig. 10.2. Select the tab for Presentations and then note the 20 choices for different types of presentations. Select the first one offered—the Auto Content Wizard. The Auto Content

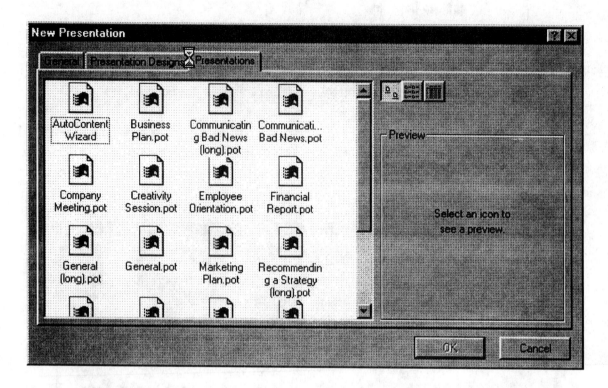

Fig. 10.2 New presentation page where selection of the Auto Content Wizard is made.

Wizard page, which appears on your screen, indicates that wizard provides ideas and organization for your presentation. Click on the Next button, and the first of five dialog boxes associated with the wizard is displayed. This first dialog box, shown in Fig. 10.3, introduces the Auto Content Wizard. The second dialog box prompts you for information pertaining to the title slide. After providing this title information, click Next, and the third dialog box provides you with a many choices for the type of presentation that you are creating. Let's hope that you do not need to Communicate Bad News. If not, select Reporting Progress. The fourth dialog box provides choices for style of the presentation and its timing. We suggest that you select the professional style. Also, your presentation should require 30 minutes or less. (Many accomplished speakers believe that 18 minutes is the ideal length for a presentation. This time is long enough to include substance, but short enough to avoid boring the audience).

The fourth dialog box gives you four choices for the type of output, and a choice of materials to hand out to the audience. For this first example, select Black and White Overheads and indicate that you will print handouts. Click on Finish and PowerPoint provides you with nine slides organized in sequence to provide the necessary information for your design review.

We will use these nine slides and prepare part of a preliminary (first) design review for the postal scale. Begin with the title slide. Note that you have already provided the title in responding to the questions raised by the Auto Content Wizard. You may want to edit the title and provide additional information in a subtitle. To edit the title, double click on the upper box to activate it; you are ready to make the changes needed to improve its appearance. Keep the title and the subtitle very short. You will have the opportunity to expound later in the presentation. Typing information onto the slides is the same as typing in a word processing program. An example of a title slide appropriate for a design briefing is shown in Fig. 10.4a.

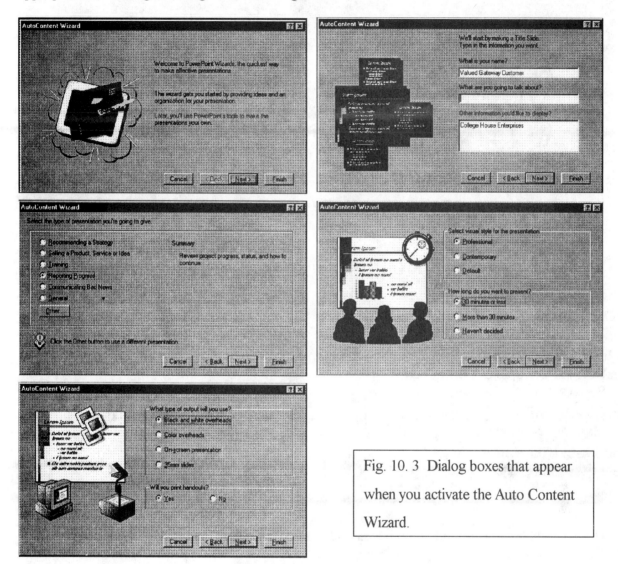

Fig. 10. 3 Dialog boxes that appear when you activate the Auto Content Wizard.

Click on the lower button on the vertical scroll bar to advance to slide 2. In the organization provided by PowerPoint, the purpose of slide 2 is to define the subject. This slide prompts us to divide the subject into topics which we plan to discuss and then to list these topics. In other words, we tell the audience in the beginning what we intend to tell them later during the presentation. It is a good idea because no one likes surprises. We show an example of this slide in Fig. 10.4b. The information contained in slide 2 was entered on the presentation outline, as shown in Fig. 10.5. You can convert from the slide format to the outline format by clicking on the outline button located to the left of the horizontal scroll bar. Under slide 2 on the outline, you respond to the prompts provided by wizard by adding bullet items to define topics that you intend to discuss. When you have completed the outline for slide 2, click the single slide button on the horizontal scroll bar and review the appearance of the slide. You may wish to edit the slide by changing the font size to better use the available space.

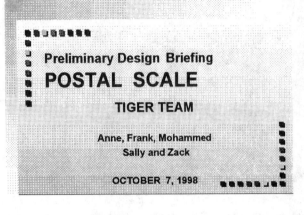

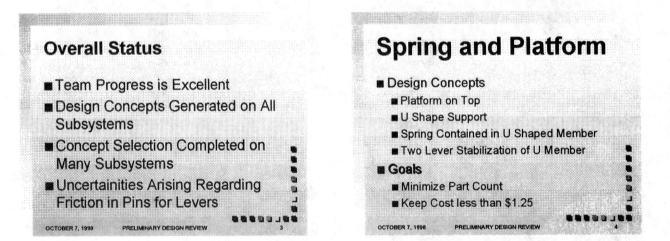

Fig. 10.4 Example showing content of four slides for a design briefing (before enhancement).

Click on the next slide button and edit the entries that the wizard provides for slide 3. The title of this slide is overall status which we believe is appropriate for a design review. In this slide, you include the following information:

- Report on the current status of the team.
- Describe the tasks that have been completed.
- Indicate if you are behind schedule on any task.
- Raise any issues that concern the team.

The example slide, presented in Fig. 10.4c, describes uncertainties regarding friction in the pins used in the lever system for the mechanical version of the postal scale.

Wizard has titled slide 4 as **Component One: Background**. This title leads us to think about reporting on the first subsystem. Therefore, we double click on the title to activate it. We edit the title, by changing it to Spring and Platform. In the large area below the title block, we describe our progress in generating design concepts and show two key criteria for selecting the concept. You will note that we have made use of bullets and sub-bullets in the descriptions. We offset one bullet relative to the other by clicking the arrow button that is located on the right side of the formatting toolbar. Try clicking on the arrow buttons and observe the changes to the position and style of the bullets.

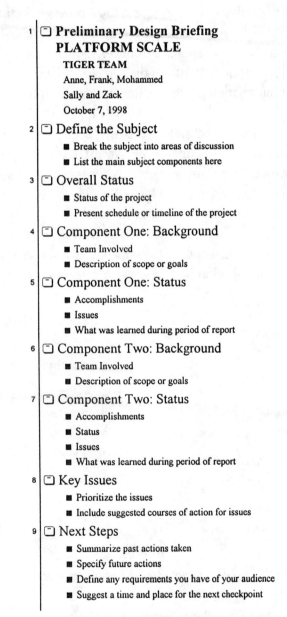

Fig. 10.5 Example of the outline page that may be used in preparing entries for the slides.

The purpose of Slide 5 in the sequence suggested by the Auto Content Wizard is to report our status on the design of Component One. We modify the title to indicate that the slide is related to the Spring and Platform subsystem. We then click on the lower block of the slide and respond to the to the prompts provided by wizard. You should provide information describing issues accomplishments, and lessons learned. These are all excellent topics to include at this stage in the design review.

In preparing a complete design review, you will cover each subsystem using two slides similar to Slide 4 and 5. The wizard has only provided slides for two components (subsystems); it is necessary to insert addition slides to cover all of the subsystems in your design. Adding new slides is easy. With slide 5 displayed on the monitor, click on the new slide button on the right side of the status bar. A new slide page is displayed on the screen that provides you with a choice of 12 different formats. A format consistent with that employed in slide 5 is highlighted. Since we want to retain this format, we click on OK, PowerPoint introduces the new slide, and changes the page count entry on the status bar to read slide 6 of 10. In the top block of your new slide, type the title, and in the lower block, type the information describing the design concepts and goals for the second subsystem.

After you have briefly reviewed the subsystems in you development, it is necessary to conclude the presentation. The last two slides in the wizard sequence provide a template for preparing your concluding remarks. We show suggestions for these two slides in Fig. 10.6. The next to last slide in your presentation is titled "**Key Issues**". This is your opportunity to raise any issues (questions), that the design team has discovered to date. Do not hide uncertainties; come forward and seek help. The design review is a formal venue for this purpose. The example shown in Fig. 10.6a indicates that we are concerned with the effects of friction at the pins connecting the levers to the U shaped member, and friction in developed between the pointer and the calibrated scale plate.

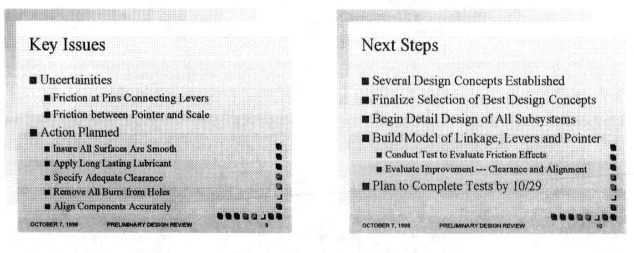

Fig. 10.6 Example of the two slides used to conclude the design review.

The final slide, illustrated in Fig. 10.6b, describes your plans for action in the near future. Wizard titles this slide as Next Steps, which is appropriate. Wizard then prompts you to:

- Summarize past actions taken by the team.
- Specify future actions that are planned.
- Define any requirements that you have of the audience.
- Suggest a time and place for the next design review.

These are excellent suggestions for the final slide concluding your design review.

You are now ready to print the output. Click on File and select Print. The print page, presented in Fig. 10.7, is adapted for PowerPoint. It contains a selection box labeled—Print What. You have several choices including: slides, handouts with 2, 3 or 6 slides per page, note pages, and an outline view. You can prepare the overheads for your presentation and the handout for the audience. Print your presentation using these options and examine the output to decide which style is most suitable for the handout intended for the audience. The black and white paper versions of the slides can be copied to give the transparencies necessary for the presentation. If you have your own laser or ink jet printer, the transparencies can be printed directly. Be certain you save the file before exiting the program.

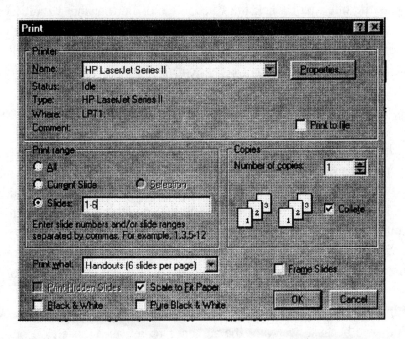

Fig. 10.7 The dialog box for printing the output from PowerPoint.

10.4 ENHANCING THE SLIDES

Suppose you decide to use color slides instead of black and white. It will cost more to use color if you take the job to a commercial copy center, but you may decide that it is worth it. How can you change the black and white slides to color slides? It is easy, and you do not have to retype all of the information into a new version of wizard. To change the color and/or the design of the slides, click on the Apply Design Template button on the right side of the PowerPoint toolbar. The design template page displayed on your screen has a menu listing of 30 templates, that start with Azure and end with Wet Sand. With the mouse, select one of the templates and note that it is displayed for your review in a small preview box. Try several templates before selecting the one you believe is the most appropriate.

If you would like to change the color of the background, go to the menu bar and click on Format, and select Custom Background. The Color page that is displayed gives you several options for the style of the background and the color employed. Click on Options and review the results until you obtain the style and the colors that you believe will be the most effective in your presentation. Be careful and stay with relatively light colors because they are more visible to the audience. Dark slides do not project well unless the room is extremely dark. Also, dark rooms are an invitation for your audience to sleep.

Suppose that you like the background style and color, but are unhappy with the color of the text and the bullets. Change them by clicking on Format and select Slide Color Scheme. The Color Scheme page has two tabs—standard and custom. The standard tab illustrates four choices of color for the titles and bullets. If none of the colors are suitable, click on the custom tab and prepare to become a painter. You have the opportunity to select the colors that you want to use in the background, shadows, fill, text and title and bullets. The colors are changed by selecting a feature on your slide such as the title text. Click the Change Color button and a multicolored hexagon appears with many choices of different shades of the basic colors. Select the color that you wish to employ for each of the features on the slide. You can have a single design for all of your slides (recommended), or you can have a different color scheme for each slide. Have fun; PowerPoint gives you the means to make great slides.

Some presenters like to use either a header or footer on their slides. A footer is helpful if you give many presentations and want to retain the slides for your files. The information in the footer can include date, occasion for the presentation, location, etc. For example, you may decide to use a footer stating October 7, 1998—Preliminary Design Review, and slide number. How do we incorporate this footer on our PowerPoint slides? Click on View on the Menu bar and select Header and Footer. The Header and Footer page is displayed with two tabs—one for the slide and the other for the notes and handouts. Let's select the slide tab and add a footer to our slides. We click on the Date and Time square, select Fixed, and type the date of the planned presentation. We mark the slide number because it will help us in timing the delivery of the presentation. We click on the footer, and type in a short message to appear in the center region of the footer. We also click on the box indicating that we do not want the footer to appear on the title slide. We delete the footer from the title slide because we want the audience to remain focused on our subject in the very beginning of the presentation. The Header and Footer page has a small preview window that permits us to review the input prior to printing. We illustrate the footer developed with this procedure in Fig. 10.8. The slide was in beautiful color on the monitor, but the figure is in black and white. Our apologies, but the use of color in printing in a low-volume, low-cost textbook is prohibitively expensive.

Still not happy with the impact that your slides will make. What about adding some clip art? Select a slide that you want to enhance with the addition of some photo or sketch which is available in the Microsoft Clip Art files. You don't know what's available. Not to worry; click on the Insert Clip Art icon button that is located on the right side of the toolbar. (The icon with the funny face on it.) There is a pause while the clip art files are loaded into the active PowerPoint program. A page entitled Microsoft Clip Art Gallery 2.0 appears on the screen. On the left side of this page, you will find a listing of clip art categories. Select the Animal category and illustrations of the selections of clip art available under this category appear in the large preview window. Click on the turtle and then the insert button.

Now examine the slide you have selected for enhancement. We have a large picture of a turtle—smack in the middle of the slide. To move the picture and change its size, we drag it by the one of the little squares either on the corners or the middle of the sides to position the insert appropriately on the slide. We show the turtle in the upper right hand area of the slide in Fig. 10.8. Note, if you push while dragging the object, it becomes smaller, but when the handles are pulled the object becomes larger. Try it, and you will soon manage to both move and size the turtle.

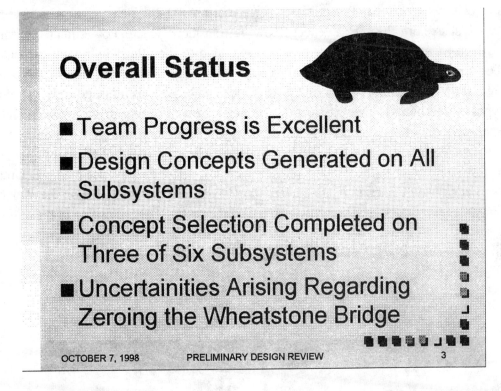

Fig. 10.8 Example of an enhance slide showing an interesting background, footnote and clip art.

10.5 REHEARSING

Rehearsing is an essential part of preparing for a design review. Fortunately, PowerPoint has a neat feature that helps in handling the visuals as you rehearse. Go to the first (title) slide and then click on the Slide Show button to the left of the horizontal scroll bar. The screen changes with all of the toolbars disappearing. The slide occupies the entire screen of the monitor and the PowerPoint presentation simulates the screen in the class room. You can practice your delivery moving from one slide to the next by using the arrow buttons on the keyboard to forward or reverse the slides.

Timing is an important part of the delivery of a design review. Many presenters make the mistake of delivering material slowly in the beginning and hurrying at the end of the presentation to meet the time constraint. The delivery should be paced to allow more than enough time at the end of the presentation. The last two slides are the most important in the overall design review. Be certain that you have sufficient time to present the conclusions properly

10.6 COMPUTER PROJECTED SLIDES

The ability to project your slides directly from the computer gives you several significant advantages which are available in PowerPoint. First, you can store a large number of slides in memory. If questions arise during your presentation, you can call on these back-up materials to improve your response. Second, you do not need a color copier to make the slides. Third, you can use special effects incorporated in PowerPoint to give your presentations some real class.

Let's first explore the special effect know as transitions where PowerPoint controls the manner in which we switch from one slide to the next. Go to the title slide in your presentation, click on Tools in the menu bar and select Slide Transition. The Slide Transition dialog box that appears is illustrated in Fig. 10.9. Click on the arrow located beside this message and a list of many different types of transitions is given. Select one of them and watch the slide switch from one view to another in the small preview window. Select the type of transition that you like the best. We think the dissolve transition is an excellent choice. Next, select the speed of the transition. Again you can review your choice in the preview window. Finally, select the advance method. Until you are a real pro at delivery of design briefings, we suggest you advance the slides on the click of the mouse. Click the OK button when you have completed the set up of the transition of the title slide.

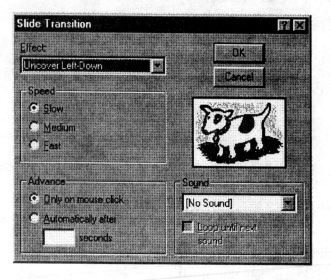

Fig 10.9 The dialog box for slide transition in PowerPoint.

Click on the Slide Sorter View button near the horizontal scroll bar and the first six slides of your presentation are displayed. The first slide has a small slide transition icon visible near its lower left corner. A new toolbar also appears at the top of the window replacing the formatting toolbar. You are now ready to select the type of transition for the remaining slides. Select slide 2, and click on the transition arrow button on the new tool bar. You can select the type of transition from the long list for slide 2. Proceed to mark any slide that you wish to transition in the Slide Sorter View, and to select the type of transition that you wish to use when presenting that slide. After you have made your choice, it is confirmed by the presence of a small transition icon just below the slide when you have the Slide Sorter View active as shown in Fig. 10.10. If you employ the same transition for all of your slides, mark all of them (Edit/Select All), and select the type of transition that you believe is the most appropriate for the presentation.

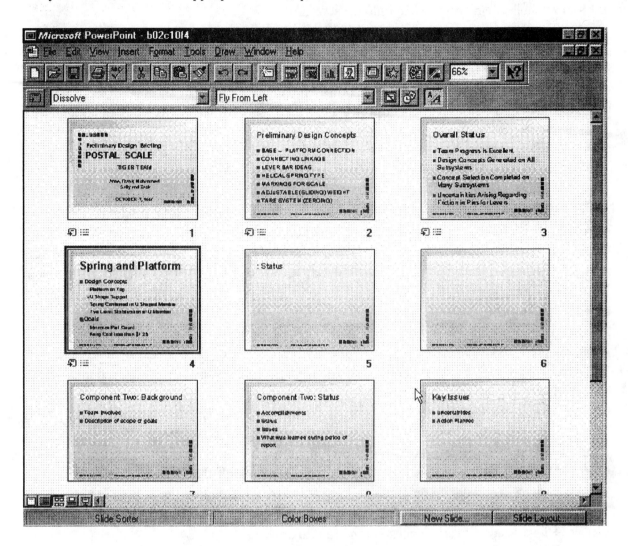

Fig. 10.10 Slide sorter view with transition and building effects icons below the individual slides.

PowerPoint also permits you to build your slides progressively for your audience. It is a bit like watching the construction of a building from the foundation to the roof. For slides, we start with the title and add one line at a time to the text in the block below the title. To modify your slide to incorporate this feature, repeat the slide transition procedure described in the preceding paragraph. The new tool bar that is evident just above the window has a box that is titled No Built Effect. Click on the arrow beside that box to reveal a wide choice of techniques for building you slides. Pick an option; fly from the left is very common. When you select an option for building the slide, a second small icon shows under the view of that slide in the slide sorter view as shown in Fig. 10.10. You can proceed to build your slides, one by one, or you can use the same style on all of them. Select all of the slides on the slide sorter view by pressing the Control + A keys.

When you have completed the transitions and the building effects, you are ready to review the dynamic nature of your electronic presentation. Real high tech. Go to slide 1 and click on the

Slide Show icon button on the horizontal scroll bar. As you advance from slide to slide with arrow keys on the keyboard or by clicking the mouse button, you will see the transitions and the lines of text for each bullet item being moved onto the screen.

If you have computer facilities available in the classroom, electronic presentations should be mandatory. Creating electronic slides and using them in a professional presentation are skills that are important for you to master. The transitions keep the audience on their toes and the building effects permits you to pace the flow of information. It takes less than an hour to learn how to prepare a very professional, dynamic presentation using PowerPoint. Take this opportunity to develop your computer graphic presentation skills.

10.7 SUMMARY

We have introduced you to PowerPoint, a graphics presentation program, that is an extremely useful tool in preparing for design reviews or any other type of presentation. We strongly suggest that you take an hour or two to acquire PowerPoint skills. You will master the skill necessary to produce professional quality overheads at the very least. If the equipment is available in the class room, you will also develop the ability to make presentations with computer projected slides.

The screen for PowerPoint has been briefly described. As with all Windows based programs produced by Microsoft, many of the buttons on the toolbars have a common purpose. The coverage given here is brief because learning is more rapid if you click, study, and understand. Hopefully, our limited coverage and examples will be helpful.

We suggest that you start learning to use PowerPoint by preparing your preliminary design review using the Auto Content Wizard. The wizard organizes your presentation, arranges the sequencing of your slides, and provides excellent suggestions for content in each slide.

After you have completed the standard presentation with the Auto Content Wizard, you should enhance the slides to improve their appearance. Procedures have been described for changing the design template, the colors of the background and the text, and by adding a footer and clip art. Clip art is particularly impressive and easy to implement.

Finally, we introduced techniques for preparing electronic presentations. An electronic presentation depends upon the availability of a computer controlled projector for the classroom. If projection equipment is available, the use of a computer in your presentation has significant advantages. Employing transitions, as you move from one slide to the other, maintains the audience interest. Utilizing build effects, where the text is presented one line at a time, has the significant advantage of controlling the flow of information to the audience. The procedures for preparing and rehearsing a computer aided presentation are described in this chapter.

EXERCISES

1. Prepare a title slide suitable for a final design briefing using PowerPoint.
2. Prepare a subject slide that provides an overview of a final design briefing using PowerPoint.
3. Prepare a status slide using the slide outline format in PowerPoint. After completing the slide in the outline format, edit the slide, and print a copy of it.

4. Prepare a pair of slides describing your progress in the development of one of the subsystems for the postal scale.

5. Prepare a slide that describes the key issues facing your team as you design the postal scale.

6. Prepare a slide showing your plans for action in the near future.

7. Using clip art, enhance a slide by inserting an animal in its lower right hand corner.

8. Prepare a transition from the title slide to the second slide. What type of a transition did you select?

9. Prepare a slide using the building effect where single lines of text are added to the slide one by one as you click the mouse button.

PART IV

PRODUCT DEVELOPMENT PROCESSES

CHAPTER 11

DEVELOPMENT TEAMS

11.1 INTRODUCTION

The development of quality, leading-edge, high-performance products requires the coordinated efforts of many skillful individuals from different disciplines over an extended period of time. This group of individuals is organized within the corporation to act in an integrated manner to successfully complete the development. The organizational structure employed varies from one corporation to another, and it also depends to a large degree on the size of the company that engaged in the development. For relatively new firms in the entrepreneurial stage, team structure is usually not an issue. These firms are small with a single product and only ten to twenty individuals are involved in the development process. Everyone on the payroll is deeply committed to this product, and communication is usually accomplished around the lunch table. On the other hand for very large corporations, where the number of employees may exceed 100,000, the company is often organized into divisions or operating groups. These large companies have many products or services that are offered by each division, and in some instances one division competes to some degree with another. For example, General Motors with its five automotive divisions produces many different models of each of its lines which compete for the same market segment.

In a large corporation, the total number of new products which are introduced each year may exceed 100. Communication is often very difficult because personnel involved in the development of a given product are sometimes separated by hundreds or even thousands of miles. Sometimes different divisions are in different countries, and business is conducted in different languages. Clearly, in a large firm, organization of the product development team and the physical location of its team members become extremely important.

In the past decade, there have been many examples [1] demonstrating that a cross disciplinary team provides a very effective organizational structure for product development. The idea of forming a cross disciplinary team to design a new product appears simple until we examine the existing organizational structure of all but the smaller of our product oriented corporations. Larger corporations are usually organized along functional lines. The organization chart presented in Fig. 11.1 shows many of the functional departments that provide the personnel needed on a typical product development team. Note, each department has a functional manager. Also, several departments are often grouped to form a division, and of course, each division has a director. Titles for the division leaders vary from firm to firm, but Director or Vice President is commonly employed.

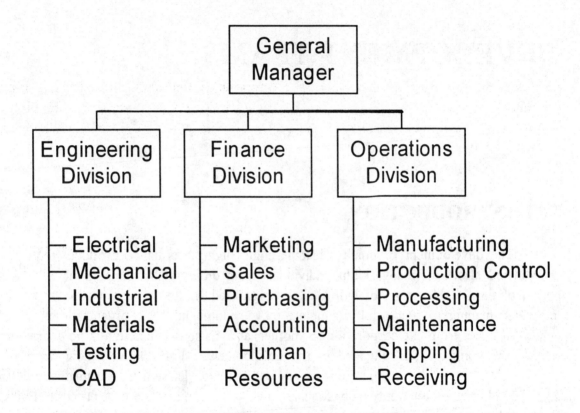

Fig. 11.1 Typical organization chart showing the disciplinary functions, three divisions and three levels of management.

The functional organization with its leadership and its ongoing operations presents a management problem when forming product development teams. As a team member, do you report to the team leader (product manager), or to the functional manager? When important decisions must be made, are they made by the product manager, by one or more of the functional managers, or by the division directors? Clearly, the establishment of a product development team within a company that is organized along functional lines creates several operational problems for management. Functional organizations may also create problems for working engineers attempting to serve several different managers. In the next section, we address these problems, and describe a popular team structure employed in many corporations which are organized along functional lines.

11.2 A BALANCED TEAM STRUCTURE IN A FUNCTIONAL ORGANIZATION

There are several organizational arrangements that are frequently used in forming teams within companies that are organized along functional lines. They include:

- Functional teaming.
- Modified functional teaming.
- Balanced team structure.
- Independent team organization.

Each of these techniques for forming product development teams in a functional organization has advantages and disadvantages. A complete discussion of these four different team structures is given in reference [2]. To introduce you to the concept of forming teaming within a functional organization, we will briefly describe the balanced team structure.

The balanced team structure is represent by the organization chart presented in Fig. 11.2. The team is formed with members drawn from the functional departments; however, in most situations, the team members are located in close proximity in a separate room or a building wing reserved for the team. The team members are usually dedicated to a single project, and their responsibilities are coordinated and focused. The team members retain their reporting relationship to the functional manager, but the contact is less frequent than in functional teaming arrangements, and the relation is less intense. Members of a balanced team often consider the product manager to be more important than their functional manager. The product manager is frequently a senior technical administrator with more experience, status and rank than the typical functional manager. The status of the product manager is dependent on the product development costs and the size of development team. The product manager reports to the general manager, and usually is equal in status to the division directors in the organization. The senior program manager with notable status and clout leads the team, and insures that resources will be available as required to meet schedules. The team members recognize the freedom from functional constraints that the balanced team structure implies. They assume ownership of the product specifications and commit fully to the success of the development activities. Communication is accomplished though the use of division and department coordinators and daily meetings. Most decisions are made by the team and only the higher level decisions require the approval of the appropriate division directors.

The primary disadvantage of the balance team structure is the dedication of the team members to a single project. Many talented team members are needed by the functional managers to support design activities across several product lines. With some of the functional resources dedicated to a single product development, other products may not receive adequate attention. The balanced team structure enhances the development capability of a given product at the cost of reducing the technical capability that can be applied across the company's entire product line. A second disadvantage is that some team member from a given department or division may not have sufficient expertise to perform adequately. With loose and distant functional supervision and review, this fact may not be apparent. The new design may not be at the leading edge of technology.

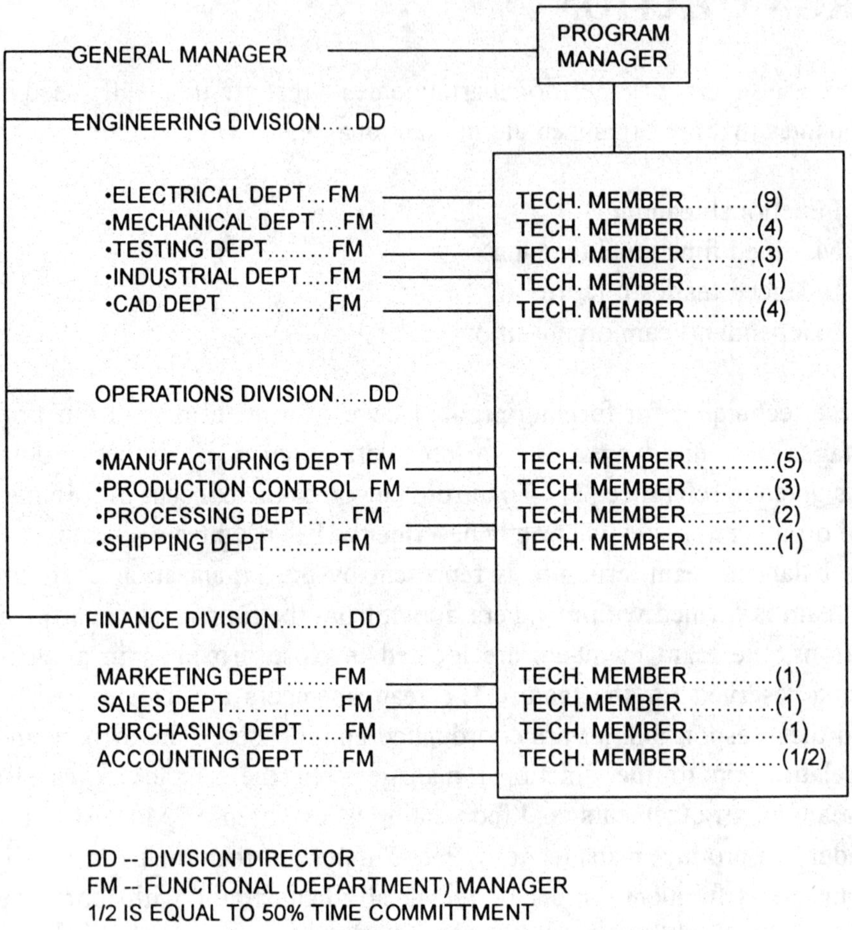

```
                                              ┌──────────────┐
                                              │  PROGRAM     │
  GENERAL MANAGER ─────────────────           │  MANAGER     │
                                              └──────────────┘
  ENGINEERING DIVISION.....DD

    •ELECTRICAL DEPT...FM ─────────    TECH. MEMBER..........(9)
    •MECHANICAL DEPT....FM ─────────   TECH. MEMBER..........(4)
    •TESTING DEPT.........FM ────────  TECH. MEMBER..........(3)
    •INDUSTRIAL DEPT....FM ─────────   TECH. MEMBER......... (1)
    •CAD DEPT...............FM ───────  TECH. MEMBER..........(4)


    OPERATIONS DIVISION.....DD

    •MANUFACTURING DEPT  FM ───────    TECH. MEMBER............(5)
    •PRODUCTION CONTROL  FM ───────    TECH. MEMBER...........(3)
    •PROCESSING DEPT......FM           TECH. MEMBER...........(2)
    •SHIPPING DEPT.........FM ──────    TECH. MEMBER...........(1)


    FINANCE DIVISION..........DD

    MARKETING DEPT.......FM ────────   TECH. MEMBER...........(1)
    SALES DEPT.............FM ───────   TECH.MEMBER.............(1)
    PURCHASING DEPT......FM            TECH. MEMBER............(1)
    ACCOUNTING DEPT......FM            TECH. MEMBER...........(1/2)
```

DD -- DIVISION DIRECTOR
FM -- FUNCTIONAL (DEPARTMENT) MANAGER
1/2 IS EQUAL TO 50% TIME COMMITTMENT

Fig. 11.2 Organization with the balanced team structure.

11.3 DEVELOPMENT PLANNING

As the development team becomes more independent and moves away from the functional departments, planning becomes more important. The team planning documents are often prepared by a small group of senior team members with talents in business, marketing, finance, engineering design and manufacturing. The planning documents incorporate detailed descriptions of the product, market analysis, business strategy for capturing market share, product performance specifications, development schedules and budgets, team staffing requirements, material selections, design strategies, technology prerequisites, manufacturing processes, tooling requirements, inspection procedures, test specifications, facility availability and distribution capabilities. An element of the plan covers techniques and personnel that are to be employed to insure quality and product reliability. Finally, the plan includes a section specifying the deliverables and another section on techniques to measure performance of the product and the productivity of the team.

The development planning documents are extremely important since they outline strategies for achieving the goals and objectives in developing a new product. Some of the documents may be used by management to monitor progress and to judge the team performance during periodic design reviews. The plan is essentially a well defined contract between the team and the division management involved in the product development. The team members are expected to sign-on, to commit to the level of effort required to achieve the goals, meet the schedules and to produce the deliverables. Management is expected to fund the program, to provide adequate staffing and to provide the necessary facilities, tooling and equipment available in a timely manner.

11.4 STAFFING

Cross disciplinary teams are necessary in staffing a product development team regardless of the structure employed the organization. Usually, at least five disciplines or functions should be represented on the team including, marketing, finance, design engineering, operations or manufacturing, and quality control. The level of participation of each team member depends on the product being developed and the stage of the development. In some cases, team members work on more than one product and have responsibilities on two or more different projects. In other instances, several engineers from the same discipline are needed to complete long and complex tasks to meet the schedule for a single product. These members are usually dedicated to a single project.

Staffing on a product development team often changes with time over the duration of the development period. Initially the team is small and staffed with senior personnel with demonstrated skills and talent. This small cadre may stay with the project for its duration. As the development proceeds, additional design engineers are required as technical strategies are converted into design concepts, then to design proposals and finally into detailed engineering drawings. Manufacturing engineers are needed in larger numbers when the early prototypes are being produced. After release of the design by engineering, the number of designers is reduced, and the emphasis is shifted to production where staffing from operations (production control, quality control, purchasing and plant maintenance) increases sharply. Clearly, the staffing of the development team is dynamic. Usually only a relatively small proportion of the total team is committed to the project on a full time basis from its initiation to its completion. In some companies, individuals are committed to the development team for the duration of the project, but at varying levels of participation. This approach enhances ownership of the project by the individual team members.

11.5 TEAM LEADERS

The experience and leadership skills required of the product manager are strongly dependent on the team structure employed as well as the size and cost of the product development. Some products can be developed with a relatively small commitment of staff (say less than 10 members). In these circumstances, the team established is often structured along functional lines, and the product manager is usually one of the functional managers. The functional managers are experts in their respective disciplines, skilled in managing personnel and knowledgeable (but not expert) in the other disciplines. They are also familiar with the company's entire product line.

Other product developments require a larger commitment of personnel (say a team numbering 10 to 50), and the balanced team structure shown in Fig. 11.2 is probably the most suitable organization. The product manager is a seasoned staff member well known for his or her expertise and respected for performing at a level that exceeds expectations. The appointment as a product manager (equal in rank and status to the functional department managers) often represents a promotion for the individual involved and a first experience in managing a technical group. The functional managers and the divisional directors provide close support through periodic reviews that monitor progress and reveal problems early in the development process.

As the size of development team becomes even larger, 100 or more members, team leadership assumes major importance. The product manager in this situation is a senior member of the company's executive management with a well established record of successful accomplishments. The program manager is senior to the managers of the functional departments, and is at least equal in status to the division directors. The authority and responsibility given the product manager is significant, and his or her career depends strongly on the success of the product in the market and the productivity of the development team.

11.6 TEAM MEMBER RESPONSIBILITIES

Members of a product development team have two different sets of responsibilities—one set to his or her disciplinary function and the other to the team. A team member, from say the electrical engineering department, provides the technical expertise in his or her specialty and ensures that the product is correctly designed with leading-edge, state-of-the-art technology. The team member acts to bring all of the important functional issues that affect product performance to the attention of the team. The team member also represents the functional department to ensure that the functional principles, goals, and objectives are maintained by the team in the design of the product.

The most important team activity for the individual member is to share responsibility. The team member shares responsibilities with the others to increase the team's effectiveness and to insure its success. Next, the team member must recognize and understand all of the product features, and fully participate in the methods and techniques employed to meet the design objectives. Individual team members must assess team progress and participate in improving team performance. The team member must cooperate in establishing all of the reporting relationships required to maintain the communication among the team and functional departments. In many of the team structures, an individual may report to three managers—the Division Director, the Department Manager and the Program Manager. If a Functional Coordinator is assigned to the team representing a certain discipline, it is necessary to keep him or her informed of your progress and plans. The matrix organization structure, inherent in forming development teams, creates complex reporting relations. Flexibility and cooperation on the part of the individual team members are required for the team structure to be effective.

Finally, the team members must be able to communicate in a clear concise manner in all three modes: **writing, speaking and graphics**.

11.7 TEAM MEMBER TRAITS

When a group of individuals work together on a team to achieve a common goal, they can be extremely effective [3]. The team interaction promotes productivity for several different reasons. First, meeting together is synergistic in that one's ideas freely expressed stimulate additional ideas by other members on the team. The net result is many more original ideas than would have been possible by the same group of individuals working independently. Another advantage is in the breadth of knowledge available in the team. The team is cross disciplinary and the very wide range of skills necessary for the product development process should be included within the team. Each member of the team is different with some combination of strengths and weaknesses. Acting together the team can build on the strengths of each member and compensate for any weaknesses. The grouping of individuals provides social benefits that are also very important to the individuals and to the corporation. There is a bonding that occurs over time, and a support system develops which is appreciated by the team members. Feelings of trust and mutual understanding, so important to everyone involved, develop over the course of the project. The team develops a sense of autonomy valued by each of its members. The team develops solutions, implements them and then monitors the results with little or no direction from management. Empowered teams assume ownership of the project. They commit to the product development process and consistently perform beyond expectations.

While cross disciplinary teams have been employed in many different companies for a decade or more, little has been done by our educational system to develop team skills. On the contrary, both the secondary school and college systems tend to encourage competitive attitudes and independence. Students compete for the few A grades that might be given in a course, and are discouraged from cooperating on the assignments. The system trains students to work independently, and team building skills are not addressed. The industrial workplace is much different. Team members compete, but not within the team. They compete with a similar development team from a rival corporation. Team members bond, cooperate and consistently help each other by sharing assignments. Competition between team members is discouraged. Recognition and rewards go to the team as a whole much more often than to select individuals.

There is a set of characteristics that describe a good team member, and another set that depicts an individual who can destroy the efforts of a team.

The characteristics of a good team member are:

1. Treat every team member with respect, trust their judgment and value their friendship.

2.   Maintain an inquiring attitude free of predetermined bias so that team members will all participate and share knowledge and opinions in a free and creative manner. Listen carefully to the other team members.

3.   Pose questions to those team members who are hesitant to encourage them to share their knowledge and experience more fully. Share your experiences and opinions in a casual, easy-going manner. Help other members to relax and enjoy the interactive process involved in team cooperation. Participate but do not dominate.

4.   Observe the body language of the other team members because it may indicate lack of interest, defensive-attitudes, hostility etc. Act with the team leader to defuse hostility and to stimulate interest. Disagree if it is important, but with good reason and in good taste.

5.   Emotional responses will occur when issues are elevated to a personal level. It is important to accept an emotional response even when you are opposed. It is also important to control your emotions and to think and speak objectively. You should be self confident in your discussions, but not dogmatic.

6.   Seating arrangements once established tend to become permanent. It is productive to break these patterns, change the seating pattern and to mix team members. The result, over time, is to enhance bonding and to stimulate new ideas generated by different combinations of members.

7.   Allocate time for self assessment so the team can determine if it is performing up to expectations. Make suggestions during these assessment periods to build team skills.

8.   Be comfortable with your disciplinary skill level and continue to study to improve your skills. Communicate effectively by speaking clearly in good English, writing well and in using modern methods in graphics.

The characteristics of a destructive team member are:

1.   No member of the team should ever participate in a conversation that is derogatory about a person on the team or in the corporation. If you can not make complimentary remarks about a person, keep your thoughts to yourself. Respect, trust and friendship are vital elements in the founding of a successful product development team. Derogatory remarks destroy the foundation necessary for respect, trust and friendship.

2.   Arguments among team members are to be avoided. Your are encouraged to introduce a different opinion, and participate in a discussion with a different viewpoint, but the discussion should never degenerate into an argument. Every member on the team is responsible for quickly resolving arguments that arise among the team members.

3.   All members of the team are responsible for their attendance at the meeting and their timely arrival. If a member is absent, the leader should know the reason beforehand and explain it to the team. We must be certain that all members appreciate and respect the reason for everyone's role on the team and the importance of every member attending every meeting.

4. Team progress is hindered when one member dominates the meeting. These people are often overly critical, intimidating and stimulate confrontations. If you have these characteristics, work very hard to suppress them.

5. The team leader must be extremely careful about intervention. If the team is moving and working effectively, the leader should be quiet because intervention in this instance is counter productive. If the team is having difficulty, intervention may be necessary depending on the problem. If a team member becomes hostile, intervention should be quick, and the disagreement producing the hostility should be dealt with immediately. If the team is seeking consensus, the process may require time and the leader should wait 5 to 10 minutes before intervening.

6. No member of the team can be opinionated including the leader. He or she is not a judge determining the right solution. The leader seeks to facilitate so that the team can reach a consensus on the correct solution. It is only when the team accepts the solution that implementation can begin.

As we compare our own traits with those listed above, we will note characteristics from both the good and bad lists. To be effective team members, we must work to enhance the favorable traits and to suppress those that are destructive to the efforts of the team.

11.8 EVOLUTION OF A DEVELOPMENT TEAM

If a number of workers or students are assigned to a new team, it is possible to observe several behavioral phases as the team melds and matures [4]. In the very beginning (Phase 1), the team members usually exhibit both excitement and concern. They are excited about working on a new project with a new group of talented people. It should be an opportunity to learn, make new friends, advance in position and/or stature and have fun in the process. At the same time, the team members exhibit signs of concern. They are worried about meeting and understanding the other team members. The tasks assigned to the team are probably extremely vague at this stage of the development, and vagueness leads to uncertainty. The skill levels and areas of expertise of fellow team members have yet to be determined. The personality of the team leader is unclear. There is worry over the new reporting relations that the new team membership implies. In this orientation phase, the team members are searching for their role in the team and evaluating their possibilities for success or failure. It is a tense time.

Dissatisfaction is the second phase in the evolution of the team. We have met our fellow team members, and the news is not good. We recognize the differences in personalities and in work habits. A few of the team members do not understand how to tell time and never arrive on time when attending team meetings. Our schedules make it very difficult for the team to meet as a whole. Some of team members lack minimal social graces. Some members are inexperienced and unable to cope with their assignments. We have met the team leader and he or she is very difficult (stressed, demanding, exacting, impatient and abrupt). The schedule calls for us to leap over mountains on a daily basis. The budget for the project is totally inadequate. Surely we are in a lose—lose situation. Our thoughts are dominated by schemes to transfer to a better team. During this phase, the progress of the team toward scheduled milestones is very slow.

With time progress is made, and the team members start learning to work together. This is the resolution phase. We agree to attend meetings and to arrive on time. Personality conflicts are smoothed over. We may never be true friends, but we manage to get along and show mutual respect. We have come to terms with the schedule, and have committed to the extra effort that it requires. The manager has mellowed, or we have learned how to handle his impatient demands. Solutions for reducing development costs have been found. There appears to be a good possibility of meeting the product development plan. The team is beginning to perform well, and more members are committed.

The team has melded into an efficient unit. Conflicts rarely arise. Lasting friendships begin to be formed. The strengths of each team member are fully utilized. The manager recognizes the talents of the team. Tasks are completed ahead of schedule and under budget. The accomplishments of the team are recognized by upper level managers. You now find yourself in a win—win situation. This is the production phase of the development team.

The project has been completed and the ribbons have been cut. The team prepares to disband. In this termination phase, we reflect on our experiences, good and bad, over the duration of the project. We examine our individual performance and evaluate various factors that improved the effectiveness of the team. We meet with the manager for a performance evaluation. The team meets to share a sense of accomplishment in achieving the goals and objectives of the development.

11.9 A TEAM CONTRACT

Teams are effective because they meet together to focus their wide range of disciplinary skills and natural talents toward the solution of a well formulated problem. The team meeting is the format for the synergistic efforts of the team. Unfortunately, not all teams are effective because some members are so disruptive they destroy the cohesiveness of the team. Bonding of team members is vital if the team is to successfully solve the multitude of problems that arise in the product development process.

Teams fail because of three main reasons. First, they deviate from the goals and objectives of the development, and cannot meet the milestones on the development schedule. Second, the team members become alienated and the bonding, trust and understanding, critical to the success of a team, never develops. Third, when things begin to go wrong and adversity occurs, "finger pointing" starts and fixing blame on someone replaces creative actions.

Clearly, many problems may arise to impair the progress of a development team. We recommend the team establish a set of simple rules for working together. We have provided some suggestions for your team to consider. Team members should meet and modify or add more rules as necessary to govern your team behavior.

- The information discussed by our team members will remain confidential.
- We will acknowledge problems and deal with them.
- We will be supportive rather than judgmental.

- We will respect differences.
- We will provide responses directly and openly in a timely fashion.
- We will provide information that is specific and focused on the task and not on personalities.
- We will be open, but will respect the right of privacy.
- We will not discount the ideas of others.
- We are each responsible for the success of the team experience.
- The team has all of the resources needed to solve the problems that arise during the development process.
- Every member of the team will contribute to its success.
- We will try to know and understand our fellow team members so as to identify ways for enhancing everyone's professional development.
- We recognize the importance of the team meeting to the success of the group, and accordingly we agree to:
- Always attend the scheduled team meetings.
- When attendance is impossible, notify the team leader and as many other members as possible in advance.
- Use meeting time wisely
- Start on time.
- Limit our breaks and return from them on time.
  - Keep focus on our goals.
  - Avoid side issues, personality conflicts and hidden agenda.
  - Share responsibility for briefing members when they miss a meeting.
  - Avoid making phone calls or engaging in side conversations that interrupt the team.

Writing a team contract has been included in this Chapter in Exercise 19. We recommend that each team prepare a contract, and each team member sign and date it.

11.10 EFFECTIVE TEAM MEETINGS

Leadership is important in ensuring that the team remains focused on the overall goals and objectives of the project and on the agenda at each meeting. Creating the correct team environment or atmosphere is essential so that the team members cooperate, share ideas and support each other to achieve solutions [5]. It is important for the team leader to assess the performance of the team as a whole at each meeting. When the team fails to make progress, it is critical for the team leader or possibly an individual member to make changes as necessary to enhance the teams effectiveness. There are several measures of team effectiveness that can be exercised by everyone on the team during the course of a typical team meeting. Corrective actions to improve team performance are not difficult to implement. We will identify some of the difficulties encountered during team meetings and suggest actions to be taken by members for enhancing its productivity.

One common problem, impairing team productivity, is when the goals and objectives of the meeting are not clear or they appear to be changing. While in many cases team meetings are planned well in advance and are supposed to follow an agenda, we observe that new topics are introduced and that the meeting drifts from one item to another. This team behavior indicates that either some of the team members do not understand the goals, or they do not accept them. Instead, they are marching to their own drummer and trying to take the team with them. Unless corrective actions are taken to focus the team on the original goals, this team meeting will fail and time and effort will be lost. A team that is focused stays on the agenda, the team members accept the goals and their discussions deal with all of the issues important in generating a wide set of solutions to the problem being considered. The discussion continues until the team members reach a consensus. This meeting is successful because a problem has been solved and all of the team members accept the solution. The team can move forward and organize its talents to address the next problem.

Leadership in the team is an essential element for its effectiveness. The most effective teams are usually democratic with a large amount of shared leadership. While there is a recognized leader, with a considerable degree of responsibility and authority, the team leader is supported by each member of the team. The team member with the appropriate expertise (relative to the problem being considered) will usually lead the discussion and effectively act as leader of the team during this period. However, in some instances, the team leader maintains strict control of the meeting and does not share the leadership role with any member of the team. Some team members resent this style of leadership and may not participate as fully as possible. The result is that complete utilization of the resources of the team does not occur. The "tightly controlled" team leader will almost always insist on sitting at the head of the table.

Attitude of the team members is an important element in the success of the team. Are the team members committed? This commitment is evidenced in their attitude. If they come to the meeting table exhibiting interest and willingness to participate, they are committed and will make a positive contribution. However, if they are bored and indifferent, they will not become engaged in a meaningful way. In fact, if they are sufficiently disinterested, they may initiate side conversations or arguments and destroy the atmosphere of trust necessary for an effective meeting. It is important to evaluate each of the team members, and to secure their commitment to the goals and objectives of the project and to the agenda for each meeting.

As the meeting progresses, it is important to assess the discussions which are occurring. For a meeting to be successful, the discussion should include everyone on the team. The discussion should stay focused on the agenda items with few if any deviations to unrelated topics. The team members must listen carefully to one another. All of the ideas presented are given a serious hearing. No one is intimidated or is made to appear foolish for making suggestions that may appear to be too radical. There is an informal and relaxed attitude exhibited by the participants. Teams tend to accomplish less when the discussion is dominated by only a few of the participants. They talk well, but they do not bother to listen to others. They may introduce a topic which is not on the agenda, and ignore suggestions for getting back on track. These members tend to intimidate others in their efforts to control and dominate the team. The result is disastrous because they essentially eliminate the contributions of the other team members.

Teams do not operate with total agreement on every issue. There must be accommodation for disagreements and for criticism. All of the ideas or suggestions made by every team member at any meeting will not be outstanding. Indeed, some of them may be ridiculous; criticism of these ideas must occur. However, the criticism should be frank and without hostility. Personal attacks must not be a part of the critique of an idea. If the criticism is to improve an idea or to eliminate a false concept, then it is of benefit to the progress of the team. The criticism should be phrased so that the team member advancing the flawed idea is not embarrassed. When disagreements occur, they should not be suppressed. Suppressed disagreements breed hostility and distrust. It is much better for the team to deal with the disagreements when they arise. The root causes of the disagreement should be ascertained, and the team should take the actions necessary to resolve them. On some occasions voting is a mechanism used to resolve conflict. This procedure must be used with care particularly if the vote indicates the team is split almost evenly in their opinion. A better practice is to discuss the issues involved and attempt to reach a consensus. For a team to reach a consensus, may require more time than a simple vote, but the results are worth the effort. Consensus implies that all members of the team are in general agreement and willing to accept the decision of the team. When the team votes, a simple majority is sufficient to resolve an issue. However, the minority members may become resentful if they are always on the short end of the voting. In a very short time, they will not accept the outcome of the voting, and they will not commit to the actions necessary to implement the decision.

The team meetings must be open. The agenda should be available in advance of the meetings and subject to change during the new business agenda item. A sample agenda is shown in Table 11.1. Team members should believe that they are important and have the authority to bring new issues before the team. They should feel free to discuss procedures used in the team's operation. Hidden agenda items detract from the harmony and trust developed through open and fair operation of the team. Secret meetings of an inside group should be carefully avoided. When the fact leaks that secret meetings are being held and that issues are prejudged by a select few, the effectiveness of the team is destroyed.

At least once during a meeting it is useful to evaluate the progress of the team. Are we still following the agenda? Is someone dominating the meeting? Is the discussion to the point and free of hostility? Is everyone properly prepared to address their agenda items? Are the team members attentive, or are they bored and indifferent? Is the team leader leading or is he or she pushing? If a problem in the operation of the team is identified during the pause for self appraisal, it should be resolved immediately through open discussion.

Table 11.1

SAMPLE AGENDA

Weekly meeting of the Tiger team
September 21, 1998
7:59 am

1. Weekly status report	Team Leader
2. Review of outstanding action items	
• Action item #12	Member responsible
• Action item #13	Member responsible
• Action item #N	Member responsible
3. Report on progress	
• Subsystem –Power	Member responsible
• Subsystem –Electrical	Member responsible
• Subsystem – Mechanical	Member responsible
• Subsystem –Control	Member responsible
• Market study results	Marketing representative
4. Identify new problems	All members participate
5. Assignment of action items	Team leader and volunteers
6. New business	All members participate
7. Summary	Team leader
8. Adjourn at 8:59 am	

Finally, the team must act on the issues that are resolved and the problems that are solved. To discuss an issue and to reach a consensus is part of the process, but not closure. Implementation is the closure, and implementation requires a plan for action. When decisions are made, team members are assigned action items. These action items are tasks to be performed by the responsible individual; only when the tasks are completed is the problem considered solved. It is important that the action items are clearly defined, and the role of each team member in completing their respective tasks is evident. A realistic date should be set for the completion of each action item. The individual responsible should be clearly identified; he or she must accept the assignment without objection or qualification. A checking system must be employed to follow up on each action item to insure timely completion. If there is a delay, the schedule for the entire product development cycle may be at risk. It is important to deal with delays immediately and for the team to participate in the development of plans to eliminate the cause of the delay. We have included a form in Table 11.2 to facilitate the assignment of action items and the follow-up on each item that the team may employ.

TABLE 11.2

ACTION ITEMS

Tiger Team
Meeting of September 21, 1998

Decisions made	Follow-up required	Responsible member	Date complete
1.	1.	1.	1.
2.	2.	2.	2.
3.	3.	3.	3.
4.	4.	4.	4.
Action at next meeting	**Preparation required**	**Responsible member**	**Date complete**
1.	1.	1.	1.
2.	2.	2.	2.
3.	3.	3.	3.
4.	4.	4.	4.

11.11 PREPARING FOR MEETINGS

Effective team meetings do not just happen. It is necessary to prepare for the meeting and to execute post meeting activities to insure success. The preparation usually involves selecting, arranging and equipping the meeting room, scheduling the meeting so that the necessary personnel are in attendance, preparing and distributing an agenda in advance of the meeting and conducting the meeting following a set of rules accepted by the team. Equally important, are actions taken by the team members following the meeting to implement the decisions that have been reached.

The meeting room and its equipment are also important to the progress made by the team. The room should be sized to accommodate the team and any visitors that have been invited. Rooms that are too large permit the members to scatter and the distance between some members becomes too long for effective and easy communication. The chairs should be comfortable, but not so soft as to promote napping. The seating arrangement should be around a table so that everyone can observe each other's face and body language. It is very difficult to engage anyone in a meaningful conversation if they have their back to you. Water, coffee, tea or soft drinks should be available if the meeting duration exceeds an hour. Smoking is strictly prohibited. The equipment that will be needed for presentations must be available, and its operation should be checked prior to the start of the meeting. Flip charts and white boards that are useful for recording the key ideas generated during the meeting are essential. Tape, pens, markers and other supplies necessary for preparing and displaying charts must also be available. A computer with projection capability is of growing importance in a well equipped meeting room. The availability of a computer during the meeting permits one to draw from a large data base and to modify the presentation in real time. Also results of analyses or experiments can be displayed in graphical form for the entire team to review. Finally, adjust the room temperature until it is comfortable. Check to ascertain that the light intensity can be controlled from low for projecting overheads to high for round table discussions.

Prior to the scheduled meeting, it is important to make careful preparations. Minutes from the previous meeting must be distributed with sufficient time for review before the next scheduled meeting. A detailed agenda, similar to the one shown in Table 11.1, is distributed which informs the team members of the topics and/or problems that will be addressed in the next meeting. Individual team members responsible for specific agenda items are identified. The details are clear to all and responsibility shared by individual members is defined. In some instances, information will be needed from corporate employees, suppliers or visitors that are not members of the team. In these cases, arrangements must be made to invite these people to the meeting so that they can provide the necessary expertise and respond to questions from all of the disciplines represented on the team.

In conducting the meeting, it is important to start on time. It is very annoying to the majority of the members to wait five or ten minutes for a straggler or two. The team leader must make certain that everyone involved understands that 8:00 am means 8:00 am and not 8:08 or 8:12 am. The objectives of the meeting should be displayed and the time scheduled for each item should be estimated. A team member should be assigned as the timekeeper, and another should act as a secretary to record notes and to prepare the minutes of the meeting. If team meetings are frequent and they are held over a long period of time (several months), the duties of the timekeeper and the secretary should be shared with others on the team. The meeting should follow a set of rules that govern the behavior of the individual members. We have assigned the preparation of a list of rules which should be followed in conducting an effective meeting as an exercise (see Exercise 17). As the meeting draws to a conclusion, the team leader should take a few minutes to summarize the outcome of the discussions. This summary gives an ideal opportunity to insure that assignments are understood, responsibility accepted and that completion dates are established. The meeting must be completed on time and everyone's schedule should be respected. If the meeting is not periodic (i. e. every Tuesday at 8:00 am), then it is important that the time and place for the next meeting be scheduled.

After the meeting, it is productive for the team leader to check to make certain that the minutes have been prepared, that they are complete and that they have been distributed. If any member was not able to attend the meeting, the team leader should brief that person on the happenings. Finally, the team leader should follow up on each action item to ensure that progress is being made, and that new or unanticipated problems have not developed. It is clear that effective meeting do not happen by accident. Many members of the team work intelligently before, during and after the meeting to make certain that the goals and objectives are clearly defined, that the issues and problems are thoroughly discussed and that the solutions developed result in assignments which are executed with dispatch.

11.12 POSITIVE AND NEGATIVE TEAM BEHAVIOR

All the members on a team and particularly the team leader must behave in a positive manner for the team to function effectively. Negative behavior, that occurs frequently, is detrimental and inhibits the progress that can be made by the team and delays the product development cycle. Group behavior is often divided into three categories including [6]:

- Leading
- Supporting
- Hindering

The leading and supporting roles are consistent with team progress and positive contributions to the development project. However, the hindering role is destructive as it impedes progress and causes conflict and concern to many members of the team.

11.12.1 Leading Roles

Any team member can play a leading role at any time in a team meeting. Of course, we expect the team leader to provide direction, but leadership is shared in a high performance development team. Leading tasks include:

1. **Initiate:** Suggest ideas or concepts for the design. Propose objectives or goals. Define a problem and give an approach for its solution. Raise issues that are important to the progress of the team. Recommend procedures for completing tasks.
2. **Provide:** Give relevant information pertaining to issues before the team. Offer suggestions, ideas and theories. Furnish facts and background knowledge. Present opinions with supporting material.
3. **Seek:** Ask for suggestions and ideas. Solicit options. Request facts relative to a team problem. Look for information leading to resolution of team concerns.
4. **Clarify:** Indicate alternatives and issues for the team. Provide examples to support another's statement. Interpreting ideas and suggestions. Define issues and terms. Offer illustrative conclusions.
5. **Test:** Check to ascertain the degree of agreement within the team. Inquire to determine if the team is prepared to reach a consensus.
6. **Summarize:** Offer conclusions or decisions at the termination of a team discussion. Gather together related ideas offered by several team members. Restate suggestions after team discussions.

11.12.2 Supporting Roles

Many team members are not always comfortable with the leading roles. Instead they can make significant contributions in a supporting role. Team member providing support engage in the following activities:

1. **Monitor:** Assist in maintaining open communication among team members. Recommend procedures for sharing ideas. Promote the participation of less active team members.
2. **Encourage:** Respect other team members by providing opportunities for their recognition. Appear friendly, warm and responsive to all members of the team. Accept others and recognize their contributions.
3. **Compromise:** When team conflicts arise, offer a compromise approach. Admit errors readily. Modify ideas or concepts in the interest of team cohesion and growth.
4. **Sense:** Share your feelings with the team. Identify feelings, moods and relationships within the team.
5. **Harmonize:** Act to defuse conflict. Reconcile disagreements. Reduce tension and stress. Encourage team members to explore their differences.
6. **Testing:** Ascertain if the team is satisfied with the procedures in place. Suggest new procedures or guidelines to improve the operation of the team.

11.12.3 Hindering Roles

Unfortunately, we sometimes find lemons in the group. These are team members who do more harm than good. Behavior indicating a member impeding the team progress includes:

1. **Uncooperative:** Stubborn resistance of ideas of others for personal motives. Disagrees for no apparent reason. Employs a hidden agenda to deliberately impede the advancement of another's idea.
2. **Degrading:** Jokes in a sarcastic manner. Hostile criticism of ideas and suggestions of others. Injures the esteem and of other team members.
3. **Dominating:** Interrupts the contribution of other team members. Asserts authority to manipulate one or more team members. Controls some team members by manipulation.
4. **Avoiding:** Absent or late to team meetings on many occasions without valid justification. Frequent changing of the topic under discussion. Uncomfortable with conflict.
5. **Withdrawing:** Avoids participation when possible in team activities. Offers few if any suggestions, ideas or recommendations. Responds to questions with abrupt and brief answers.
6. **Distracting:** Ignores the team meeting and begins side conversations with other team members.

Most of us are capable of playing all three roles at any time. We can help or hinder the progress of the team. It is important that we pause occasionally toward the end of a team meeting to assess the roles played by each team member. We have provided an assessment form in Exercise 20 to facilitate collecting the data from your assessment.

11.13 SUMMARY

We have presented arguments for the importance of development teams in industry. Experience has shown that effective development teams are vital if a corporation is to develop highly successful products and introduce them in time to win a major share of the market.

We have described functional organizations that exist within most corporations, and have given reasons why this organizational structure often interferes with rapid low-cost product development. The balanced team structure commonly employed in establishing development teams within large functionally organized corporations has been described.

We have not discussed the importance of location of team members in either enhancing or inhibiting communication, but you should be aware of its importance. The rule is simple. To enhance communication, minimize the distance between those that are to communicate with each other. The ideal situation is to place the entire development team in the same room.

Development planning and staffing of the development team has been covered. The success of the team depends strongly on generating a comprehensive plan. The plan is used initially to justify the development budget and its schedule. During the course of the development, the plan provides guidance for monitoring the progress and for judging both the team and product performances. Team staffing is cross disciplinary with adequate representation from both the engineering and finance divisions. Engineers interact with the business personnel from marketing, sales and purchasing. Responsibility is shared between disciplines and functions.

We briefly discussed team leadership in industry, and related the organizational structure of the team with the level of management provided by the team leader. This is important because of the authority that goes with the level of management. Higher level managers carry more authority, many decisions can be made more rapidly, communication is more effective and team progress is often enhanced if the team leader has status and clout associated with executive management.

Next, team member responsibilities and team member traits were discussed in considerable detail. Team members work in a matrix organization and often report to two or more managers. This is sometimes difficult particularly when you get mixed signals from different managers. It is important to be flexible and always remain cooperative with the program manager and the functional managers. Team member traits are very important to your career. Please evaluate your own traits in an honest self assessment. This is a critical first step in building team skills.

Most design teams evolve over the duration of a development project. We have described five phases usually experienced by the team in this evolution. Not all of the phases provide pleasant experiences. Early in the evolution, a team encounters the dissatisfaction phase with very slow progress, low productivity and member gloom. To minimize the time in this phase, we suggest the team prepare and execute a contract that provides guidelines for member behavior.

Finally, we have emphasized team meetings because the progress of the team is very dependent on effective meetings. Also, you will spend more time than you can imagine in meetings. From the outset, learn the techniques for a successful meeting. We provide reasons for the failure of some meetings and the success of others. The role of the team leader and the behavior of the team members are described. The meeting room is more important that most engineers imagine. We described features and furniture in a room conducive to productive meetings. We also emphasized positive and negative behavior on the part of the individual team members and gave explicit examples of both types of behaviors.

EXERCISES

1. Write an engineering brief describing why team structure is so important in the product realization process that takes place in a large corporation. Add a second paragraph indicating why it is much less important in a very small company.

2. Prepare an organization chart, like the one shown in Fig. 11.1, for the University or College that you are attending. Describe the logic, as you see it, upon which the organization is based.

3. List the advantages and disadvantages of the balanced team structure.

4. Outline a development plan that you would prepare if you were a senior team member representing the engineering function on a newly formed development team. The product to be developed is a new trail bike.

5. Write an engineering brief describing staffing required for the development team for the trail bike. Give the reasons for changing the staff as the bike evolves during the development process.

6. Describe the experience and status of the product manager for the balanced team structure.

7. Describe a matrix organization structure. Explain why this organization impacts an engineer assigned to a product development team with a balanced structure. If you were this engineer, how would you handle your functional manager?

8. Write a letter to a prospective employer explaining why you are a good team member.

9. Write a letter, with the best intentions, to a friend explaining why you think he or she is not an effective team member.

10. You have been a member of a functional team for four months. The team is not doing well and is beginning to miss milestones in the schedule for the development plan. You are to meet with your functional supervisor for your formal six month performance review. Write a script for a two person play, with your supervisors questions and comments, and your answers and comments covering what you expect to occur during your review.

11. You (a male/female figure) are attending weekly team meetings and are always seated next to Sally/Bill who is young and very attractive. Sally/Bill is apparently interested in you because she/he frequently involves you in side conversations that are not related to the ongoing team discussions. Write a plan describing all of the actions you will take to handle this situation.

12. Brad, the team leader, is a great guy who believes in leading his team in a very democratic manner. He encourages open discussion of the issue under consideration for a defined time

period, usually 5 or 10 minutes. At the end of the time period, he intervenes and calls on the team to vote to resolve the issue. Write a critique of this style of leadership.

13. You together with Zack and Mary are senior members on a development team with 14 other members. The three of you are really very knowledgeable and experienced. You get along very well and have developed a habit of gathering together at the local watering hole the evening before the schedule team meeting. During the evening you discuss the issues on the agenda and reach some decisions between the three of you. Write a brief describing the consequences of continuing this behavior.

14. Sue has a wonderful personality, and as a team leader is skillful in promoting free and open discussion by all of the team members. After the team reaches a consensus she calls for volunteers to implement the required action. When two or more people volunteer, she assigns them as a group to handle the action item. When no one volunteers, she assigns the least active person on the team to the action item and moves promptly to the next item on the agenda. Write a critique of this leadership style.

15. Efficient Ed prepares a meeting agenda that includes 16 major items. He schedules each item on the agenda with a 15 minute discussion period without indicating which team member will lead the discussion on each topic. The meeting is scheduled to begin at 8:00 am and to adjourn at 11:59 am. Write a description of what you imagine would happen during the meeting. If you were the team leader, how would you handle the 16 issues that had to be resolved in a short time frame?

16. You are the leader of a team of eight members meeting for the first time to develop a new hair dryer. During this initial meeting the following events occur.
    • The temperature the meeting room is 88 °F.
    • The room is set up for a seminar speaker.
    • The meeting is scheduled to begin at 9:30 am and at 9:35 am only five of the eight members have arrived.
    • During the meeting Horrible Harry begins to verbally abuse Shy Sue.
    • At 9:50 am Talkative Tom begins to describe his detailed positions on the world situation and is still going at 10:10 am.
    • The team breaks for coffee at 10:15 am and has not returned at 10:30 am.
    • Sleepy Steve begins to snore.
    • Procrastinating Peter refuses to accept an action item after leading the discussion on the issue for 15 minutes.
    Describe the approach that you would follow in dealing with each of these problems.

17. Prepare a list of ground rules that should be followed by all the members of a team in conducting an effective meeting.

18. You are a member of a team with an ineffective leader. What can you do to improve the productivity of the team? In responding to this question consider the irresponsible actions on the part of the team that are listed in Exercise 16. Remember in your response that you are not the leader and the most that you can do is to share the leadership role from time to time.

19. Meet with your team and discuss the list of rules that will be used in guiding the conduct of the team during the semester. Prepare this list of rules in the form of a contract similar to

that shown in Section 11.9. As a final step, each member of the team is to sign the contract indicating his or her commitment to the rules of conduct.

20. Record your assessment of the leading, supporting and hindering roles that are played by each member of your team. We have provided you with a form shown below to facilitate this process. Use a score from 1 to 5 in each cell with 1 indicating little activity and 5 very significant activity. Place the name of each team member in the column headings

TEAM MEMBERS' NAMES						
SCORES	1 -5	1 -5	1 -5	1 -5	1 -5	1 -5
LEADING ROLES						
1. Initiate						
2. Provide						
3. Seek						
4. Clarify						
5. Test						
6. Summarize						
SUPPORTING ROLES						
1. Monitor						
2. Encourage						
3. Compromise						
4. Sense						
5. Harmonize						
6. Test						
HINDERING ROLES						
1. Uncooperative						
2. Degrading						
3. Dominating						
4. Avoiding						
5. Withdrawing						
6. Distracting						

REFERENCES

1. Smith, P. G., D. G. Reinertsen, <u>Developing Products in Half the Time</u>, Van Nostrand Reinhold, New York, 1991.
2. Cunniff, P. F., Dally, J. W., Herrmann, J. W., Schmidt, L. C. and Zhang, G., <u>Product Engineering and Manufacturing</u>, College House Enterprises, Knoxville, 1998.
3. Barczak, G. and Wilemon, D., "Leadership Differences in New Product Development Teams," Journal of Product Innovation Management, Vol. 6, 1989, pp. 259-267.
4. Lacoursiere, R. B. <u>The Life Cycle of Groups: Group Development State Theory</u>, Human Service Press, New York, 1980.

5. Barra, R. Putting Quality Circles to Work, McGraw Hill, New York, NY, 1983.

6. Foxworth, V., "How to Work Effectively in Teams," Class notes ENME 371, University of Maryland, College Park, September 5, 1997.

7. Clark, K. and Fujimoto, T., Product Development Performance, Harvard Business School Press, Boston, 1991.

8. Wheelwright, S. C. and Clark, K. B., Revolutionizing Product Development, Free Press, New York, 1992.

CHAPTER 12

A PRODUCT DEVELOPMENT PROCESS

12.1 INTRODUCTION

The design and development of new products and services are essential for the welfare of most industries operating in the world. The only industries that can operate today, disregarding the marketplace and continuing to ignore to the consumers' needs, are those protected by government regulations (i.e. the U. S. Post Office). For private enterprise to be successful, their product line must be competitive in every category. The products must be attractive, functional, efficient, durable, reliable, affordable, and most importantly they must satisfy the needs of the customer. How do we systematically design products that will be a success in the marketplace while generating profits for the company? Over the years most companies have evolved a product development process that is intended to provide a steady stream of new successful products. During the past decade, there have been many changes, and new approaches to product development have been advanced by several of the more successful companies. A general outline of a nine phase product development process which is followed by many companies in the U. S. today is listed below:

1. Identify the customer needs.
2. Establish the product specification.
3. Define alternative concepts for a design that meets the specification.
4. Select the most suitable concept.
5. Design the subsystems and integrate them.
6. Build and test a prototype and then improve it with modifications.
7. Design and build the tooling for production.

8. Produce and distribute the product.
9. Track the product after release developing an awareness of its strengths and weaknesses.

A graphic describing the product realization process, which is derived from references [1, 2], is depicted in Fig. 12.1. The idea stage of the development involves interpreting the information derived from customer interviews and benchmarking. Then the product is defined in the product specification. The product specification establishes objectives for all of the important parameters controlling the performance of the product.

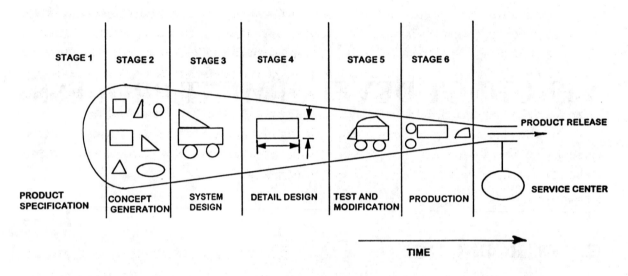

Fig. 12.1 Stages followed in a product development process showing design activities that occur from the initiation of a development to release of a product to market. After [1, 2].

In the second stage of the process (concept generation), the development team considers many different ways to design each subsystem in the product. The ideas flow and different design concepts are explored. The better ideas are differentiated from inferior concepts. Models may be constructed and tested to determine initial feasibility of the newer concepts. The number of design ideas initially increases with time during the concept generation phase, and then the number decreases during the concept evaluation process.

The third and fourth stages, in the pictorial of Fig. 12.1, deal with design. System design encompasses the entire product line. Functions of the product are considered that lead to the division of the design of the product into various subsystems. Interfaces between the subsystems are defined and the subsystems are integrated. Only after the subsystems have been integrated, is the detail design initiated. Engineering drawings are made of the component parts. Final selections of materials are made. Manufacturing and process control personnel begin to design tooling and to plan the production sequences. Purchasing agents begin to negotiate price with qualified suppliers. Quality assurance procedures are developed and delivery schedules for parts to be procured are established.

Stage 5 in the pictorial, entails building one or more prototypes. Building these prototypes insures that the parts fit together properly and that the unit can be assembled without difficulty.

Testing the prototypes verifies performance and gives a basis for design modifications leading to product improvements or refinements.

Stage 6 of the development process, includes procuring the final tooling, establishing the production procedures, and writing the production routings. The product is manufactured at increasing production rates until the inventory is sufficient; then the product is release to the market.

After product release, service centers are contacted to investigate the reasons for early product returns. Failures of the product are diagnosed, corrections to the design are made, and product modifications are implemented at the earliest opportunity.

The illustration in Fig. 12.1 implies that product development is a sequential series of steps. However, it is important to recognize that we must attempt to conduct many of these stages of development concurrently. The phrase **"concurrent engineering"** has been defined to convince managers that the various stages of development are conducted in parallel and not in series. By overlapping the phases, beginning certain aspects of stage 2 and possibly 3 before the completing stage 1, we can significantly reduce the development time and reduce the time to market.

In this chapter, we will describe many aspects of the of the product development process. Hopefully, you will gain an understanding of the activities that take place within a typical corporation as they develop a new product for the marketplace.

12.2 THE CUSTOMER

We develop a product so that it may be sold at a profit. If a product is to be sold, someone must like it well enough to spend his or her money to purchase it. How do we design a product that the customer will like? The answer is to talk to the customer. Yes, design engineers do go into the field and talk with the customers one on one. These talks (interviews) are conducted at the customer's location for 30 minutes to an hour. Another approach for learning about the customer's needs, is to quietly watch the customer use the product over an extended period (several hours). By observation, we determine all of the ways the product is used and record the customers' reactions in each application. Interviews conducted by two members of the design team with a user of the product and observation periods with the customer represent very cost effective methods of establishing what improvements the customer prefers in a new product. Studies [3] indicate that 30 to 40 interviews usually reveal the needs of the customer sufficiently well to provide direction for developing a new improved product.

Focus groups, similar to those used in the news programs on television to determine political preferences, are also effective in determining the customers' needs. The focus group (eight to ten customers) is assembled at an off-site location to discuss the product. A moderator with skills in group dynamics leads the discussion. The focus session, which is about two hours in duration, is video taped. The development team reviews and analyzes the tape. They gain insight regarding the needs of the customer, and the value that customers place on features that may be included in the development of a new product. Focus groups have the advantage that they encourage cross communication among the customers which often reveals information that is not discovered in the individual interviews. However, the disadvantage of this approach is higher costs because of the need to hire a moderator, tape the session and rent the meeting room in a hotel.

Who is the customer? Clearly the person buying the product in the marketplace is a customer. However, usually there are others involved with the product that we need to consider as stakeholders if not customers. To illustrate this point, let's consider Black & Decker® (B&D), a company that designs, manufactures and sells moderately priced power tools. Clearly, Harry and Harriet, happy home owners, are customers because they go to the store and buy power tools. B&D will certainly interview Harry and Harriet to obtain their input, but they will also interview the store managers at retail outlets like Home Depot, Walmart, Lowes, etc. These store managers are stakeholders because they share in the success of the product. They are concerned with attractive packaging, the size of the package, the lead time for the delivery, the market introduction date, price, etc. The retail store managers needs are as important as the needs of the individuals buying the product. They control the entry to the marketplace because they place the purchase orders with the company's distributors.

The service centers are another rich source of data that must be carefully considered before beginning to write the product specification. The service centers receive the products that are returned for a variety of different reasons. They listen to the complaints of the users and they repair products that have failed in service. They know in great detail the strengths and the weaknesses of the product line. They know the importance of each flaw to the customers because they are exposed to the remarks made when a faulty product is returned for either refund or repair.

Suppose you are assigned to a development team, and your first task is to interview six previous customers all located in the same city. You schedule a 30 minute interview period with each, leaving a 30 minute period for travel between their locations. What questions do you ask? We suggest that you try the following questions:

- What are the features about the product that you like the most?
- What characteristic about the product that you do not like?
- Has the product every failed?
- If so, what component failed?
- Do you like the appearance of the product?
- How do we compare to the competitor?
- How would you improve the product?
- What features would you incorporate in a new product?
- What are the important product characteristics that you consider before purchasing.
- Show us how you use the product in your applications.

Your job in the interview is to listen. Talk only enough to encourage the customer or stakeholder to relate his or her experiences with your product or one of your competitors' products. Keep the flow of information moving from the customer to you. Ask the customer to demonstrate the use of the product as a way to trigger his or her memory. Keep your own ideas for product improvement to yourself. You want the learn about the customers ideas, and you should not try to sell your ideas or to influence his or her thoughts.

It is a good idea to make notes or to use an audio tape to record the conversation. If you use an audio tape, make certain that the customer is aware that you are taping the session and agrees, on record, to the use of the tape recorder. If you interview six customers in the course of a day, notes and the audio record will help immensely in developing a complete listing of their significant comments at the end of the day. The audio tape has an advantage because it provides, by the emotions evident in the voice of the customer, a record of the importance of the issue being discussed.

12.3 INTERPRETING THE CUSTOMER NEEDS

If the development teams interviews 30 to 40 different customers, they probably will have generated several hundred significant comments, favorable and unfavorable, about the features and characteristics of the product. Some comments will indicate satisfaction, others will be critical, and still others will be superfluous. What are we to do with all of this information?

In examining the listings of the customer's comments, you will attempt to identify their needs. What does the customer need from the product? At this stage of the process, do not worry if changes suggested by the customers are feasible or about the changes that you intend to make in redesigning the product. Keep the listing of the needs positive by **avoiding** wording like—it should not—. Arrange the listing of the comments and the needs in two columns as shown in Table 12.1.

Table 12.1
Listing of customers' comments converted to customers' needs.

Customers' comments	Customers' needs
I hate it when the tool breaks when I drop it.	Impact resistant tool.
I can't quit working even in the rain.	Waterproof tool.
I keep losing the key to the Jacob's chuck.	Keyless chuck.
I want to use the tool all day	Continuous power available.

Only four of the customers' comments are illustrated in Table 12.1. Well organized and properly conducted interviews with about 40 customers will usually generate a hundred or more significant comments that the team considers in establishing the customers' needs. It will be necessary for the development team to place priorities on these comments by listing the most important first and the least important last. Judgments by the team are necessary in prioritizing the comment list. Three rules often followed in ranking the importance of the customers' comments are:

1. Frequency—if many customers describe the same problem.
2. Emotion—if some of the customers show emotion when describing a feature of the tool.
3. Price—if the customer demonstrates a willingness to pay more for a new feature.

The final step in identifying the customers' needs is to expand the table shown above by expressing the need as a product feature. We accomplish this by adding a third column to our listing as demonstrated in Table 12.2. Note the three step process moves from the customers' comments, which are often vague and ill defined, to a need defined in engineering terms and then to the product feature involved in the suggestion for product improvement.

Table 12.2
Adding the product feature required to meet the customers' needs.

Customer's Comments	Customer's Needs	Product Feature
I hate it when the tool breaks when it's dropped.	Impact resistant tool.	Toughened case. Shock proof components.
I can't quit working even in the rain.	Waterproof tool.	Waterproof switch and motor.
I keep losing the key to the Jacob's chuck.	Keyless chuck.	High gear ratio chuck.
I want to use the tool all day without worrying about the cord.	Continuous power available.	High capacity batteries. Interchangeable batteries.

12.4 THE PRODUCT SPECIFICATION

Once the customers' and stakeholders' needs are established and prioritized, we must convert the vague statements gleaned from the customer interviews into meaningful engineering terms. A meaningful engineering term is a number or a statement that gives a well defined design objective for each product feature or attribute. If the customer states that the tool vibrates too much, we must conduct experiments and measure the vibration level. Only then can we set a realistic vibration level in the product specification that will satisfy the customer. We need metrics (measurable quantities) describing each product feature that can be employed as targets to include in the product specification. We establish these metrics by testing not only our own products but also those of the competitors. In this extensive testing program (called benchmarking in industry), we establish the market norms for the product. If our new product is to be successful, we must improve on these norms to produce a superior product.

Benchmark testing provides the basis for writing the product specification. In preparing the product specification, we attempt to provide metrics associated with each feature or subsystem in the product. We establish a target value for a certain feature in the specification with a lower limit that is marginally acceptable and a higher limit that is essentially the best that we hope to achieve. Design is a compromise particularly when several metrics are to be satisfied simultaneously. We design to meet all of the subsystem targets; however, with some features, it may not be possible to achieve the highest level of performance while maintaining target values for cost and development time.

There is a systematic procedure for generating a product specification, where the vague statements made by the customer are converted to well defined metrics that are then included as design targets in the product specification. This procedure called **quality functional deployment (QFD)** is a set of planning and communication tools that are used by a development team to design and produce products that closely meet the customers' needs. The products designed also conform to the capabilities of the company to manufacture, procure and assemble the product on time and at a predictable cost. Quality functional deployment is outside the scope of this book but the interested reader is referred to the excellent book by Dr. D. Clausing [4].

12.5 DEFINING ALTERNATIVE DESIGN CONCEPTS

Let's suppose we begin an assignment on a development team the day after review of the product specification has just been completed. Our next task is to develop design concepts. A design concept (an idea) determines the approach taken to develop a specific feature or subsystem in the product. For example, in the development of an electronic version of a postal scale, we must generate an electrical signal that is uniquely related to the weight of the letter or package. Clearly, it is necessary to utilize a sensor of some type to convert the force (weight) to an electrical signal. Which sensor is the most suitable?

- Resistive
  - Potentiometer
    - Rotary
    - Linear
  - Strain gage
    - Metal foil
    - Semi-conductor
- Capacitive
  - Flat plate
  - Cylindrical
- Inductive
  - Linear variable differential transformer
  - Eddy current
- Electro-optical
  - Photo cell, cadmium sulfide type
  - Pin photo diode
  - Phototransistors

We have presented many different design approaches (concepts) for converting the weight applied to the postal scale to an electrical signal. Some of these ideas may be well known and others may be new but wild. Hopefully, one of the design concepts will be new and feasible. In the initial stages of generating design alternatives, we consider all ideas acceptable (the good, the bad and the very bad). We will evaluate all of them and select the better ideas later in the concept development process.

While several different types of postal scales are already available in the market, suppose your team is to develop an improved version. In this phase of the product development process, it is important to generate as many design ideas as possible to consider in designing each product feature or subsystem. Do not worry about their feasibility in this first effort to generate ideas. We will sort the good ideas from the crazy ones later in the development process.

Is there an organized method that we can follow to generate design concepts? **Yes!** There are so many methods to follow that complete books have been written to adequately describe all of them. We include three of these books as references [5, 6, and 7]. In this treatment, we will describe a few of the more popular techniques for generating design concepts including functional decomposition and both creative and rational techniques. We will use the postal scale as a product that we are redesigning and provide examples of each method where appropriate.

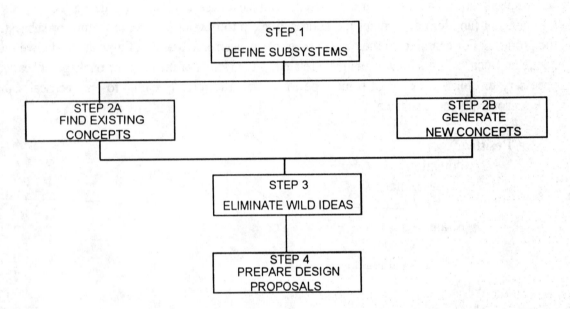

Fig. 12.2 A block diagram showing a systematic approach for concept generation and subsequent evaluation.

A systematic approach to the concept generation phase of the development process, is shown in Fig. 12.2. The first step in this approach is to define the subsystems involved in the operation of the product. In our case, the design of a postal scale will depend on the overall approach. If we decide to develop an electronic type of postal scale, we will have several subsystems including:

1. The platform supporting the letters and/or packages.
2. The force transfer mechanism from platform to transducer.
3. The transducer and sensor.
4. The power supply.
5. The signal conditioning circuits.
6. The digital display.
7. The containment case.
8. The tare (zeroing) system.

We have examined our concept of an electronic postal scale and have subdivided it into eight different subsystems. This is not a unique list. In fact, your team may decide to divide the many functions of the scale differently. The uniqueness of the list is not important providing it is complete. We divide the product into its major parts to simplify the design problem. It is easier to think about several smaller design tasks involving a more limited number of concepts than it is to handle a giant task involving a very large number of concepts.

The second step is to generate design concepts for each of the subsystems identified above. We approach the concept generation in two different ways. First, look to others for ideas. You are not the first design team in the world to design a postal scale or any other product. Someone else was first. If you go to the library and search the literature, you will find scores of papers or books describing many different types of balances, self-indicating scales, spring scales, electronic scales, etc. that have been built over the past century or even earlier in history. A careful review of this literature, and a search of the patents will enable you to develop a listing of design concepts that already exist for each of the eight subsystems listed above. Don't worry about using another persons idea. We do it all the time. We call it by a special name—"reverse engineering". We perform a detailed examination of the competitor's product, test it, adopt the good features, and reengineer the features that we believe to be inferior. We also try to improve on the good features by making them better. Providing we do not violate a patent, there is no requirement that the design be original. However, the new product must be superior to the competing products, feature by feature, when we compare them.

Okay! We have a complete listing of the existing design concepts used in previous designs of our postal scale machine. Let's next generate some new ideas from within the design team. There are two approaches to follow. First, we suggest that each team member think about the subsystems; understand their purpose, function, and the way that they work. Then after developing a thorough understanding, the individual team members should list as many design ideas as possible for each subsystem. The independent thinking by team members is important early in our attempts to generate alternative design concepts. Write your ideas on index cards, with one idea on each card, and then give the cards to the team leader. Keep a copy of the card for yourself because you will use it in the next part of the new concept generation task.

The second approach in generating alternative design concepts involves a collective effort of the team. One of most widely used methods for generating ideas is brainstorming. The development team, five or six people, is a good number of participants in a brainstorming group. The team leader moderates the group keeping it on task and avoiding any tendency to begin a round table discussion of any idea. The team leader must formulate the questions posed to the team that initiates its response. A typical question is—give us ideas for the mechanism connecting the platform to the transducer. The questions should be focused on a subsystem, but the inquires should not be so narrow as to limit the response from the team.

The team spends the next few minutes quietly thinking about the question. If ideas occur to you during this period for quiet thought, write them on an index card. You may also use the index cards that you prepared when you considered the design concepts independently. The team leader goes around the table in sequence, and each team member briefly presents one idea to the team. After presenting the idea, he or she hands the index card to the team leader. The brainstorming method is very effective when a few simple rules are followed:

1. No criticism is allowed during the session. The critique comes later.
2. The team responses should be short and to the point. Detail is deferred until later.
3. Wild ideas are OK because our goal is to generate as many ideas as possible.
4. Improvements and combinations of ideas are sought.

Brainstorming should be fun with everyone participating. Begin with prepared ideas from your index cards, and then add spontaneous ideas as they occur to you from the synergism of the group. You should, if possible, try to embellish the ideas of other team members with improvements. Humor helps during the brainstorming process because it breaks down barriers and improves the bonding of the team members. As the session moves from discussion based on the index cards to the generation of spontaneous ideas, the team needs a recorder to write the ideas on the blackboard or a flip chart (a very large pad of paper visible to the entire team). The visual display of the emerging ideas keeps the team on track and facilitates more thoughts on improving the ideas already identified.

There are many questions that the team leader can put to the brainstorming team to trigger ideas. Let's consider several of these questions. The questions are underlined and the descriptions that follow are to provide an example that relates to the postal scale:

1. Adapt.
   - Can we copy a previous design of the spring assembly?
   - Can we emulate the appearance of the packaging of the competitor?
   - Are there other fundamental methods for displaying the weight?
   - Is the design of the lever system that was abandoned five years ago suitable?
2. Substitute
   - What other material can be used for the case?
   - What other means can be used to power the electronics?
   - What new additive can be used with the lubricant to reduce friction?
   - Is there an improved process for heat treating the spring?
3. Modify
   - Can we change the color of the case to make the product more attractive?
   - Can the shape of the platform be changed to improve the scale's appearance?
   - Can we change the motion of platform to improve stability of the weight?
   - Can we change the size of the spring to reduce its displacement at high loads?
4. Rearrange
   - Can we change the location of the display?
   - Can we change the sequence of development in the schedule?
   - Can we interchange components like the levers in different scales to reduce inventory?
   - Can we introduce pilot prototypes to the distributors earlier?
5. Combine
   - Can we combine the circuit card with the digital numerals to eliminate a part number?
   - Can we combine the lock washer with the screw to reduce assembly time?

- Can we blend glass fibers in the plastic to improve the toughness of the case?
- Can we combine your idea with mine to reduce the size of the scale?

6. Eliminate
   - What feature can we eliminate to reduce the cost of the product?
   - What process can we eliminate in manufacturing to reduce cycle time?
   - What practice can we eliminate to reduce inventory?
   - What fasteners can we eliminate to reduce assembly time?

7. Reverse
   - Can we reverse roles in the interview process to improve communications?
   - Can we reverse the position of the scale in the packing carton to reduce packing time?
   - Can we reverse the color of the printing and the background color of the case?

8. Magnify
   - Can we make the case stronger?
   - Can we make the platform longer?
   - Can we add more time for the tooling in our schedule?
   - Can we provide more value with an added feature?

9. Reduce
   - Can we omit the felt covering the feet on the case?
   - Can we reduce the diameter of the spring?
   - Can we reduce the weight of the scale?
   - Can we reduce the number of parts required to produce the scale?

The questions listed above are not exhaustive. You can certainly think of many more. They should be suggestive of the types of questions that can be raised at a brainstorming session. Do not worry that different questions may lead to the same answers. The idea is to learn answers to as many questions as you can imagine. Also, we want to share those answers with all of the team members.

A brainstorming session should run for 20 to 30 minutes, or it should be concluded when the team has exhausted all of their ideas. The team leader takes the index cards and recorders notes, and is responsible for sorting and classifying the teams contributions. These team generated ideas are added to the list of concepts that were determined from our literature and patent searches. Together they give us relatively complete lists of design ideas for each subsystem. At this point in the process, the lists—one for each function or subsystem—are composed of many ideas with very brief descriptions.

A follow-up meeting of the team is conducted to consolidate the lists. Wild ideas were encouraged in the brainstorming session to create a fun-like, open and creative environment. The following session is a return to realism. Its purpose is to remove the crazy ideas from our list. The team reduces the list so that it includes only three or four of the most feasible ideas for each subsystem. Reducing the lists means that someone's idea must be eliminated. It is important that the team reach consensus in rejecting ideas. The procedure to cull ideas should not generate ill feelings. Good reasons for excluding an idea should be given in good humor. Considerable time is needed for thorough discussion. No one should feel belittled or the butt of a joke by the idea elimination process.

When the lists of the design concepts have been reduced so that they include only feasible ideas, we expand on the ideas. The ideas are represented by short written statements on index cards or notes without significant detail. Discussion of the concepts occurred in the meeting to reduce the number of alternatives, but we have not prepared detailed written descriptions of the concepts. To proceed with the evaluation of the remaining concepts, it is necessary to develop brief design proposals which include more detailed information and drawings or sketches showing the key features of each idea. Individual team members should develop a design proposal for his or her concept. The proposal should describe the idea in a few paragraphs, present a sketch if required, a list of advantages and disadvantages, and, if possible, a design analysis. The proposal should contain much of the information necessary for the team to select the superior idea from among the few remaining design concepts.

12.6 SELECTING THE SUPERIOR CONCEPT

We have reduced the number of design concepts originally generated by a preliminary meeting that weeded out the obviously weak and deficient design ideas. We have also expanded the ideas in the design proposals, providing the team with drawings, descriptions and perhaps even analyses. But, hopefully, we still have too many good ideas. We must now select the best from the good, and that is not always easy to do. Fortunately, we have the technique developed by Stuart Pugh [8] to help us in the selection of the best concept. The Pugh selection technique is a team activity that involves the preparation of an evaluation matrix. Keep calm, it's not a math matrix, and it is very easy to evaluate. Let's consider the nine steps in the Pugh selection technique that are:

1. Select the alternative design concepts to be evaluated.
2. Define a criteria upon which the evaluation is based.
3. Form the Pugh matrix.
4. Choose a datum or a reference for concept comparisons.
5. Score the concepts.
6. Evaluate the ratings.
7. Discuss the good and bad elements of each concept.
8. Select a new datum and repeat the Pugh matrix.
9. Revise select concepts as needed and evolve a superior concept.

Step 1 concerns the selection of the design concepts to be evaluated; however, we have already selected the most feasible three or four concepts based on previous team discussion. The essential element in successfully completing this step is that all of the team members understand each concept in considerable detail. If all of the team members are to participate effectively in the Pugh selection process, then every one must understand each concept in great detail. For example, consider that we have generated four concepts for the sensor to be employed in the electronic version of the postal scale, which include:

A. A cylindrical potentiometer.
B. A metal foil strain gage.
C. A linear variable differential transformer.
D. A cadmium sulfide photocell.

Do you understand how each of these four concepts can be implemented in the design? The sensor is a critical part of the transducer that converts the weight on the platform of the postal scale to an electrical output signal. Can the concept be implemented in the limited time allotted for prototype evaluation in this course?

Step 2 defines the criteria that will be used in the selection process. This criteria may change in evaluating concepts for different features or subsystems, but it is fixed for any one feature or subsystem. All of the concepts being considered must be judged against the same measure. What are the criteria that should be used to evaluate the four concepts listed above for supporting the weight of the person standing on the scale? In the instructions provided to you in Chapter 2 for the design of the postal scale, we specified the performance expected in terms of the accuracy and resolution of the weight measurement, and we listed some constraints on the size of the scale. Also, the time available for testing, the visibility of the display, design innovation, cost, and the quality of the product were listed as design criteria. Finally, the importance of safety and avoiding all hazards is vitally important. Clearly, the criteria used to judge the concepts must take into account all of the requirements imposed by the course instructors as well as any additional criteria that your team imposes. Let's list some items that you may wish to include in the criteria:

1. Cost of materials
2. Availability of materials or components
3. Simplicity
4. Ease of manufacturing
5. Ease of assembly
6. Ease of disassembly
7. Impact on the environment
8. Weight
9. Power requirements
10. Friction effects on accuracy

Your team may want to add addition items to include in this list of criteria. For instance, do you believe that stability of the electrical signal output or the ease of maintenance should be included as a criteria?

Step 3 involves the development of the Pugh matrix as shown in Fig. 12.3. The matrix is a simple table with rows and cells. You could use a spreadsheet program to prepare the matrix or a sheet of engineering paper as illustrated, in Fig. 12.3. The concepts A, B, C are placed in the column headings, and the criteria 1, 2, 3,....10 are placed in the row headings.

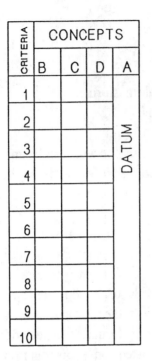

Fig. 12.3 Format for the Pugh chart used for concept selection.

In step 4, you choose the datum or reference against which your comparisons will be made. The datum is one of the concepts that you are evaluating. We initially select what is believed to be one of the better concepts, and this design approach serves as our reference. Suppose that we consider Concept A (the cylindrical potentiometer) to be the datum. We then compare the remaining three concepts B, C and D to the cylindrical potentiometer as the concept for the sensor.

In Step 5, you score the concepts and enter the results in the Pugh matrix. The scoring is simple. We consider each criterion separately and compare a selected concept (say B) against the datum concept (say A). The evaluation is a team activity. If the team reaches a consensus and decides that concept B is superior to concept A for criterion 1, then we score the intersecting cell in the Pugh matrix with a plus sign (+). On the other hand the team may decide that concept A is superior to B, and mark the intersecting cell with a minus sign (−). If the team decides that there is not a significant advantage of one concept over the other, the intersecting cell is marked with the letter S indicating that the concepts are essentially the same. It is important to recognize that the team discussion leading to the scoring is of great benefit. The team members interchange information in reaching a consensus. This heightens everyone's awareness and understanding of the design issues involved in implementing the concept. The scoring continues with each concept compared to the datum for every criterion. An example of a completed Pugh matrix is shown in Fig. 12.4.

Let's use the matrix presented in Fig. 12.4 as an example to illustrate the scoring technique. Note, this scoring was performed by one of the authors without the significant benefit of team discussion. In your team's evaluation, items may be scored differently because you are considering different information or the concept may be modified during the discussion. The matrix is not unique since it is totally dependent on the information available to the team at the time of the evaluation.

CRITERIA	CONCEPTS			
	B	C	D	A
1	S	–	S	DATUM
2	S	–	S	
3	–	–	S	
4	+	+	+	
5	S	S	S	
6	S	S	S	
7	S	S	S	
8	+	–	+	
9	S	S	S	
10	+	+	+	

Fig. 12.4 Example of a completed Pugh chart where four concepts have been compared.

Let's consider the author's scoring for the 1st criteria in Fig. 12.4. I was comparing concepts B, C and D against concept A on the basis of cost. First, recall the concepts:

A. A cylindrical potentiometer.
B. A metal foil strain gage.
C. A linear variable differential transformer (LVDT).
D. A cadmium sulfide photocell.

Second, recall the criteria:

1. Cost of materials
2. Availability of materials or components
3. Simplicity
4. Ease of manufacturing
5. Ease of assembly
6. Ease of disassembly
7. Impact on the environment
8. Weight
9. Power requirements
10. Friction effects on accuracy

To begin, consider the first criterion cost. I estimated the cost of the datum (the cylindrical potentiometer). A catalog for electronic components showed a wide variety of potentiometers both cylindrical and linear. Prices for cylindrical potentiometers ranged from a few dollars to more than $40. A standard "stock pot" is priced at $3.18. A metal foil strain gage suitable for the sensor beam is priced at about $25 for a package containing five gages or $5 per gage. The price of a linear variable differential transformer is difficult to determine because they are not available from electronic supply houses. The LVDT is a displacement transducer sold by a few firms supplying specialty instruments and transducers to engineering companies. One of the few companies producing LVDTs is Schaevitz Engineering. You may contact them on the Internet at www.schaevitz.com and request prices by e mail. The quote I received by e mail for a LVDT suitable for our postal scale was $35. Cadmium sulfide photocells are available from many electronic supply houses. The cost of these semiconductor devices ranges from a couple of dollars per unit to about $10 depending on size, sensitivity and range. Other photo transducers fabricated from silicon are in the same price range.

Now that the approximate cost for implementing each concept has been determined, we proceed to score the matrix. There is a relatively small difference in the cost of the sensors for concepts A, B and D; therefore, we mark cell B1, and D1 with the letter S. The LVDT, concept C, is relatively expensive; thus, it is scored with a (−).

We repeat the scoring for each criterion, and mark each cell in the matrix with +, −, or S. Each score represents a team consensus, and the discussion leading to that agreement gives each member of the team an opportunity to share information in depth about every concept as it is judged against every criterion. The matrix should not be considered fixed. If a criterion is not effective in the selection process, it may be eliminated from then matrix. Concepts may be modified during the rating process, or elements from one concept can be incorporated in another. The Pugh matrix is not a strict scoring device. Instead, it is a technique that leads to the interchange of information between team members which gives additional opportunities to modify and improve the design concepts.

Step 6 evaluates the ratings, by examining the scores for concepts B, C and D when compared to concept A. In Fig. 12. 4 for concept B (strain gages), we have three plus signs, one negative sign and six Ss. Do not cancel the negative with one of the positives because the negatives will be examined later in the Pugh process. For concept C (the LVDT), we have two plus signs, four negative signs and four Ss. For concept D (the photocell), we have three positive signs, no negative signs and seven Ss. Concept C is clearly inferior to concept A, but concepts B and D appear to have positive ratings relative to the datum concept (A). We have evaluated the ratings and summarized the pluses and minuses, but we defer the final selection.

Step 7 is inserted in the Pugh selection process to formalize the discussion of the advantages and disadvantages of each concept. Let's examine concept B and focus on the minus sign in cell B3. That minus score is due to the fact that the low signal output from the strain gage signal conditioning circuit will increased the complexity of the design because of the necessity for an amplifier to increase the signal output. Can we retain the same concept but reduce the complexity? The amplifier cannot be eliminated, but it is possible to reduce the complexity by using a modular design where an instrument amplifier is a single integrated component. The minus sign in the Pugh matrix permits the team to focus its attention on a weakness in a concept, and to improve the design by eliminating the weakness.

Sometimes minus signs lead us to abandon a concept. Let's consider cells B10, C10 and D10, all of which contain a plus sign indicating that we anticipate friction problems if we implement concept A (the potentiometer). The friction due to the wiper moving over the resistor will detract from the accuracy of the scale. If the friction force is large, it will prevent us from achieving the specified accuracy requirements. Is it possible to change the concept and eliminate friction effects while retaining all of the advantages of the potentiometer? If we cannot determine a technique for eliminating friction effects, concept A will not be feasible and will have to be abandoned.

Another minus sign for cell C1 indicates that the cost of the LVDT is prohibitive. It is approximately ten times more expensive that the other three types of sensors. Is it possible to reduce the cost of this sensor? Suppose Shirley Sharp suggests that the sensor is made of three coils of small diameter magnet wire. Magnet wire is not expensive and winding a few coils is easy. If you design and manufacture the LVDT, the costs become comparable to the motion other sensors. This approach eliminates the cost disadvantage effects, but increases the time required for the team to procure a sensor. As a consequence this new approach will change the scoring on the Pugh chart for some of the other criteria. If we make modifications to improve a concept relative to one criterion, it probably will be necessary to repeat the scoring of the Pugh chart.

Step 8 selects a new datum and repeats the Pugh procedure. We recommend this additional step if there are many concepts that remain after the first evaluation which represent uniquely different approaches. For the example given here, a second round of the Pugh procedure is recommended. The strain gage (concept B) and photocell (concept D) sensors are nearly identical in scoring the Pugh matrix. With only two concepts to consider, the criteria could be improved and more detail regarding the characteristics of the sensors could be brought to bear in the comparisons.

Step 9 revises the selected concept to evolve to a superior concept. Suppose that we have selected concept B (strain gages) for the sensor after eliminating the negative mark for simplicity. However, we can consider improvements by working to eliminate the reasons for the negative scores or even the scores of S. Keep asking questions about the factors detracting from the merits of an idea. New approaches emerge—negative scores or scores of S can change to positive scores. Answers to your questions often lead to design modifications that eventually provide the superior concept. When we finally have superior concepts for every feature or subsystem, we can proceed with the detailed design.

12.7 DESIGN FOR ????

No, we have not made a typo in the section heading. When we design, we must consider so many different aspects, that we replaced the listing of them with ???? in the section title. For example, in the design of any system, subsystem, or component, we must consider the following major factors:

- Performance
- Appearance
- Manufacturing
- Assembly
- Maintenance
- Environment
- Safety

12.7.1 Detail Design

Let's begin our discussion of design with the four Fs, namely form, fit, function and finish. The four Fs refer to detail design of piece parts involved a some system. This is the easiest part of the design which is the reason for considering it first. Form refers to shape. Is the part shaped correctly to perform its function? If it is too thin, it may fail by breaking or by deforming excessively. If it is too large or too thick, it will weigh too much and cost too much. If it has a sharp external corner, it represents a safety hazard because sharp corners are cutting edges. If it has a sharp internal corner, high stresses, which may lead to premature failure, will develop.

Fit is involved when we assemble two or more pieces to form a subsystem. Will the pieces fit together without gaps or discontinuities? If holes are drilled in a given component to accommodate a bolt pattern, are they in alignment so that the bolts fit without interference? Fit becomes critical when we deal with rotating shafts with bearings. Tolerances required for several of the diameters in a bearing housing are very tight and require precise machining of the component parts if the bearing is to fit properly. If the fit is too tight, the bearing will overheat and seize. If the fit is too loose, vibration will occur which detracts from performance and shortens the life of the bearing and other components.

Function refers to the ability of the system to perform satisfactorily with its components. If a part fails in service, then the system cannot function. Suppose you design a crankshaft for an auto that resonates at an engine speed of 3800 RPM. Your crankshaft is great. It has the right form, fits properly and has the correct surface finishes on the bearing journals—but the crankshaft vibrates at or near highway cruising speed. The design fails because the performance of the auto has been severely compromised by the vibrations induced in the engine by the crankshaft. The crankshaft has not functioned correctly.

Finish refers to the surface finish of the part. Is the surface rough, smooth or polished, or does it matter? Some surfaces are important because of appearance. The sheet metal on our auto is very smooth so that the paint that is applied will produce a high gloss finish. Other surfaces are not important. The aluminum block on an auto engine is produced by die casting, and its outside surface reflects the finish of the die. Since rarely look at the engine until it dies, a relative rough surface finish is acceptable. Some surfaces are polished or ground to enhance their performance in bearing applications. When designing a part, we must know how it will be used in service. With this understanding, we can specify the appropriate surface finishes on the detail drawing for that part.

12.7.2 Design for Manufacturing

In designing a piece part or a subsystem, it is essential that you consider how the parts are to be fabricated. Manufacturing methods are probably as important as function in the design of parts that are to be produced in large quantities. For a component that is mass produced, we must carefully consider manufacturing methods such as die casting, injection molding, forging, casting, drawing, machining, stamping etc. in the detail design of the parts. We have good and bad design rules to follow depending on the production methods to be employed in manufacturing [9].

In the development of the prototype, manufacturing methods remain important, but they are less critical. For the design of the prototype scale that we are developing in this course, you probably will be much more limited in your manufacturing capabilities. Hopefully you will have a model shop with power saws and safety qualified operators to cut wood and plastic according to your drawing. For sheet metal, you may have a shear and brake for cutting, bending and folding the thin sheets of metal employed in the design. You may also have hand tools such as hammers, saws, chisels, wrenches, etc. available for your use. Small power tools such as the drill, bayonet saw and belt sander, illustrated in Fig. 12.5, may also be available in the student workshop. Any component that you design must be fabricated with the tools available to you unless the components are procured from a retail outlet in finished form.

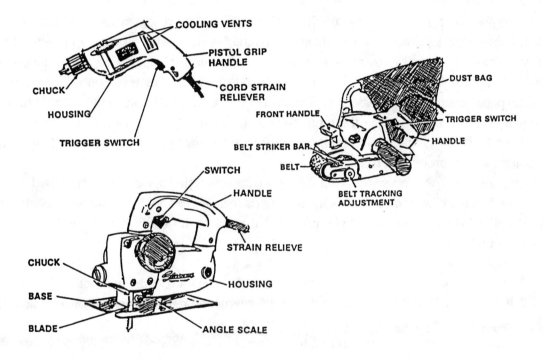

Fig. 12.5 Power hand tools available in the student workshop.

As an example, let's consider the base of the postal scale that serves to support the entire assembly. We will assume that your base is a flat, rectangular plate, 4 by 8 inches in size fabricated from a piece of ½ inch thick plywood. The case is fabricated by cutting it from a larger sheet of plywood with either a power or a hand saw. Each member of the postal scale is limited in size so that they can be packed for storage in the shoe paper box. (Remember the constraint that was placed on the design in Chapter 2). We can make the pieces required for the base and case from plywood cut to the appropriate lengths with a saw. Alternatively, we can make the members from aluminum sheet which can be cut with a shear and bend to shape with a brake. However, fastening the aluminum pieces together to form the base will pose many more problems that working with wood which can be nailed, screwed or glued together to form strong joints. It appears that we will not encounter any problems manufacturing the pieces needed for the base and the case. Will we encounter any problems when we try to assemble all of these pieces in short period of time?

12.7.3 Design for Assembly

Products are systems containing several subsystems, and each subsystem contains several components. In the design or selection of each component, we must consider the methods and procedures used to assemble them in the construction of the subsystems. We also must be concerned with the assembly of the various subsystems to form the complete system (product). There are several levels of consideration in design for assembly. The first and most important is whether or not the components can be assembled. Don't laugh. The author has seen designs which were impossible to assemble and had to be scrapped. Second, the assembly operation should be easy and not require special operator skills. Finally, the time required for the assembly should be minimized.

Assembly time is very critical to the success of very complex products that are mass produced. For example, Ford Motor Company leads the big three auto producers in the U. S. in assembly. They can assemble an auto with 24.72 people-hours compared to 26.32 and 27.76 people-hours for Chrysler and General Motors respectively. Nissan required only 17.84 people-hours at its plant in Smyrna, Tennessee. It is estimated that General Motors could save about $3.5 billion each year if it could reduce the time required for assembly to that currently achieved by Nissan. Clearly, savings in billions of dollars can be made by employing design for assembly in the development of mass produced products.

One approach to design for assembly is to minimize the number of parts used in the product. As designers, we tend to use many more parts than are needed based on assembly theory. Boothroyd et al [9] have developed three criteria that are used to justify the use of a part in the design of a product.

1. The part must move relative to other parts in the operation of the product.
2. The part is fabricated from a different material or must for some other reason be separated from adjacent parts.
3. The part must be separate from adjacent parts to permit either assembly or disassembly of the product.

Using these three criteria, we can develop a theoretical part count for any given product. With very rare exceptions, we greatly exceed the theoretical minimum part count when we design. Nevertheless, these rules are valuable because they stimulate the development team to consider combining parts and to design unique parts to serve dual functions. Part counts are often reduced by 50% or more by applying the design for assembly criteria.

How does this affect you in the design of your postal scale? After all you are designing a one of a kind prototype to learn about the product development process. Well, you are stuck with an assembly problem, because you have a very limited time to assemble, test and disassemble your postal scale. If you need more than 45 minutes to assemble and 15 minutes to disassemble, then you will have to use test time for assembly, and you may not have enough time to accurately calibrate and adjust your scale.

There is a lot of work involved in the assembly of the prototype. Can we shorten the time by conducting several tasks at the same time. After all we have five or six team members, and they should all be involved in the assembly operation. The team should divide the assembly operation into several assembly tasks that can be conducted in parallel. Many of the subsystems can be assembled simultaneously, and the overall time required to complete the assembly can be shortened by working tasks in parallel rather than series. You are not expected to work like a pit crew at the Indianapolis raceway, but every one should stay productively busy.

There are many aspects to design for assembly, and this treatment is only a brief introduction to a very important topic. For a much more complete coverage see the excellent book identified in reference [9].

12.7.4 Design for Maintenance

Have you ever changed the oil filter on your auto? On some cars you have to raise the body, crawl under the car, and reach up to unscrew a horizontally oriented filter. When the filter is loosened, it leaks, oil runs down the side of the engine, and onto you. The oil spill is difficult if not impossible to contain. Clearly, the design of this engine, and the placement and orientation of the oil filter did not accommodate the need for periodic and scheduled maintenance. Other cars have the oil filter on the other (front) side of the engine. The filter is replaced from the front without going under the car. The designers have also oriented the filter nearly vertically so that the oil does not spill when the filter is loosened and the seal is broken. This second filter is an example of design for maintenance. The designer recognized the need for periodic maintenance and developed the product to allow easy access, and rapid, inexpensive replacement of the required parts.

In the design of the postal scale, you should not have to perform any maintenance. However, you should recognize that many commercial products do require some periodic maintenance, and that it is a very real design criterion for many product developments.

12.7.5 Design for the Environment

Customers and society at large is concerned with preserving the environment. Society enjoys new products and enhanced performance but they clearly want to minimize the impact on the environment. In some countries in Europe, it is necessary for a company to show the methods to be utilized in recycling the product and disposing of any waste. In the U. S., most companies are sensitive to environment issues. Emissions are minimized, waste water is cleaned before being discharged into steams, toxins are eliminated from processes, waste paper is used in containers, recycled plastic and metal is used in most components.

We are at the early stages of design for the environment. As world population grows and third world countries become more affluent, preserving the environment and conserving resources will become even more important. Today many products are designed to be partially recyclable. Today about 75% of the weight of an auto is comprised of components that may be recycled. These percentages will grow with time as disposal becomes more difficult and concern over the environment assumes major importance.

A very new approach to conservation is remanufacturing. Remanufacturing involves rebuilding an older model of a product rather than building a completely new model. In the remanufacturing process critical components are replaced and some components are updated to enhance performance. Personal computers and copy machines are examples of products that can be remanufactured. Of course, auto components have been remanufactured for several decades because it costs much less to rebuild an internal combustion engine than to build a new one.

12.8 PROTOTYPE BUILD AND EVALUATION

There are several important reasons for building a prototype. The first is to determine if all of the component parts fit together. Frequently we find errors, and some of the parts do not fit properly when we attempt to assemble the subsystems. It is much easier to change the drawings and to fix the fit problems when we build the first prototype. Imagine trying to assemble 10,000 units of product using parts that are in error. The prototype, often called the first article build, is vital to insuring parts that fit perfectly. The assembly of the first article also reveals difficulties that may arise at a later date on the production line when the product is mass produced. The idea is to find the problems and fix them immediately while the development is in the prototype stage. Finding problems later is much more expensive. Also, a problem discovered after the product has been introduced to the market requires a recall and harms the reputation of the company for producing a high quality product.

The prototype also enables engineers to test the product. These tests permit us to measure the performance parameters and to verify the analysis used in the design phase to predict performance. If any deficiencies in performance are revealed in the test program, design modifications are made to the prototype and the tests are repeated. The prototype is a vehicle for design change. It is instrumental in enhancing the performance of the product. In some instances, prototype development is nearly a continuous process with one prototype following another. At select times in this process, the company will freeze a design and introduce a new product based on these prototype experiences.

You are building a prototype of a postal scale. As you build the first article, you will probably find some mistakes. Do not be depressed. Drawing errors and other analytical mistakes are common. Find the errors and fix them immediately. Make the prototype work as early as possible in the semester. When you fix a problem, find the mistake and learn from it. You should make certain that you revise the drawing to reflect the revisions made in correcting errors. An early test of your prototype, prior to the last week of class is strongly encouraged. The workshops and testing facilities will be ready for your early evaluations of the prototype.

12.9 TOOLING AND PRODUCTION

In many instances we design and develop products for a very large market with sales of 100,000 units or more each year. Also, the life cycle of the product may range from five to ten years, so that the total production over the product life is huge. The production of large quantities requires extra care in the design phase. It is widely accepted that 70 % of the final product costs are determined by the design [10]. Decisions made in the design affect the cost of producing the product over its entire life. If one million units are produced over the life cycle of the product, selecting a fastener with a cost premium of only 2 cents adds $20,000.00 to the production costs.

Designing a component that requires an extra machining operation could easily add more than several hundred thousand dollars to the life cycle costs for high volume products. Clearly, when designing components intended for use in high volume products, we need an excellent understanding of costs, manufacturing process design rules, and assembly theory.

We trust that the procurement of the materials and the components for your postal scale will provide a small lesson in cost control. For lessons in manufacturing processes and assembly theory, we defer to later courses in the curriculum or independent study on your part. The references at the end of the chapter will help you start the study.

12.10 MANAGING THE PROJECT

When a development team works on a project, they are charged with the responsibility for completing the project according to a schedule and a budget. The team leader should manage time and costs. Development costs can be extremely large. Chrysler spent about one billion dollars to develop the Concorde auto. The project took 3.5 years and involved a total of about 2250 people [2]. As you move up in a corporation, your position will likely involve some degree of management. You will need to learn about scheduling and project costs relatively early in your career. Let's start now with scheduling.

You already know something about scheduling because the University imposes a schedule on you from day one. Classes start on September xx or January xx and end on December yy or May yy. The final exam schedule is fixed before the semester starts. You have a schedule of classes—M, W and F at 9, etc. The instructor has imposed some deadlines for design reviews that you must meet in this course. You will need to incorporate these deadline dates into the Gantt schedule that we will develop later in this section.

Scheduling starts with listing the tasks that you need to formulate in order to develop the prototype postal scale. The list of tasks should be in reasonable detail to enable you to realistically assess the time that you can allocate to each task. Let's begin by setting up this list, presented in Table 12.3, that shows all of the tasks that must be completed during the development of an electronic scale. Note, the listing is in sequence according to time except for the design reviews.

Your team should review the list and the time estimates and make those modifications necessary to conform with your design. The time estimates are particularly important. We have scheduled 4 weeks for the team to generate, evaluate and select design concepts. That is a significant block of time (five or six people-weeks). Do you need that much time, is it the correct amount of effort, or will it prove to be insufficient? Carefully estimate the time required to complete a task, and then check to see if your estimates are correct. Estimating time for task completion is very difficult. Errors in underestimating time lead to delays in design releases and product introductions. These delays are costly to the corporation, and they are serious detriments to career advancement. Many of us even after many years of experience still underestimate the time needed to properly complete a task. Take this opportunity to start developing your time estimating skills.

TABLE 12.3
Task List for Developing a Postal Scale

TASKS	TIME (weeks)
CONCEPT DEVELOPMENT	
Receive and understand specification	1
Concept generation	2
Concept evaluation and selection	2
DETAIL DESIGN	
Spring element	1/3
Strain gage --- bridge circuit	1/3
Cover	1/3
Base	1/3
Amplifier and digital display	1/3
Drawing reviews	1/3
DESIGN REVIEWS	
Preliminary review	1
Final review and design release	1
MANUFACTURING	
Procure materials and components	1
Manufacture piece parts	1
Inspect piece parts and rework as required	½
Prepare assembly kit	½
ASSEMBLE AND TEST	
Pre-assembly	1
Final assembly and test	1
TOTAL	14

After completing the listing of tasks, we construct a Gantt chart as shown in Fig. 12. 6. The Gantt chart shows a listing of the tasks in the rows down the left side. The time line, in this case expressed in terms of calendar months during the fall semester, is displayed along the X axis. We draw bars to represent the time required to complete a specific task. The left edge of the bar corresponds to the time when the task is scheduled to begin. The right edge of the bar indicates the completion date. Of course, the length of the bar shows the number of weeks scheduled to complete the task. When the tasks are completed the bars are darkened. In the example presented in Fig. 12.6, the bars for the concept development tasks are darkened indicating that these tasks are complete. A vertical line is shown in Fig. 12.6 to indicate a current date—in this example, sometime in mid October. The vertical line intersects the horizontal darkened bar for concept evaluation and selection. This configuration indicates that the team is ahead of schedule because the bar is darkened for a week ahead of the current date.

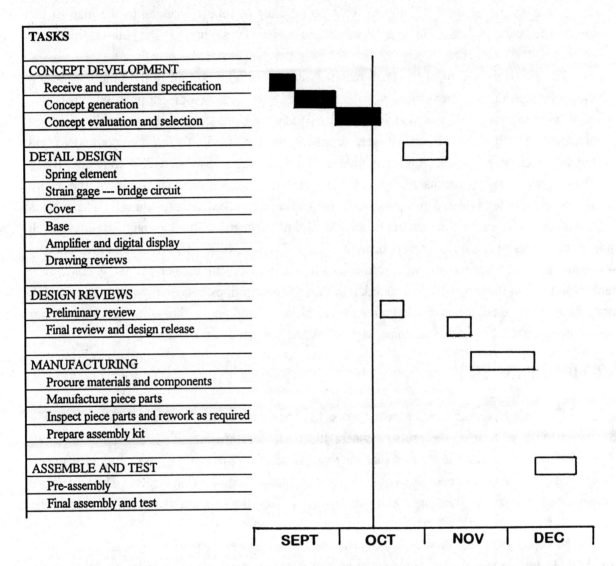

Fig. 12.6 Example of a Gantt chart for the development of an electronic scale during a semester.

The placement of the bars in Fig. 12.6 is typical of what is called series scheduling. We schedule one task after the completion of the previous task. However, it is possible to overlap the bars along the time scale. Overlapping the bars (parallel scheduling) implies that we will start a new task before the current task is completed. In many instances, we can effectively pursue two tasks simultaneously, and important time savings can be achieved whenever parallel scheduling is possible. In procuring materials, we often encounter delays of weeks or even months in delivery of components necessary to build the prototype. In procuring these long-lead-time components, parallel scheduling is mandatory because it is necessary to place the purchase order long before the design is complete.

In Fig. 12.6, we have provided your team with an example Gantt chart for developing an electronic scale over a 14 week semester. You should consider your team's schedule and change the beginning and completion dates for each task to correspond with your team's plan. Prepare the Gantt chart early in the semester, and revise it periodically to reflect changes to the plan as the design of the scale evolves. Be sure to add the deadline dates imposed by your instructor as milestones on the Gantt chart.

The physical location of the team members is extremely important in the time required to develop a new product. Communication so essential to the entire development process is markedly affected by the distance. If two people are located only 20 meters apart, the probability of their communicating with each other once a week drops to only 10% [11]. To facilitate communication during the development process it is imperative that team members be located close together. Black & Decker, the largest manufacturing company in the State of Maryland, places the entire development team for a related group of products in a single room called a fusion cell. The room is large with desks for each team member located around its perimeter. Shoulder high screens to each side of the desks give a little privacy, but not much. The idea of the room arrangement and the office furniture is to facilitate communications and to eliminate private conversations. A conference table is situated in the center of the room that is used for all meetings concerning the program. The concept is effective because every team member is close at hand and within eyesight. You do not need e-mail, a phone, or a FAX machine when everyone is located in the same room.

12.11 SUMMARY

A complete product development process has been outlined in this chapter. We very briefly described the beginning of the process, where the customers are identified, and interviewed to determine their needs and the amount that they would be willing to pay for new product features. The vague customers' comments are converted into meaningful engineering targets. The targets are incorporated into an engineering specification that defines the goals and objectives which the designers are expected to meet in the development of the product.

Methods for generating alternative design concepts were covered in detail. The most essential element of this process is the synergism of the team members in generating new ideas for design. The approach is to divide the product into subsystems, and then to generate as many design ideas for each subsystem as possible. From these many concepts, we select the best one and then improve it by introducing modifications that eliminate its disadvantages. A selection technique based on the Pugh chart has been described in considerable detail with an appropriate example.

The four Fs of design were introduced to show the importance of form, fit, function and finish. We then briefly discuss several important aspects of design for manufacturing, assembly, maintenance and the environment. These are all extremely important subjects and the treatment here is only an introduction. Several references are cited for independent study.

Reasons are given for the importance of building the prototype. The prototype is critical to insuring the success of a development. It verifies that all of the parts fit, that they can be assembled with ease, and that the system performs in accordance with the product specification. The prototype is a vehicle for design change—it affords a critical opportunity to correct the design errors prior to the introduction of the product to the market.

Finally, we introduce a method for managing the project. The method involves a listing of all of the tasks that must be accomplished to complete the prototype development and a careful and intelligent estimate of the time necessary to complete each task.. The data from the listing is then used to construct a Gantt chart that enables the team to track the completion of each task relative to a predefine time schedule.

REFERENCES

1. Wheelwright, S. C. and K. B. Clark, <u>Revolutionizing Product Development: Quantum Leaps in Speed, Efficiency and Quality</u>, The Free Press, New York, 1992.

2. Ulrich, K. T. and S. D. Eppinger, <u>Product Design and Development</u>, McGraw-Hill, New York, NY, 1995.

3. A. Griffin and J. R. Hauser, "The Voice of the Customer," <u>Marketing Science</u>, Vol. 12, No. 1, 1993.

4. Clausing, D. <u>Total Quality Development</u>: <u>A Step by Step Guide to World Class Concurrent Engineering</u>, ASME Press, New York, NY 1994.

5. Cross, N., <u>Engineering Design Methods: Strategies for Product Design</u>, 2nd ed., Wiley, New York, NY, 1994.

6. Dym, C. L., <u>Engineering Design: A synthesis of Views</u>, Cambridge University Press, New York, 1994.

7. Pahl, G. and W. Beitz, <u>Engineering Design: A Systematic Approach</u>, Ed. K. Wallace, Springer-Verlag, New York, 1977.

8. Pugh, S., "Concept Selection---A Method that Works," <u>Proceedings of the International Conference on Engineering Design (ICED)</u>, Rome, March 1981, pp. 479-506.

9. Boothroyd, G, Dewhurst, P. and W. Knight, <u>Product Design for Manufacture and Assembly</u>, Marcel Dekker, New York, NY 1994.

10. Anon, Company Literature, Munro and Associates, Inc., 911 West Big Beaver Road, Troy, MI 48984.

11. Allen, T. J. <u>Managing the Flow of Technology: Technology Transfer and the Dissemination of Technological Information within the R & D Organization</u>, MIT Press, Cambridge, MA, 1977.

12. Kalpakjian, S., <u>Manufacturing Processes for Engineering Materials</u>, Addison Wesley, Reading, MA, 1984.

13. Carter, D. E. and B. S. Baker, <u>Concurrent Engineering: The Product Development Environment for the 1990s</u>, Addison-Wesley, Reading MA, 1992.

EXERCISES

1. Consider a household refrigerator as a product. List the customers that you should consider in any redesign. Also give reasons for including each person or organization as a customer.
2. Interview a family member, Mom, Dad, Sister or Brother about some product that they own and use frequently. Try to list all of the favorable comments as well as their complaints. Are they willing to pay for product improvements? If so, how much?
3. Working together with your fellow team members divide the postal scale that you are designing into sub systems. Briefly describe each sub system and define all of the interfaces that exist between sub systems.
4. Working with your team conduct a brainstorming session to generate design concepts for each of the sub systems defined in exercise 3. List all of the concepts generated, the good and the bad, for each sub system.
5. Define the criteria that you plan to use to evaluate one of your subsystems in the Pugh selection process.
6. Construct a Pugh matrix and evaluation several design concepts using the criteria defined in exercise 5.
7. When the Pugh analysis of exercise 6 is complete, discuss the implications of the positive and negative scores that appear in the Pugh matrix.
8. Write an engineering brief describing design for manufacturing.
9. Write an engineering brief describing design for assembly.
10. Write an engineering brief describing design for maintenance.
11. Write an engineering brief describing design for the environment.
12. Why do we invest funds and time in building prototypes before we release a product to the market?
13. What was your reaction when your auto or a friend's auto was recalled to replace a design defect?
14. Discuss the implications of mass production on the development process.
15. Prepare a task list for your development of the prototype of a postal scale.
16. Prepare a Gantt chart for the postal scale development.

PART V

COMMUNICATIONS

CHAPTER 13

TECHNICAL REPORTS AND INFORMATION GATHERING

13.1 INTRODUCTION

We hope that you like to write, because engineers often have to prepare several hundred pages of reports, theoretical analyses, memos, technical briefs, and letters during a typical year on the job. We trust that you will learn how to effectively communicate — by writing, speaking, listening, and employing superb graphics. Communication, particularly good writing, is extremely important. Advancement in your career will depend on your ability to write well. We recognize that you will be taking several courses in the English Department and other departments in Social Sciences, Arts and the Humanities which will require many writing assignments. The courses should help you immensely with the structure of your composition and the development of good writing skills. Most of the assignments will be essays, term papers, or studies of selected works of literature. There are several differences between writing for an engineering company and writing to satisfy the requirements of courses like English 101 or History 102.

In school, you write for a single reader, namely the instructor. He or she must read the paper to grade it, subsequently determining how well you are doing in class. In industry, your report may be read by many people, within and outside the company, and with different backgrounds and experience. In class, the teacher is the expert, but in industry, the writer of the report — you —are supposed to be the one with the knowledge. Many of those working in industry do not want to read your report as they are busy: the phone is ringing continuously, meetings are scheduled back to back, and many other things need to be accomplished. They read — if not skim — the report only because they must be aware of the information that it contains. They want to know the key issues, why these issues are important, and who is going to take the action necessary to resolve them. In writing reports in industry, we cut to the chase. There is no sense in writing a 200 word essay when the facts can be provided in a 40 to 50 word paragraph that is brief and cogent. An elegant writing style, so valued by our colleagues in the arts and humanities, is usually avoided in engineering documentation. In writing an engineering document, do not be subtle; instead be obvious, direct and factual.

In the following sections, we will cover some of the key elements of technical writing, including an overall approach, report organization, audience awareness, and objective writing techniques. Then we will describe a process for technical writing that includes four phases: composing, revising, editing and proofreading.

13.2 APPROACH AND ORGANIZATION

The first step in writing a technical report is to be humble. Realize that only a very few of the many people who may receive a copy of your report will read it in its entirety. Busy managers, and even your peers, read selectively. To adapt to this attitude, organize your report into short, stand-alone sections that attract the selective reader. There are three very important sections, namely the title, summary, and introduction, located at the front of the report so they are easy to find. At the front of your report, they attract attention and are more likely to be read. In college, you call the page summarizing your essay an abstract, but in industry, it is often called an executive summary. After all, if it is prepared for an executive, perhaps a manager will consider it sufficiently important to take time to read it. Follow the executive summary with the introduction and then the body of the report. A common outline to follow in organizing your report is:

- Title page
- Executive summary
- Table of contents, list of figures and tables
- Nomenclature—only if necessary
- Introduction
- Technical issue sections
- Conclusions
- Appendices

The title page, the executive summary and the introduction are the most widely read parts of your report. Allow more time polishing these three parts, as they provide you with the best opportunity to convey the most important results of your investigation.

13.2.1 Title Page

The title page provides a concise title of the report. Keep the title short—usually less than ten words; you will have an opportunity to give detail in the body of the report. The authors provide their names, affiliations, addresses, and often their telephone and fax numbers and e-mail address. The reader, who may be anywhere in the world, should know how to contact the authors to ask questions. For large firms or government agencies, a report number is formally assigned and is listed with the date of release on the title page. An example of a title page of a report prepared for a government agency is shown in Fig. 13.1.

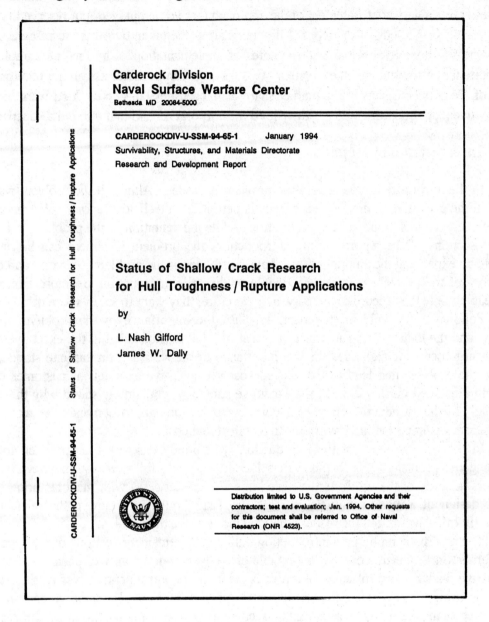

Fig. 13.1 Illustration of the title page of a professional report.

13.2.2 Executive Summary

The executive summary should, with rare exceptions, be less than a page in length (about 200 words). As the name implies, the executive summary provides, in a concise and cogent style, all the information that the busy executive needs to know about the content of the report. What does the big boss want to know? Three paragraphs usually satisfy the chief. First, briefly describe the objective of the study and the problems or issues that it addresses. Do not include the history leading to these issues. Although there is always a history, the introduction is a much more suitable section for historical development and other background information related to the issues or problems. In the second paragraph, describe your solution to the problem and/or the resolution of the issues. Give a very brief statement of the approach that you employed, but reserve the details of the approach for the body of the report. If actions are to be taken by others to resolve the issues, list the actions, those responsible, and the dates of implementation. The final paragraph indicates the importance of the study to the business. Cost savings can be cited, as can improvements in the quality of the product, gains in the market share, enhanced reliability, etc. You want to convince the executive that your work was worthwhile to the corporation and that you did a superb job.

13.2.3 Introduction of Report

In the introduction, you restate the problem or issues. Although the problem was already defined in the executive summary, redundancy **is** permitted in technical reports. We recognize that we have many readers, so important information — like the definition of the problem — is placed in different sections of the report. In the introduction, the problem statement can be much more complete by expanding the amount of information provide for the problem. A paragraph or two on the history of the problem is in order. People reading the introduction are more interested than those reading only the executive summary, and therefore, they want to know more detail.

Following the problem statement, establish the importance of the problem to both the company and the industry. Again, this is a repeat of what was included in the executive summary, but your arguments are expanded for the importance of the study. You can cite statistics, briefly describe the analyses that lead to the alleged cost savings, give a sketch of customer comments indicating improved quality, etc. In the executive summary, you simply stated why the study was important. In the introduction, however, you develop arguments to convince the reader that the investigation was important and worth the time, effort and cost.

In the next segment of the introduction, you briefly describe the approach followed in addressing the problem. You begin with a literature search, followed by interviews with select customers, a study of the products that failed in service, interviews with manufacturing engineers, and the design of a modification to decrease the failure rate and improve the reliability of the product. In other words, tell the reader what you did to solve the problem.

The final paragraph in the introduction outlines the remaining sections of the report. True, that information is already covered in the table of contents, but again, we repeat to accommodate the selective reader. Also, placing the report contents in the introduction gives us the opportunity to provide a one or two sentence description about each section. Perhaps we can attract some reader interest in one or more sections, thus convincing a few people to study the report more completely.

13.2.4 Organization

The organization of the report was suggested by the bullet list shown at the beginning of this section. However, you should recognize that an organization exists for every section and every paragraph in the report. In writing the opening paragraph of the section, you must convey the reason for including the section in the report, and the importance of the content covered in the section. You certainly have a reason for including the section in the report, or you would not have wasted your time in writing it. By sharing this reason with your readers, you convince them that the section is important.

A paragraph is written to convey an idea, a thought, or a concept. The first sentence in the paragraph must convey that idea, clearly and concisely. The second sentence should describe the significance or importance of the idea. The following sentences in the paragraph support the idea, expand its scope, and give cogent arguments for its importance. State the idea in the first sentence and then add more substance in the remaining sentences. If written properly, a speed reader should be able to read the first two sentences of every paragraph and glean 80 to 90% of the important information conveyed in the report.

As you add detail to a section or a paragraph, develop a pattern of presentation that the reader will find easy to read and understand. First, tell the reader what will be told, then provide the details of the story, and finally summarize with the action to be taken to implement the solution. Do not try this approach in the theme that you have to write for English class, because the instructor will clean your clock for being redundant. Technical writing is not theme writing, however. In technical writing, we reiterate to the reader to convey our message. Often the readers must take corrective actions for the benefit of the company and the industry.

We frequently include analysis sections in engineering reports. In these sections, we state the problem and then give the solution. After the problem statement and the solution, we introduce the details. The reader is better prepared to follow the details (the tough part of the analysis) after they know the solution to the problem.

13.3 KNOW YOUR READERS AND YOUR OBJECTIVE

Another important aspect to technical writing is to know something about the readers of your report. The language that you use in writing the report depends on the knowledge of the audience. For example, we are writing this book primarily for engineering students. We know your math and verbal skills from the SAT scores required for admission. We know that you have very good language skills, and even better math skills. You are bright and articulate. We believe that most of the chapters in this book are in line with your current abilities. However, there are a few sections in the book that deal with topics usually found in the business school. Profits and costs are important in product development, and although they may not be as interesting as a new finite element program they need to be clearly understood.

As you write a report for a class assignment, you know that the instructor is the only reader with which you need to communicate. You also know that he or she is knowledgeable. In that sense, the writing assignments in a typical class are not realistic unless they are modified. Suppose we asked you to write an assembly routing (a step by step set of instructions for the assembly of a certain product) for the production of your new product in a typical factory in the U. S. What language would you use as you write this routing? If the plant is in Florida or the Southwest for example, Spanish may be the prevalent language. You should also be aware that a significant fraction of the factory workers in the U. S. are functional illiterates, and many others detest reading. In this case, it might be a good approach to use fewer words and many cartoons, pictures and illustrations.

Know the audience **before** you begin to write, and adapt your language to their characteristics. Consider four different categories of readers:

1. Specialists with language skills comparable to yours.
2. Technical readers with mixed disciplines.
3. Skilled readers, but not technically oriented.
4. Poorly prepared readers, who read with difficulty if at all.

Category 1 is the audience that is the least difficult to address in your writings, and the audience in the last category is the most difficult. Indeed, with poorly prepared readers, it is probably better to address them with visual presentations conveyed with television monitors and video tapes.

A final topic, to be considered in planning, is the classification of the technical document that you intend to write. What is the purpose of the task? Several classifications of technical writing are listed below:

- Reports: trip, progress, design, research, status, etc.
- Instructions: assembly manuals, training manuals, and safety procedures.
- Proposals: new equipment, research funding, construction, and development budget.
- Documents: engineering specifications, test procedures, and laboratory data.

Each classification of writing has a different objective and requires a different writing style. If you are writing a development funding proposal, you need persuasive arguments to justify the costs of your proposed program. On the other hand, if you are writing an instruction manual to assemble a product, arguments and reasons for funding are not an issue. Instead, you would be trying to prepare complete and simple descriptions that precisely explain how to accomplish a sequence of tasks involved in the assembly.

13.4 THE TECHNICAL WRITING PROCESS

Whether you are writing a report as part of your engineering responsibilities in industry, or as a student in college, you will face a deadline. In college, the deadlines are imposed well in advance and you have a reasonable amount of time to prepare your report. In industry, you will have less time, and you will have to allow for the time required to obtain approvals from management before you can release the report. In both situations, you have some limited period of time to write the report. The idea is to start as soon as possible. Waiting until the day before the deadline is a recipe for disaster.

Many professionals do not like to write, and consequently suffer from writer's block. They sit at their keyboard hoping for ideas to occur. Clearly, you do not want to join this group, and there is no need for you to do so. There is a technique for you to follow that helps to avoid writer's block.

13.4.1 Define Your Task

Start your report by initially following the procedures described in the previous section. Understand the task at hand, and classify the type of technical writing that you have to produce. Define your audience, in order to establish, before starting to write, the sophistication of the detail and the language that you may employ. Prepare a skeletal outline of the report. Start with the outline presented previously, and add the section headings for the body of the report. This outline is very brief, but it provides the organization of the report. That organization is important because it has divided the big task (writing the entire report) into several smaller tasks (writing the report a section at a time).

13.4.2 Gather Information

It is impossible to write a report without information. You must generate the information that goes into the report. You can use a variety of sources: interviews with peers, instructors, or other knowledgeable persons who that are willing to help; complete literature searches at the library and then read the most suitable references; and Internet and searches for information. Take notes as you read, gleaning the information that you can use in your report. Be careful, however, not to plagiarize. While you can use material from published works, you cannot copy the exact wording; the statements must be in your own words. If the report has an analysis, go to work and prepare a statement of the problem, execute its solution, and make notes about the interpretation of the solution.

13.4.3 Organize Data

When you have collected most of the information that is to be discussed in your report, organize your notes into different topics that correspond to the section headings. Then incorporate your topics in an initial outline, which will grow from a fraction of a page to several pages as you continue to incorporate your notes in an organized format.

13.4.4 Compose Document

You are now ready to sit down and write. Writing is a tough task that requires a great deal of discipline and concentration. We suggest that you schedule several blocks of time, and reserve these exclusively for writing. The number of hours that you should schedule will depend on the rate at which you compose, revise, edit and proofread. Some authors can compose about a page an hour, but most students need more time.

In scheduling a block of time for composing, you should know your productive interval. Most writers take about a half an hour to come up to speed, and then they compose well for an hour or two before their attention and/or concentration begins to deteriorate. The quality of the composition begins to suffer at this time, and it is advisable to discontinue writing if you want to maintain the quality of your text. This fact alone should convince you to start your writing assignments well before the deadline date.

While you are writing, avoid distractions. Since writing requires deep concentration, you must remember your message, the supporting arguments, the paragraph and sentence structure, grammar, vocabulary, and spelling. Find a quiet, comfortable place, and focus your entire concentration on the message and the manner in which you will present it in your report.

Naturally, some sections of a report are easier to write than others. The easiest are the appendices, because they carry factual details that are nearly effortless to report. The interior sections of the body of the report that carry the technical details are also easy to prepare, as the writing describing the detail is less concise and less cogent. Do not become careless on these sections; they are important; however, each sentence does not have to carry a knock-out punch.

The most difficult sections are the introduction and the executive summary. It is better to write these two sections after the remainder of the first draft of the report is completed. We suggest that the introduction be written before the executive summary. Although the introduction contains much of the same information as the executive summary, it is more expansive. The introduction can be used later as a guide in preparing the executive summary.

13.5 REVISING, EDITING AND PROOFREADING

Writing is a difficult assignment. Do not expect to be perfect in the beginning. Practice will help, but for most of us, it takes a very long time to improve because writing is such a complex task. Expect to prepare several drafts of a paper or report, before it is ready to be released. In industry, several drafts are essential, since you often will seek peer review, and receive mandatory manager reviews. In college, you have fewer formal requirements for multiple drafts, but they are a good idea if you want to improve the document and your grade.

13.5.1 First and Subsequent Drafts

The first draft is focused on composition, and the second draft is devoted to revising the initial composition. Devote the first draft to getting your ideas down in reasonable form and in the correct sections of the report. In the second draft, seek to revise the composition and do not concern yourself with editing. Concentrate on the ideas and their organization. Make sure the message is in the report and that it is clear to all of the readers. Polish the message later in the process.

Several hours should elapse between the first and second drafts of a given section of the report. If we read a section over and over again, we soon lose our ability to judge its quality. We need a fresh, rested brain for a critical review. In preparing this textbook, the author composes on one day and revises on the next day. Revising is always scheduled for the early morning block of time, when the brain is rested and concentration is usually at its highest level.

Let's make a clear distinction between composing, revising, editing and proofreading. Composition is writing the first draft, where we formulate our ideas, and we then organize the report into sections, subsections and paragraphs. Unless you are a super talented writer, your first draft is far from perfect. The second draft is for revisions where you focus on improving the composition. The third draft is for editing where we correct errors in grammar, spelling, style, and usage. The fourth draft is for proofreading where we eliminate errors.

13.5.2 Revising

As you revise your initial composition, you should be concerned with the ideas and the organization of the report. Is the report organized so that the reader will quickly ascertain the principle conclusions? Are the section headings descriptive? Sections and their headings are helpful to the reader and the writer as they help the reader organize his or her thoughts. Additionally, they aid the writer in subdividing the writing task and keeping the subject of the section in focus. Are the sections the correct length? Sections that are too long tend to be ignored, or the reader gets tired and loses concentration before completing them.

Question the premise of every paragraph. Are the key ideas presented together with their importance, before the details are included? Is enough detail given, or have we included too many trivial items? Does the paragraph contain a single idea or have we tried to pass two or even three ideas in the same paragraph? It is better to use a paragraph for every idea even if the paragraphs are short.

Have you added transition sentences or transitional phrases? The transition sentences, usually placed at the end of a paragraph, are designed to lead the reader from one idea to the next. Transitional phrases, embedded in the paragraph, are to help the reader place the supporting facts in proper perspective. You contrast one fact with another, using words like **however** and **although**. You indicate additional facts with words such as **also** and **moreover**.

13.5.3 Editing

When the ideas flow smoothly, and you are convinced that the reader will follow your ideas and agree with your arguments, you can begin to edit. Run the spell checker, and eliminate the typos and the misspelled words which the spell checker finds. Search for additional misspelled words, because the spell checker does not detect the difference between certain words like **grate** or **great**, or say **like** and **lime** and between **from** and **form**.

Look for excessively long sentences. When sentences become 30 to 40 words in length, they begin to tax the reader. It is better to use shorter sentences, where the subject and the verb are close together. Make sure that the sentences are actually sentences with a subject and a verb of the same tense. Have you used any comma splices (attaching two sentences together with a comma)? Examine each sentence, and eliminate unnecessary words or phrases. Find the subject and the verb, and attempt to strengthen them. Look for redundant words in a sentence, and substitute different words with similar meaning to eliminate redundancy. Be certain to employ the grammar check incorporated in the word processing program. It is not perfect, but it does indicate a significant number of grammatical errors.

13.5.4 Proofreading

The final step is proofreading the paper to eliminate errors. Start by running the spell checker for the final time. Then print out a clean, hard copy to use for proofreading. Check all the numbers and equations in the text, tables and figures for accuracy. Then read the text for correctness. Most of us have trouble reading for accuracy, because we read for content. We have been trained since first grade to read for content, but we rarely read for correctness.

To proofread, you read each word separately. You are not trying to glean the idea from the sentence, so do not read the sentence as a whole. Read the words as individual entities. If it is possible, arrange for some help from a friend; one person reads aloud to the other, and both have a copy of the manuscript. The listener concentrates on the appearance of each word and then checks the text against the spoken word to verify its correctness.

13.6 WORD PROCESSING SOFTWARE

Word processing software is a great tool to use in preparing your technical documents. The significant advantage of using word processing software is the ability to revise, edit and make the necessary changes without excessive retyping. You can mark, delete, and copy or move text. You can easily insert words, phrases, sentences and/or paragraphs.

While working with word processing software, it is very difficult to revise, edit, and proofread on the screen. Better results are obtained if you print a hard copy and can view the entire page. With the entire page, you can see several paragraphs at once and can check that they are in the correct sequence. Use double spacing when printing the first few drafts to give adequate space between the lines for your modifications.

After you have completed the revisions on the hard copy, make the required modifications to the text using word processing software, and save the results on your own floppy disk. We recommend that you keep only the most recent version (draft) of the document. If you save several versions of the document, it is necessary to keep a log book of the changes to each version. If the writing takes place over several weeks, it is easy to lose track of what changes you made and which version of the hard copy goes with which electronic copy. We find it easier to keep one electronic file (on your floppy disk) of the most recent draft with a hard (paper) copy to serve as a back-up.

Word processing software has several features that are helpful in editing. The spellchecker program finds most of our typos and provides suggestions to correct many of the words that we misspell. The search command permits you to systematically examine the entire document so that you can replace a specific word with a better substitute. The thesaurus is available on command to help you with word selection, but you must be careful when using it. Make sure you understand the meaning of the word that you select; do not try to impress the reader by using long or unusual words. Short words that are easily understood by the reader are preferred in technical writing.

Formatting the report is another significant advantage of word processing software. You can easily produce a document with a professional appearance. The formatting bar permits you to select the type of font, the point size (the height of the characters), the pitch (the number of characters per inch) and emphasis such as bold, underline or italic.

You can also format the page with four different types of line justification which are commonly available. We using word processing software to prepare this textbook, and applied the "justify" alignment for both the right and left margins. The word processing software automatically added the spaces in each line and aligned both margins. In designing your page, use generous margins. One inch margins all around are standard unless the document is to be bound; then the left margin is usually increased from 1.00 to 1.25 inch.

Tables and graphs can be inserted in the text. Take advantage of the ability to introduce clip art into the text or to transfer spread sheets and drawings produced in other software programs into the word processing program. Position the tables and figures as close as possible to the location in the text where they are introduced. Avoid splitting a table between two pages, even if it means leaving an unused space at the bottom of a page. If the table is longer than one page, consider placing it in an appendix. Identify figures with a suitable caption. Place the caption directly beneath the figure, and write it to embellish the message conveyed to the reader.

13.7 SEARCHING FOR INFORMATION

13.7.1 In the Traditional Library

In design, you do not have to reinvent the wheel in order to use it. The same is true for information. You do not have to go into the laboratory and recreate knowledge. It is acceptable practice to use the knowledge generated by others. In fact you do not have a choice in the matter. Demands are placed on you to generate information to use in your technical reports and design briefings. This information may exist in your brain, but more likely it does not. The specific information that you need exists in the library[1]. The difficulty you encounter in either the traditional library or the electronic library is finding the information that you need, as we have information overload. There is so much information available to us that we need to be experts to sort among the excess to find that which suits our needs.

To survive the information overload, we need to develop search procedures. These search procedures differ considerably depending upon whether we are searching in the traditional library or the electronic library. Let's begin by developing search routines for the traditional library and then consider the electronic library later.

The volume of scientific and engineering literature is huge and complex. Today we have nearly 100,000 technical journals which periodically publish new papers describing recent research and development. Add to that new books, reports from both government and private companies, proceedings of technical meetings and patents, it is little wonder that it is so difficult to find narrowly defined facts in the information haystack. Where do we begin?

---

[1] We will make reference to two different types of libraries in this section. The first, is the traditional library with its stacks of books and periodicals. The second is the new electronic library with information compiled at many web sites and available to you at your computer through the Internet and the World Wide Web.

13.7.1.1 Encyclopedias

If you are at a complete loss, begin with an encyclopedia. The encyclopedias give an overview, a brief description of the topic, some history and references that are useful for further inquiry. A list of useful encyclopedias is:

- Encyclopedia Britannica
- Encyclopedia Americana
- Harper's Encyclopedia of Science
- McGraw-Hill Encyclopedia of Science and Technology
- Van Nostrand's Scientific Encyclopedia

This list of encyclopedias offers information on general and scientific topics. It is not a complete list since there are many encyclopedias devoted to more specific areas of science such as astronomy, biology, chemistry, physics, mathematics, etc.

13.7.1.2 Library of Congress

If you are knowledgeable about the subject that you are investigating, then you can begin a more focused search. Card catalogs have given way to the computer which does an electronic search based on author, subject or keywords. The electronic search is very quick, and if the library has books on your topic, call letters and numbers will be provided to help locate the book in the stacks. The call letters assigned to engineering subjects by the Library of Congress are:

- TA General engineering including civil engineering.
- TB Hydraulic engineering.
- TD Sanitary and municipal engineering.
- TE Highway engineering.
- TF Railroad engineering.
- TG Bridge engineering.
- TH Building construction.
- TJ Mechanical engineering.
- TK Electrical engineering.
- TL Motor vehicles. Aeronautics and Astronautics.
- TN Mining engineering. Mineral industries. Metallurgy.
- TP Chemical technology.
- TR Photography
- TS Manufacturers
- TT Handicrafts. Arts and crafts.
- TX Home economics.

The call letters and numbers provide a guide for locating the material in the stacks. Browse the surrounding books that are on the shelves. If the subject is only a few years old, there is a good chance that you will find a book or two that provides excellent coverage on your specific topic. If the subject is older (a decade or more), you will probably find several books available to you. Browse through these books, and note the different treatments by the various authors. Select the book that covers your topic in sufficient detail and provides the depth of coverage suitable for your purposes.

13.7.1.2 Textbooks

Books are one of the best sources of information for your technical reports or design briefings. The authors have studied the literature and have prepared a section or chapter that organizes and describes the topic with clarity. There is a great temptation to copy sections and/or paragraphs verbatim (word for word), and insert them into your report. Avoid the temptation — it's plagiarism. It is okay to use the **material** from any book or paper, but the **wording** must be yours. To avoid problems with plagiarism, read the reference; make notes outlining the organization, the ideas, and the supporting arguments. Then compose your text from your notes. Do not refer back to the book as you compose.

Most textbooks contain an extensive list of references of the material covered in each chapter. Sometimes these references are listed at the end of the chapter, and other times, they are listed at the end of the book. If you need more detail on a given topic, these references provide an excellent starting point for your search. They will identify the book or periodical and the authors who developed the information you seek. A difficulty with the references from books is that they are usually several years old. The information in these older books will be useful, but it will not be up-to-date. You will therefore, need to fill in the gap to determine the current state of the art. To accomplish this, stay with the same periodical and even the same authors. Libraries bind the periodicals annually, and usually the December issue of a given journal has an author and a subject index. Look in this index for the author's names who you know are experts on your topic. Chances are they have continued to research the same subject. Find one of their most recent papers, and examine their references. You will find additional authors and more periodicals that are current. Continue looking up additional papers, finding more authors and more periodicals. Soon you will have an abundance of information covering the past two decades showing a complete development of the topic of interest.

13.7.1.3 Science Citation Index

A common problem in performing an information search for a college assignment is that you find **too much** material and become overwhelmed. You have to weed out the leading material from the average material; otherwise, a section of your report will turn into a chapter. How do you judge the merits of one paper against another? One approach is to use a citation index. The **Science Citation Index (SCI)** lists the number of times a specific paper by a known author has been cited in the references of papers written by other authors. If one of the papers you have found has been cited frequently in the citation index, then it is well recognized in the field as being a significant reference. On the other hand, if one of your references has not been cited in the citation index, then it has been largely ignored by the research community. The SCI keeps score as the research community judges the merits of all writers' contributions to the literature. The SCI is available in the library, on CD-ROM or on the World Wide Web (WWW).

You have performed a computer search, found several books in the stacks, updated your sources with select papers from the correct periodicals, and found the well recognized sources from the SCI. It is now time to organize your materials, outline your sections, and begin to compose.

13.7.2 Electronic Libraries On the World Wide Web

During the past decade or so, an electronic library has evolved that differs in many ways from the traditional library. There are no books on the shelf and no stacks with periodicals on the fourth floor. The information resides in digital format in countless computer memories (web sites) distributed worldwide. You access this information from a computer that is connected to the Internet. The Internet is a network of computers that are connected together by wires, fiber optic cables, or microwaves. It is similar to a system of roads that lead to the information which resides at web sites. The roads, in this analogy, connect cities; whereas, the wires and fiber optic cables connect computers. The advantage of the Internet is that you basically have access to the world's greatest libraries, all from your personal computer. All that is necessary is a computer, a modem to connect to a phone line, an Internet Service Provider (ISP) and Internet software. For readers still in college, the computer center, at either the college or the university, provides Internet access. All that is needed is access to one of the college computers, an account number, and a password.

13.7.2.1 Basic Internet Service

The Internet, a system of many nets, provides the means of transporting the information from one site to another. The information resides on the World Wide Web (WWW) at individual web sites. At each web site, you will find one or more pages of information. There are a vast number of web sites[2], so there is no shortage of sources.

Each page shows information in the form of text, pictures, video clips, and audio clips, among others. Some of this information may be useful, while other information will be worthless for your application. Interesting are the linkages that exist between different pages on different web sites. You can obtain facts from several different web sites, by clicking on keywords or phrases (called "hot links") and your browser (to be described below) transfers to another information source at a new web site. These web sites may reside in any country, and the information may be accessed 24 hours a day, 365 days of the year. There are no "hours of operation" on the Web; the information is always available for you.

Another advantage of the links embedded in web sites is that they permit you to quickly expand the information search. The Web, with the appropriate links, connects pieces of information from different sources, viewpoints and regions of the world. Geography and time constraints are

---

[2] The 4th edition of The Internet for Dummies, published in 1997, quotes 600,000 as the number of web sites, but the number is doubling every two months. By the time this book reaches publication, the number will certainly exceed several million.

removed. The information exists, and the browser will quickly find it for you. Your problem is in organizing the information and presenting it in a form suitable for your audience.

When seeking information on the Web, you will need to use a browser which is software that locates the web pages and then displays them on the screen of your monitor. Two of the most common browsers are Netscape and Internet Explorer. The screen for Internet Explorer is shown in Fig. 13.2.

The top row shows the title of the Web page. The second row gives the menu items such as File, Edit, View, Go, Bookmarks, etc. Click on the menu titles, and explore the options available to you on the drop down menus. The third row shows a string of buttons that you click to move forward or backward, print, search or stop your search. The address of the web site that you intend to visit is typed into the box occupying the fourth row. Such an address is known as the Uniform Resource Locator (URL) and often looks like **http://www.title.com**. Type the address of the web page into the address box, press the Enter key, and your search is initiated. Review the web page that appears on your screen for the information which you seek. Also, look for the "hot links" in the text of the web page. Underlined blue text or blue borders on photos indicate links to other web pages that may contain more useful information. Pointing at them with the cursor will change the arrow to a small hand. Click on the link, and the browser will connect you to another source of information. There is, however, no guarantee that each link will be helpful.

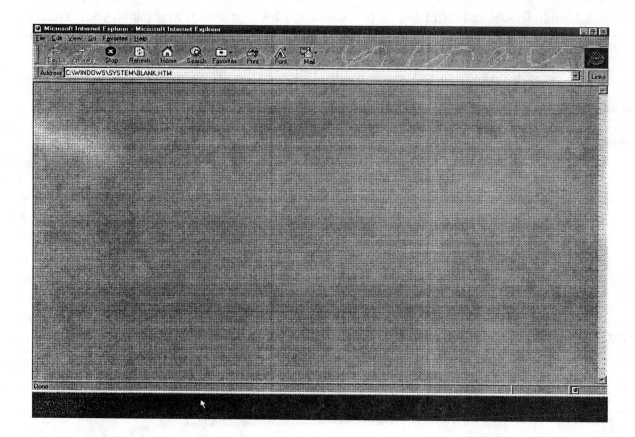

Fig. 13.2 Display for Internet Explorer.

For those readers who are curious, the "http" that is used in all of the addresses stands for HyperText Transfer Protocol. The use of http permits Web users to receive hypertext material that is linked to hypertext material on other web pages.

13.7.2.2 Advanced Search

Sounds cool! But what if you haven't developed an Internet address book and cannot locate the first web site? Fortunately, several software programs, called search engines, are available to help you search. Two of the most popular programs are **Yahoo!** and **AltaVista**. (Yes, the exclamation mark belongs after the word Yahoo. Again for the curious, the word Yahoo stands for Yet Another Hierarchically Organized Oracle.

The choice between these two programs depends entirely upon whether you prefer to work from a directory or an index in your search. **Yahoo!** uses a directory based search routine, similar to using the table of contents in a book to search for a topic. **AltaVista** uses an index based search routine that employs keywords to locate the information. The best way to learn to search is to begin using one search engine through to completion. Start with **Yahoo!** by typing the following address into the address block of either Netscape or Internet Explorer:

http://www.yahoo.com

Click on Reference and then Libraries to see how much information is available to you. You may also be interested in clicking on Education, then Universities and exploring the information available.

Next try **AltaVista** as a search routine by typing the following address into Netscape or Internet Explorer:

http://www.altavista.digital.com

The database for **AltaVista** is established by a web "scooter" that continuously searches all of the web sites, cataloging keywords and addresses of every page. This data is transmitted to the index in **AltaVista** to keep it up to date as new web pages are added to the Internet. When you employ **AltaVista** for a search, you type keywords that describe the topic. Use lower case for most words, except for proper names which begin with a capital letter, like Bill Clinton. The difficulty with this type of search is the large number of matches found. Often you will find thousands of pages relating to the keywords describing your topic. Look over the first few pages of the listings to determine if you were successful in locating the information you were seeking. If not, try using additional keywords, in an attempt to narrow the search.

Some other addresses that you may find useful for searches are:

http://www.lycos.com	An automated index.
http://www.webcrawler.com	An automated index.
http://www.infoseek.com	An index with a limited directory.
http://www.excite.com	An index with concept search and reviews.

13.7.2.3 Specific On-line Libraries

Okay, the search engines are running and able to seek information using either indices or directories. Let's return to the massive electronic library. We previously mentioned the library available to you if you call Yahoo!; however, there are many other libraries. Since you have a wide selection, let's review a few of the more popular offerings to give you a start.

The WWW Virtual Library is one of the oldest libraries. It maintains a collection organized into about 100 categories. Each category is maintained by a volunteer, and its organization resembles a collection of reference journals published by different societies and companies. Its address is:

http://www.w3.org/hypertext/DataSources/bySubject/Overview.html

EINet Galaxy is another library that is maintained by a for-profit company. The organization of Galaxy is like a library with its card catalog sorted by subjects. Engineering and Technology is a main subject heading with most of the disciplines of engineering listed as sub-topics. Its address is:

http://galaxy.einet.net/galaxy.html

If you prefer to use an electronic library with an organization like the one found at many universities, then you will probably be most comfortable with the Clearinghouse for Subject-Oriented Internet Resource Guides. Let's call it Clearinghouse for short. The Clearinghouse comes from the University of Michigan, and its emphasis is on information generated by other universities. You find Clearinghouse at:

http://www.lib.umich.edu/chhome.html

The addresses of other library-like sources that may be useful in one of your searches are:

http://www.nova.edu/Inter-Links/
http://www.uwm.edu/Mirror/inet.services.html
http://gnn.com/
http://duke.usask.ca/%7Elowey/encyclopedia/index.html
http://www.mid.net/NET/
http://library.microsoft.com
http://www.bizweb.com/

We suggest that you try using the Web as an information source. It is available to you at little or no cost, and the hours are much better than those posted by the Engineering library. Finally, you will not be frustrated to find the item you wanted is either missing or checked out to someone on sabbatical leave in West Africa. True, you may be frustrated by the information overload, but you will not be not by the absence of a suitable reference.

13.8 SUMMARY

Writing is a difficult skill to master, and most engineers experience a number of different problems early in their careers. Unfortunately, the writing experiences in college do not correspond well with the writing requirements in industry. In school, we write for a knowledgeable instructor. In industry, we write for a wide range of people, with different reading abilities, that varies from assignment to assignment. In both college and industry, we write to meet deadlines imposed by others. While writing may not always seem to be fun, at least early in your career, there are many techniques that you can employ to make writing much easier and more enjoyable .

The first technique is to organize the report, and we have suggested an outline for a typical report. Gather information for the report from a wide variety of sources, and generate notes that you can use to refresh your memory as you write. Sort the notes and transpose the information on them to expand the outline for your report.

As you organize the outline, but before you begin to write, determine as much as possible about who will be reading your document. They may be technically knowledgeable of your subject or they may be functionally illiterate. The language that you use in the report will depend on the reader's ability to understand. Also be clear about the objective of your document. Is it a report, an extended memo, a proposal or an instructional manual? Styles differ depending on the objective, and you must be prepared to change accordingly.

There is a process to facilitate the preparation of any document. It begins with starting early and working systematically to produce a very professional document. Divide the report into sections and write the least difficult sections (appendices and technical detail portions) first. Defer the more difficult sections, such as the executive summary and the introduction, until the other sections have been completed.

The actual writing is divided into four different tasks, namely composing, revising, editing and proofreading. Keep these tasks separate, and compose before revising, revise before editing, and edit before proofreading. Multiple drafts are necessary with this approach, but the results are worth the effort.

Word processing software does not substitute for clear thinking, but they are extremely helpful in preparing professional documents. They save enormous amounts of time in a systematic editing process. They enable a mix of art, graphics, and text neatly integrated into a single document. Word processing software has a thesaurus and word search features that are helpful in editing. They essentially turn a computer into a print shop, so that you have a wide latitude in the style and appearance of your professional documents.

Searching for information is an essential part of writing a good technical report, although you will usually complete your search **before** beginning to write. We described techniques for using the traditional libraries to seek specific information, and then we introduced the electronic libraries that reside on the World Wide Web. We briefly described the Internet, and the Web. Browsers and search engines that will be useful to you as you learn to move about on the Web were described. Finally, addresses for search programs and a number of different electronic libraries were provided.

REFERENCES

1. Eisenberg, A., *Effective Technical Communication*, 2nd ed., McGraw Hill, New York, NY, 1992.
2. Goldberg, D. E., *Life Skills and Leadership for Engineers*, McGraw Hill, New York, NY, 1995.
3. Elbow, P. *Writing with Power*, Oxford University Press, New York, NY, 1981.
4. Levine, J. R., C. Baroudi, and M. L. Young, *The Internet for Dummies*, 4th Ed., IDG Books Worldwide, Chicago, IL, 1997.
5. Hoffman, P. *Netscape and the World Wide Web for Dummies*, 2nd Ed., IDG Books Worldwide, Chicago, IL, 1996.

EXERCISES

1. Prepare a brief outline of the organization of the final report for the scale development.
2. Prepare an extended outline of the final report for the scale development.
3. Write a section describing one of the subsystems in the weighing machine.
4. Write a section covering the calibration of the scale during the prototype test period.
5. Write the Introduction for the final report on the postal scale development.
6. Write the Executive summary for the final report on the postal scale development.
7. Revise the Introduction that another team member wrote.
8. Edit the Introduction after it has been composed and revised by others.
9. Proofread the Introduction after it has been composed, revised, and edited by others.
10. Go to the Engineering Library and develop a one-page description of the plastics used in the injection molding process.
11. Using the Internet and the Web, develop a one page history of the Ford Motor Company.
12. Using the Internet and the Web, develop a one page history of the largest manufacturing company in the State in which you reside.
13. Write a brief description of performing a patent search using the Internet and the Web.
14. Write a brief description of performing a patent search using the Engineering Library.
15. Using a computer reference such as Microsoft Bookshelf®, find a suitable quotation for this class.

CHAPTER 14

DESIGN BRIEFINGS

14.1 INTRODUCTION

We communicate by writing, speaking and with a number of different visual methods. All three modes of communication play a role as we try to convey our thoughts, ideas and concerns to others. All three modes are important. Let's focus our attention on speaking in this chapter. We use speech almost continuously in our daily life—why do we need to study about design briefings? There are several reasons. We usually speak with our friends and family in an informal style. We know them and feel comfortable with them. They know us. They are concerned with our well being, genuinely like us and are usually interested in what we have to say. A professional presentation is different. It is a formal event that is usually scheduled well in advance. The audience may include a few friends, but usually they are strangers. The time we have in which to convey our thoughts is very limited, and the audience may not be interested in our message. A person or two in the audience may be managers who control our advancement in the company. There are many reasons for tension headaches as one prepares for a professional presentation.

The design briefing is extremely important to both the product development process and to your career. Information must be effectively transmitted to your peers, management, and any external parties involved in the project. Clear messages, that accurately define problems which the development team will effectively address, are imperative. On the other hand, ambiguous messages are often misunderstood, hinder the definition of the problem and lead to delays in implementing timely solutions. Clearly, such messages should be avoided. The design briefing provides an opportunity to review the status of a specific product development process. It permits peers to share their ideas with you, and it affords management an open forum for assessing the quality of your work and the progress made by your development team. Since the professional presentation is critically important, let's become knowledgeable about some great techniques for accurately delivering our messages to a mixed group of strangers and acquaintances.

14.2 SPEECHES, PRESENTATIONS AND DISCUSSIONS

To begin, let's distinguish between three types of formal methods of oral communication—the speech, the presentation and group discussion.

14.2.1 Speeches

The speech is the most formal of the three. In 1996, the year of the Presidential election, we had many examples of speeches. We were besieged with political addresses that clearly illustrated the key features of speeches. They are given to large audiences in huge rooms, stadiums or coliseums. The setting is usually not appropriate for visual aids although the speaker often uses a teleprompter. The audience is diverse with a wide range of concerns. They are of widely different ages with different interests and persuasions. The speech is carefully scripted, and the speaker usually reads or closely follows the script. Ad-libbing is carefully avoided. Communication is one way—from the speaker to the audience. Questions are usually not appropriate and most generally not appreciated. Time is strictly enforced unless you are the President of the United States giving the State of the Union address. Fortunately, engineers are rarely called upon to make speeches; hence, we will not dwell on this topic.

14.2.2 Presentations

Professional presentations differ from speeches. Presentations are made to smaller audiences in smaller rooms that are usually equipped with electric power, light controls and projection devices. We depend on visual aids, demonstrations, simulations and other props to help convey our message during a typical presentation. The audience is knowledgeable (about our topic) and usually has many common characteristics. The presentation is very carefully prepared because of its importance. It has order and structure, but it is not scripted. The flow of information is largely from the speaker to the audience; however, questions are permitted and often encouraged. The speaker is considered the expert, but discussion, comments and questions give the audience an opportunity to share their knowledge of the topic. Time is carefully controlled and is often insufficient from the speaker's viewpoint. The professional presentation is a vitally important mode of oral communication that you must quickly learn to master.

14.2.3 Group Discussions

Group discussions are also very important to the design engineer. The audience is smaller with much more in common. The group may all be members of a development team. The topic being discussed will be narrowly focused. The speaker (central person) serves as a moderator, and he or she is an expert on the topic being considered. However, members of the discussion group (audience) may be as knowledgeable as the speaker (moderator). The moderator works in two modes. He or she may act as a presenter giving brief background information to frame an issue that leads to discussion from any member of the group. The moderator may also direct questions to a group member known to be the most knowledgeable on the issue being addressed. The speaker (moderator) controls the flow of the information, but the flow is clearly two ways—from speaker to the group and from one group member to another. In group discussions, time is difficult to control and the content and the range of coverage is strongly dependent on the skill of the moderator. Group discussion is very important in industry because the leader of a development team will often use this method of communication to identify problems and to initiate an effort directed toward their solution. We will not address discussion group methods in this textbook; however, we recommend that you watch *Washington Week in Review* on the PBS television network to gain some insights. While the participants are journalists, we can adopt many of their clever techniques to guide engineering discussions.

14.2.4 Design Briefing

A design briefing is a type of professional presentation. It is the method that you will use in speaking before the class and will be an approach emphasized in this course. Indeed, you probably will be required to make presentations describing your product development on at least two occasions. The first will be the preliminary design review, and the second will be the final design review. In these design reviews, you will employ professional visual aids to assist in delivering the information required to describe the status of your project with conciseness and clarity.

14.3 PREPARING FOR THE BRIEFING

A design briefing is far too important to take casually. You should prepare very carefully to insure that you will accurately report the status of your teams' development and to identify problems or unresolved issues. There are three aspects that you should consider in your preparations: identify the audience; plan the organization of the presentation and your coverage; and prepare interesting and attractive visual aids.

14.3.1 Identify the Audience

In a typical college classroom, it is much too easy to identify the audience. You have your classmates, the instructor and perhaps a visitor from industry. In an industrial setting, the audience will be more difficult to identify because it will be much more diverse. The size and diversity of the audience depends on the magnitude of the development project. A small briefing will have 10 to 20 people in attendance with most participants from within the company. A large briefing may include 50 to 100 representatives who are both internal and external to the company. The characteristics of the audience are important because you must adapt the content of the presentation to the audience. Classify your audience with regard to their status, interest and knowledge before you begin to plan the style and content for your presentation.

The status of the audience refers to their position in the various organizations they represent. Are they peers, managers or executives? If they are a mix, which is likely, who is the most important? If you are preparing a design briefing for high level management, it must be concise, cogent and void of minor detail. Executives are busy, stressed and always short on time. High-level managers are impatient and rarely interested in the technical details that engineers love to discuss in their presentations. Recognize these characteristics and adjust the content and the timing of your presentation accordingly. The executives are interested in costs, schedule, market factors, performance and any critical issue that will delay the development or increase projected costs of either the development or the product. Usually you will have only 5 to 10 minutes to convey this information at an executive briefing.

First—and even second level—managers are more human. They usually are more interested in you and the designs that you are creating. If you prepare the presentation for this lower level of management, you can safely plan for more time (10 to 20 minutes). These managers are likely to be engineers, and they will share your knowledge of the subject. In fact, they probably will be more experienced, expert and knowledgeable than you. They will want to know about the schedule and costs because they share responsibilities with the higher-level management for meeting these goals. However, they are also interested in the important details, and you can enter into discussion of subsystem performance with them. The first—and second—level managers usually control the resources for the development team. If you need help, make sure that they get the message in enough detail to provide the assistance required. Do not hide the problems that your team is encountering; managers do not like to be surprised. If you have a problem, make sure they understand it and are able to participate in its solution. If you hide a problem from management and it causes a delay or an escalation in costs, the manager will take the heat and you will be in one very unpleasant dog house.

14.3.2 Organization

Design briefings for peers are often less tense, and you will have more time (20 to 30 minutes). You will address schedule and cost issues because everyone needs to know if you are meeting the milestones on the Gantt charts. However, the main thrust of your presentation will be on the design details. If you are working on one subsystem that interfaces with other subsystems, you will cover your subsystem in sufficient detail for all to understand its geometry, interfaces, performance, etc. It is particularly important that the interface issue be fully addressed in the presentation. If four subsystems are to be integrated in a given product, many details about all four subsystems must be addressed to ensure that the integration goes smoothly. Suppose, for example, that we are developing a power tool with a motor that draws a current of 10 amps, and plan to employ a switch to turn the power on/off that is rated at 6 amps. The switch will function satisfactorily in the short term tests with the prototype, but it will malfunction in the field with extended usage sometime after the product has been released to the market. Clearly, the two persons responsible for the interface between the motor and the switch did not communicate properly. The integration of the two subsystems failed.

The most important purpose of design briefings with peers is to make certain that all of the details have been addressed. We strive to integrate the subsystems without problems arising when the prototype is assembled and tested.

Peer reviews in the absence of management are very beneficial in the development process. The knowledge of most of the team members is about the same although some members are more experienced than others. The topic is usually a detailed review of one subsystem or another. The audience is fresh and capable of providing a critical assessment of the technology that you employed. They can check the accuracy of your analysis, comment on the choice of materials, and discuss methods to use for manufacturing and assembly and relate their experiences with similar designs on previous products. Peer reviews afford the opportunity for the synergism that makes fine products even better. They also provide a forum for passing on important lessons from the more experienced designers to the beginning designers in a friendly, stress-free environment.

The subject of the presentation is self-evident in a design review. We are developing a product, and the content the design briefing will deal with one or more issues regarding the development. There is a choice of topics and considerable latitude regarding the content. Although we have given advice in previous paragraphs about matching content with the interests of your audience, you must understand the importance of knowing who your audience will be. Executive reviews serve a different purpose than peer reviews, and the content and the time allotted for the presentation is adjusted accordingly.

For every type of review, there are two absolute rules that you must follow. First, you must know the material absolutely cold. It only takes an audience about two seconds to understand that you are faking it. Some presenters fail to rehearse adequately, and the audience often misinterprets this lack of preparation for insufficient knowledge. In either event, when they realize that you are not the expert you allege to be, your presentation fails. You have **lost** the opportunity to effectively communicate your message. Second, be enthusiastic. It is alright to be calm and cool, but do not be dead on arrival. You must command the attention of your audience, or they will quickly turn you off. You control the attention of the audience only if you are enthusiastic and knowledgeable in presenting your material (story).

14.4 PRESENTATION STRUCTURE

There is a well accepted structure for professional presentations. If you are familiar with PowerPoint, a graphics presentation program marketed by Microsoft, you are already familiar with an excellent structure for a design briefing. The structure, or outline, for a professional presentation is shown in Table 14.1.

Table 14.1
Recommended structure for a professional presentation

- Title
- Overview
- Status
- Introduction
- Technical Topics
    - First Topic
        - Background
        - Status
    - Second Topic
        - Background
        - Status
    - Final Topic
        - Background
        - Status
- Summary
- Action Items

Let's examine each of these topics individually and discuss the content that will be included on the visual aids that accompany your presentation. Recognize that the visual aids (35 mm slides or overhead transparencies) control the flow and the information which you will present. For this reason, we will address the topics listed above in the context of information to be placed on the presentation slides.

14.4.1 Title Slide

The title slide[1] obviously carries the title of the presentation. However, it also states your affiliation and the names of the team members that contributed to the work. Remember, since you may be reporting on the work of the entire team, it is necessary for you to acknowledge their contributions on the title slide and in your opening remarks. A brief title—ten words or less—is a good rule to follow in drafting the title for your presentation. It is also a good idea to use descriptive words in the title. For example, you could use **Preliminary Design Briefing: POSTAL SCALE DESIGN—64 OUNCE RANGE**. Nine words and you have told us the topic (a design briefing or review), the type of briefing (preliminary), the product being developed (a postal scale), and the capacity of the scale. A lot of information is successfully conveyed in nine words.

Below the title, the team name together with the name of each team member involved in the development is listed. It is essential that you share the ownership of the material covered in your presentation. Often the work of several people is covered in one design briefing by a single member of the team. It is not ethical to make the presentation without acknowledging the contributions of others. As students, we typically use first names in identifying team members, but in industry, the complete names of the team members are used. If necessary, division and department affiliations are also stated.

It is also a good idea to date the title slide and to cite the occasion for the presentation. If you make many presentation, this information is useful when referring to your files, allowing you to easily reuse some slides after selective editing and/or updating. An example of a title slide prepared using PowerPoint® is shown in Fig. 14.1.

14.4.2 Overview Slide

The second slide is an overview. The overview for a presentation is like a table of contents for a book. You list the main topics that you intend to cover in the presentation. In other words, you tell the audience what you plan to tell them in the next 10 to 20 minutes. The number of topics should be limited. In a focused presentation, you can cover about a half of dozen topics. If you press and try to cover a ten or more topics, you will encounter difficulty maintaining the attention of the audience throughout the presentation. Constrain the tendency to tell all, and instead, focus on the important issues. In a design review, the topics usually are organized to correspond with the subsystems involved in the product under development. For a component redesign, we concentrate

---

[1] We will refer to slides in this discussion; however, the information described is also true for overheads.

on the issues guiding the component modification being considered. An example, of an overview slide for a preliminary design review for a postal scale, is illustrated in Fig. 14.2.

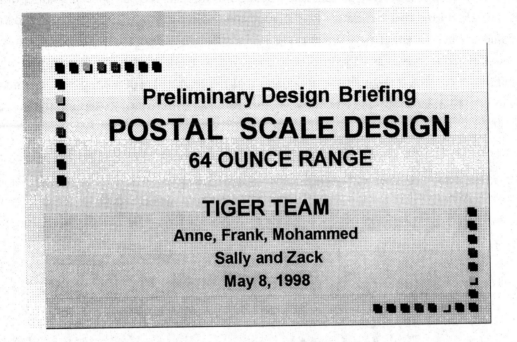

Fig. 14.1 Illustration of a title slide that is informative and descriptive.

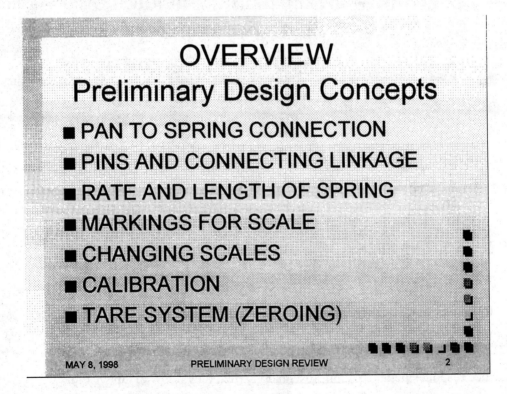

Fig. 14.2 The slide with the overview indicates the content in your presentation.

14.4.3 Status Slide

The third slide is used to present the product development status. You should report on the progress made by the team to date. It is a good idea to incorporate a Gantt chart showing the development schedule either on the status slide or a separate slide. The progress of the team on each task defined on the chart is immediately conveyed to the audience. If there are problems or uncertainties, this is the time to air the difficulties that the team is encountering. An example slide, shown in Fig. 14.3, illustrates uncertainties that the team has with regard to zeroing the Wheatstone bridge circuit in the design of the transducer for the postal scale.

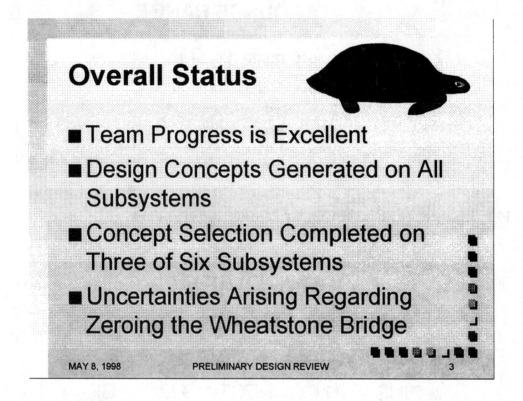

Fig. 14.3 Status slide indicates overall progress and accomplishments while signaling potential problems.

14.4.4 Introductory Slide

The fourth slide covers the introductory material such as background, history or previous issues. It is important that the audience understand the product, the main objectives of the development, the role the team has in that development and the most pressing of the current issues. This introductory slide permits you to set the stage for the body of the presentation that follows. The audience responds better to your presentation when they know in advance the background and the topics that you will cover. They are better prepared mentally to receive your message.

14.4.5 Technical Topics

The body of the presentation usually deals with five or six technical topics. In most instances, the topics selected correspond to the more critical subsystems involved in the design of the product. In the redesign of a component, the topics deal with the issues arising when considering new design concepts. Avoid trying to cover a large number of topics; it is better to report on a limited number of topics thoroughly than to rush through a dozen topics with incomplete coverage. We suggest two slides for each topic. The first slide describes the progress made by the team in generating design concepts and indicates the criteria employed in selecting the best concept. The second slide covers the status of developing the selected concept. The type of information reported on this slide includes accomplishments, outstanding issues, lessons learned, etc. An example of the status of the early stages of a postal scale redesign is presented in Fig. 14.4.

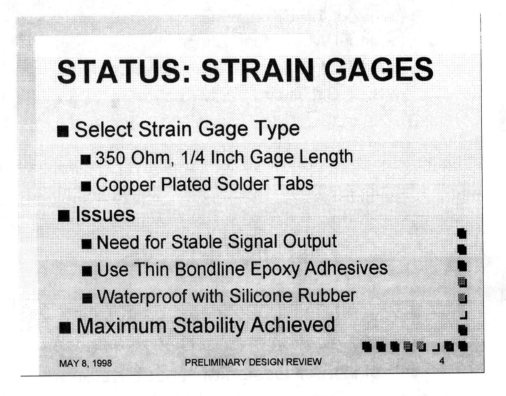

Fig. 14.4 Example of information presented in describing the redesign of a postal scale.

14.4.6 Concluding Slides

After the technical topics have been covered, you must conclude the presentation with a decisive pair of slides. Suggestions for your two concluding slides are given in Figs. 14.5a and 14.5b. The next to last slide in the presentation, shown in Fig. 14.5a, is titled "Key Issues." This is your opportunity to identify very important issues (questions) that your team has discovered. Do not hesitate. This is the time to introduce the uncertainties and to come forward and seek help. Design reviews are not competitive. We seek to help regardless of our status or our role as a member of the audience. Managers will arrange support for your team if it is required. Peers will

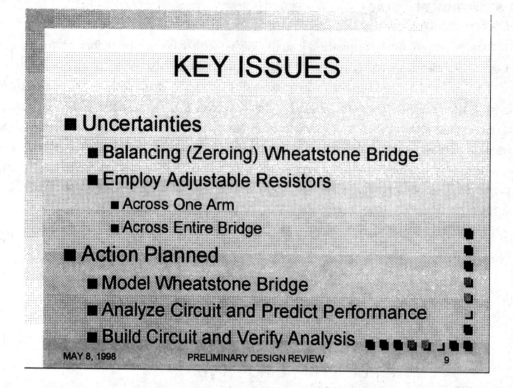

Fig. 14.5a The "key issues" focuses the team and management on your most significant problems.

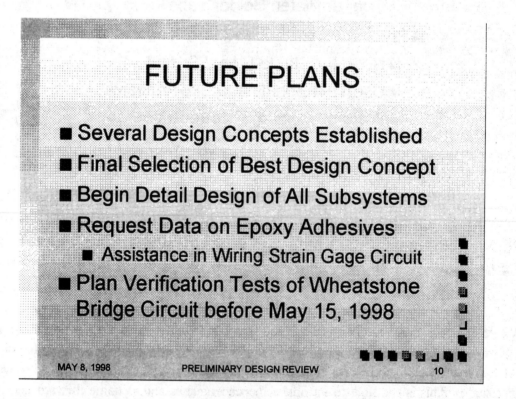

Fig. 14.5b The "future plans" slide gives you the opportunity to describe your approach for solving the problems that the team is encountering.

make suggestions and introduce fresh approaches to unresolved problems that the team may find useful. The design review is a formal process in the product development in which everyone participates. Take advantage of the review to reveal uncertainties and to seek help whenever your team needs assistance.

The final slide is to present your plans for the future. Management wants to know how you intend to solve the issues that you have revealed and how much time is required for the solutions. Your peers will be interested in the technical approaches that you intend to follow. You should also define any requirement that you have of anyone in the audience. If you need help from your peers or more resources from the manager, be certain it is clearly indicated on this final slide. You should clearly identify those individuals, with resources, who have agreed to provide the requested support. If action items are listed, clearly identify the responsible individual and estimate the time to completion. The planning incorporated in the final slide is very important. It is the prescription for your team's get-well program.

14.5 TYPES OF VISUAL AIDS

Because they control the content and the flow of the information conveyed to the audience, visual aids are essential for a technical presentation. They provide visual information that reinforces the verbal information. You communicate using two of a person's five senses. Take time to carefully select the type of visual aid that will most effectively convey your message.

In selecting visual aids, we are usually limited—first, in our ability to produce the visual aids within the time and financial constrains and second with regard to the availability of equipment and the room in which the presentation is to be made. Is the room equipped with a 35 mm slide projector, an overhead projector or a computer-controlled projector? Another important consideration is the control of the light intensity in the room. Most rooms have light switches used to control the overhead lights. However, some rooms do not have blinds and bright sunshine will cause problems with the visibility of 35 mm slides when they are projected.

14.5.1 Computer Projected Slides

Computer projected slides are an excellent choice if the room can be darkened. Computer projection of slides has several advantages: the presentation is easy to prepare using PowerPoint; the slides are in color and you can take advantage of advanced visual effects (slide transition and slide building feature) that enhance the impact of the presentation. It is also easy for you to copy your presentation onto a floppy disk, use the computer attached to the projection system to load your program and employ a mouse to click through your slides line by line. Finally, low-cost hand-outs, with four six slides per page, can be prepared which help the audience follow your presentation. The only disadvantage of computer projected slides is the dark room, required for projection, eliminates eye contact with the audience or even reading their body language. You lose your visual contact with the audience that is so important in reading their response to your presentation style.

14.5.2 Overhead Transparencies

Projecting overhead transparencies is probably the most common method of presenting visual material. Overhead projectors are relatively inexpensive (a few hundred dollars compared to a few thousand dollars for the computer controlled projectors), and for this reason they are available in most rooms. Overhead projectors are also the best choice if the room cannot be darkened. The projectors are sufficiently bright to provide quality images without extinguishing the lights. Bright rooms have an advantage because they help you keep the audience alert. Overhead transparencies are also easy to make on a copy machine. If you are willing to sacrifice the benefits of color, utilize inexpensive black and white transparencies. If you want to use color in the presentation because it is more effective, the costs increase significantly to use a color laser printer or a color copier. However, color ink-jet printers can be used to produce good quality color transparencies at a reasonable cost.

14.5.3 35 mm Slides

The 35 mm slide presentation is a third option for visual aids; however, the equipment required is not always available. Projectors for 35 mm slides are common, but in an engineering college, they usually are much more difficult to find than the overhead projector. A slide presentation is the preferred option if you have many photographs to project because the quality of colored 35 mm slides is excellent. The slides are inexpensive and easy to prepare if you have a camera and the time necessary for the film processing. It is possible to prepare colored slides from your PowerPoint files, but conversion from the digital format to the 35 mm format requires special equipment. Again a main disadvantage is that the room must be darkened to view the image from a 35 mm projector. Also, the relatively long distance required from the projector to the screen precludes the use of very small conference rooms for presentations.

Although ineffective, sometimes speakers will employ two or three different types of projectors. Such a tactic should be avoided as the switch from one projector to the other is disruptive. If possible, use only one type of projector. If you must use more than one type of projector to properly present your message, try to minimize the number of times you switch from one device to the other.

14.5.4 Other Types of Visual Aids

There are several other visual aids that are sometimes used in professional presentations. Video tapes are common; unfortunately, the smaller size of TV monitors often makes viewing difficult for a large audience. However, if you want to show motion or group dynamics, video clips are clearly the best approach. Video cameras are readily available, and after some practice, you can become a reasonably good video producer.

Motion picture films, particularly of older material with historical interest, are very effective in developing a long-term prospective. Finding an 8 mm or 16 mm motion picture projector may be difficult—allow time in your schedule for making the necessary arrangements.

Hardware, or other materials to be used in the product development, is sometimes passed around the audience during a presentation. The hands-on opportunity for the audience is a nice touch, but you pay for it with the loss of attention by some members of the audience during the inspection period. We recommend that you defer passing materials to the audience during your presentation. Instead, after the presentation, invite the audience to inspect exhibits that are placed on a strategically located table. This approach maintains the attention of the audience throughout the presentation while giving those interested in the hardware time for a much more thorough examination.

14.6 PRESENTATION DELIVERY

Excellent presentations require well prepared slides and a smooth, well-paced delivery. There are many aspects to the delivery part of the presentation process including: dress, body language, audience control, voice control and timing.

14.6.1 Attire

Let's start with dress. Broadly speaking, there are four levels of dress. The highest level, black tie for men and formal gowns for women, is a rare event and is never appropriate for design briefings. The next level, business attire for men and women, is sometimes appropriate for design briefings. A conservative Eastern company may have a dress code requiring business attire; whereas, a less formal Western company would encourage more casual attire. Casual attire should not be confused with sloppy attire. Casual attire is neat and tasteful, but without suits, ties and white shirts for men and suits for women. Although becoming more common in the workplace, sloppy attire (old jeans and a sweatshirt) is strictly taboo. We suggest that you select either business or casual yet tasteful attire when you dress for the presentation. We recognize that most students these days prefer a relaxed, collegiate style of dress, but resist the impulse to dress that way. If you look like a bum or a homeless street lady, who will believe your message?

14.6.2 Body Language

Posture is another important element in the presentation. In the words of former President Reagan, "stand tall." Your body language signals your attitude to the audience—you are the presenter and in control. Make sure you are calm, cool and collected. Nervous gestures with your hands, rocking on the balls of your feet, scratching your head, pulling on your ear, etc. should be avoided.

14.6.3 Audience Control

Before you begin your presentation, take control of the audience. One approach is to pause a moment before projecting the title slide and immediately make eye contact with the entire audience. How do you look at everyone in the house? You scan the audience from left to right and then back looking slightly over their heads. Occasionally drop your eyes and make eye contact for a second or two with one individual and then another. Pause long enough for the audience to become silent (10 to 20 seconds). If someone is rude and continues to talk, walk toward them and politely ask for their attention. When you have everyone's attention, project the title slide and begin your delivery.

If you have rehearsed, you will not need notes. The slides carry enough information to trigger your memory. If they do not, you have not rehearsed long enough to remember the issues. Scripting the presentation is not recommended. People who script will eventually start to read their comments which is a deadly practice. Rehearse until you are confident. Make notes to help you rehearse, but do not use them during the presentation.

14.6.4 Voice Control

When you begin to speak, be certain everyone can hear you. If you are not sure of how far your voice carries, ask those in the back of the room if they can hear. If you use a microphone with an amplifier and speakers, be careful; the tendency is to speak too loudly. Try to control the loudness of your voice within the first minute of the presentation when you introduce the topic with the title slide. The audience will read that slide with anticipation, and they will bear with you as you adjust the volume of your voice.

Do not mumble. Speak slowly, clearly and carefully enunciate each word. It is not necessary to speak as slowly as Vice President Gore, but it is better to present your material too slowly than too rapidly. If you can improve your enunciation, you will be able to increase the rate of your delivery without loosing the audience.

Don't run out of breath while speaking. Learn to complete a sentence, pause for breath (without a gasp) and then continue with the next sentence. There is nothing wrong with an occasional pause in the presentation, provided the pause is short.

14.6.5 Timing

If you forget some detail, do not worry. Skip it, and move to the next point that you are trying to make. If the detail is critical, count on one of your team members to raise the issue in the question and answer period. We all forget some of the material that we intended to present and introduce some additional items on an ad-hoc basis. Usually, the audience will never be aware of our omissions or additions.

Studies have clearly shown that people can listen faster that the presenter can speak. The trouble with following someone who speaks very rapidly is not their rate of speech, but instead, their enunciation. Many people who speak rapidly tend to slur their words, and we have trouble understanding poorly pronounced words. The trouble in listening to Al Gore is to stay with his message; while you are waiting for his next word, your mind goes off on a mental tangent. You anticipate that your mind will return to the speaker in time for the next word, but unfortunately our minds are sometimes tardy. We often miss keywords or even complete sentences before our attention returns to the speaker.

Listening is a skill, and many of us have not developed it properly. As speakers, we have the responsibility to keep the listener, even the poor ones, on the topic. We use several techniques to maintain the attention of the audience. Employing the screen is probably the most effective tool for this purpose. As you project the slide, walk to the screen and point to the line on the slide that corresponds to the topic that you are addressing. If a few members in the audience are coming back from their mental tangents, you reset their attention on the current topic.

When referring to the image projected on the screen—and good speakers use this technique—position yourself correctly. We show the proper position in Fig. 14.6. Stand to the left side of the screen being careful not to block anyone's view. Face the audience and maintain eye contact. When you have to look at the slide, turn your head to the side and read over your shoulder. The 90° turn of your head and the glance at the slide should be quick because you want to continue to face the audience. Under no circumstance, should you stand with your back to the audience and begin reading from the slide as if it were a script.

As you develop the discussion, point to key phrases to keep the audience focused on the topic. Engage both of their senses. Deliver your message through their eyes and ears simultaneously. When you point to a line on the slide, point to the left side of the line. People read from left to right—so start them from the initial point on the left side.

As you speak, modulate your tone and volume. Avoid both a monotone and a sing-song delivery. A continuous tone of voice tends to put the audience asleep. A sing-song delivery is annoying to many listeners. If you have a high-pitched voice, try to lower the pitch because a high-pitch tone is bothersome to many people.

If a question is asked, answer it promptly, unless you intend to cover material later in the presentation that addresses the question. The answer should be brief, and you should try to avoid an extended dialog with some member of the audience. If a member of the audience is persistent, simply indicate that you will speak with them off-line immediately after the presentation. The timing of your presentation can be destroyed by too many questions. Some questions are helpful because they permit audience participation, but excessive questions cause the speaker to loose control of the topic, the flow and the timing of the presentation. The format of the communication changes from a presentation to a group discussion. Group discussions have a purpose; however, they are different than a design briefing.

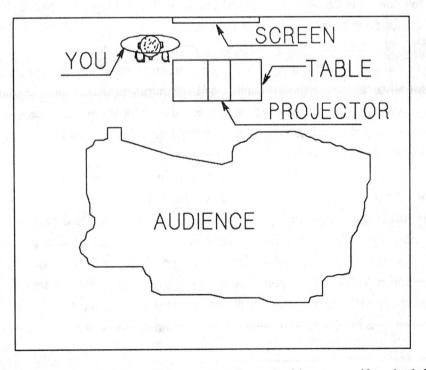

Fig. 14.6 Typical room arrangement for a presentation. Position yourself to the left of the screen and face the audience.

14.7 SUMMARY

The design briefing is extremely important both to a successful product development and to your career. Information must be effectively transmitted to your peers, the management and to external parties involved in the project. Clear messages, that accurately define the problems which the entire development team can effectively address, are imperative. The design briefing provides the opportunity to inform management and our peers of our progress and problems.

There are three types of formal communication: the speech, the professional presentation, and the group discussion. The characteristics of these three types of communication are described. For the engineer, the professional presentation is the most important and the formal speech is the least important. The main emphasis of the chapter is on the professional presentation with a focus on the design briefing.

Preparation for a presentation is essential. Accomplished speakers often spend the better part of a day preparing for a 20 minute briefing. There are three important elements in the preparation. First, know the characteristics of your audience, and be certain that the content that you include corresponds to their interests. Second, organize the presentation in a manner that is acceptable to the audience. Ensure that the organization is efficient so that you use the allotted time effectively. Prepare high-quality visual aids (slides) that will help you convey your message and rehearse until you command the subject material.

A structure for the presentation is recommended that is commonly employed in design briefings. This outline includes four front-end slides—title, overview, status and introduction. The body of the presentation is devoted to technical details with from eight to twelve slides. The two closure slides are for a summary and the action items.

The type of visual aid that you select is important because the visual transmission of information will markedly affect the outcome of the presentation. Computer projected slides have some significant advantages, but the availability of the equipment and a suitable (darkened) room often precludes their use. The overhead transparencies are the most common medium. However, if you have a presentation that includes many photographs and a large room in which to deliver the presentation, 35 mm slides are the most suitable medium.

The delivery is the make or break part of the presentation. We have provided you with two pages of details about how to do this—and do not do that. However, the best way to learn to deliver a design briefing is by practicing your presentation. On your first two or three attempts, have a friend video tape the event. Then review your delivery style, technique and appearance during a trail presentation. You will identify many problems of which you were not aware. Another suggestion is to use one of your free electives to take a speech course. The experience gained in a typical communications course will not help much in crafting the content to include in a design briefing, but it will help you to develop very important delivery skills.

REFERENCES

1. Eisenberg, A., *Effective Technical Communications*, 2nd ed., McGraw Hill, New York, NY, 1992.
2. Goldberg, D. E., *Life Skills and Leadership for Engineers*, McGraw Hill, New York, NY, 1995.
3. Wilder, L. *Talk Your Way to Success*, Simon and Schuster, New York, NY, 1986.

EXERCISES

1. Write a brief description of the characteristics describing your classmates in this course. Focus on the characteristics that will influence the language and content in your design presentations.
2. Prepare an outline for a preliminary design review. Include in the outline the titles of all of the slides that you intend to use.
3. Prepare the title slide for your preliminary design review.
4. Prepare an overview slide for your preliminary design review.
5. Prepare a status slide for your preliminary design review.
6. Prepare the "Key Issues" slide for your preliminary design review.
7. Beg, borrow, but do not steal a video camera. Video tape a practice presentation by a teammate. Critique his or her presentation with good taste.
8. Locate all of the projectors that you are permitted to borrow for the presentations to be made in this class.
9. Practice the delivery of your presentation employing the most suitable projector that you can borrow.
10. Prepare a group of slides for a design briefing that utilizes the slide transition and slide building features of PowerPoint.

PART VI

ENGINEERING AND SOCIETY

CHAPTER 15

ENGINEERING AND SOCIETY

15.1 INTRODUCTION

It should be made clear from the beginning of this chapter that engineering and society are closely coupled together. Engineering is an integral part of society, and consequently, we must treat engineering as a social enterprise as we continue to develop sociotechnical systems.

When engineers develop a new product or improve an existing system, we change society. The magnitude of the change differs from one product to another, and the effect on society may be small or very large, but no matter what, there is an effect. When a few companies introduced a random orbit sander a few years ago, the time, effort, and cost of producing a flat, smooth, scratch-free surfaces was reduced. These sanders helped several hundred thousand workers in the finishing business producing high-quality finishes on furniture, cabinets, counters and floors. The results also benefited many millions of customers who purchase furniture each year. This new product had a modest, but beneficial, impact on society. There was, however, a small negative effect in that minor emissions to the environment occurred during manufacturing of the product. Overall though, society has clearly gained by the introduction of this product.

When transistors were first developed from new ultra pure semiconductors, the impact was revolutionary. While the first transistor was developed in the mid 1950s (the integrated circuit was introduced in 1959, and the microprocessor was released in 1971), the technological revolution in microelectronics is still underway. Micro-electronic based products are markedly affecting the way in which society lives today. We call it the "information age". New products, which are introduced almost daily, are driven by technology. They will continue to change the way we work and play in a major way for at least another generation.

We hear a lot about new products and new life styles. Television, radio and even newspapers advertising keep us well informed of the availability and prices of both old and new products. However, we rarely hear much about the actual technology involved even when the product is new and revolutionary. The media and the public usually are not interested in learning about the technology employed in developing a new product; this is unfortunate because it would be easier to introduce new or modified systems if the public were more technologically literate.

Engineers are, in part, to blame for this apparent wall between the public and the understanding of technology. We often speak in such technical terms that we leave those rare individuals interested in our work in a confused state. As engineers, we must recognize our responsibilities to effectively communicate to society. We can begin by learning how to explain why and how our products work in terms easily understood by the general public. We must also be more enterprising in understanding society and their changing views relative to technology. Lastly, we need to think on a more global scale. Engineers develop sociotechnical systems which are usually large and complex and which rely on hundreds of products—and many different types of services—to function. Even though, as an engineer, we may be dedicated to developing a single product that is only a small part in a much larger system, we must understand the entire system. The superior designs are developed by those teams that have a thorough understanding of all aspects of the complete sociotechnical system.

Let's consider an example of a sociotechnical system—the commercial air transport system. Do we have technical products involved in this system? **YES**, literally hundreds of advanced state-of-the-art products. Do we provide technical services for the airlines? **YES**, again. Does an engineer working for General Electric in the development of a new jet engine need to understand what Boeing is doing in the design of a new aircraft or modifications of an existing aircraft? Does that same engineer need to know about the sound levels that the public will accept as the plane lands and takes off? Is the engineer required to understand the causes of each and every airline accident? Again, **YES** to all.

From these simple questions, it should be evident that our sociotechnical systems are extremely complex. It will require significant effort for you to understand the detail of the interaction of the many products incorporated in these systems with the public. Nevertheless, the most successful products will be developed by engineers that **understand** and **appreciate** the improvements that both the customer and society seeks.

Finally, as an engineer, you must understand that the systems we develop are adaptive. They are dynamic and change with time in response to several different forces. First, society changes. What society and the customer wanted in the 1950s is a far cry from what either will accept today. The customers, a segment of society (more on this fact later), typically expect that the new model of the product will be much improved over the older model and cost less. Others in society, not necessarily the customers, will expect (perhaps demand) that you manufacture the product without degrading any aspect of the environment.

An example is in order to distinguish between the customer and society. Let's consider the automobile as the product. The customer of an automobile demands style, comfort, mobility, convenience, affordability, reliability, performance and value. Society requires (through federal regulation) safety, fuel efficiency and reduced pollution. The fact that the requirements from the customer and society often conflict should be recognized. Conflicting requirements challenge engineers to create improved designs that advance the state of the art in all areas of technology.

The governments (Federal, State, County and City) are also involved in the relationship between engineers and society. Various government agencies continually issue regulations that affect both the sociotechnical systems and the product. Consider the commercial air transport system again. Obviously, safety is a very serious issue. The Federal Aviation Administration (FAA) is responsible for issuing regulations to insure safe operations of commercial aircraft. 1996, however, was not a good year for the FAA or our commercial airlines because several very serious accidents occurred with significant loss of life. As a consequence of the fire in the cargo bay, which caused the crash of a Value Jet Boeing 737 aircraft in the Florida Everglades and the loss of all aboard, the FAA is now requiring the airlines to retrofit all planes with a fire suppressant system. Regulations from government agencies drive a portion of the changes to any sociotechnical system.

The understanding of the relationship between engineering and society is relatively new. If we examine the five volume tome, *A History of Technology*, [1] published by Oxford University Press over the period from 1954-58, we find a chronological, deterministic catalog of technological development. However, the editors almost completely ignore the relation between engineering and society. In his review of this tome, Hughes [2] states that the editors belatedly conclude their five volume treatise with an afterthought essay on technology and the social consequences. Today, the relationship is more clearly recognized although we have not yet managed to include adequate treatment of this very complex relationship in the typical engineering curriculum. We will attempt to show a few examples of the relationships between engineering and society in this chapter. For a more complete treatment, we encourage you to read reference [3].

15.2 ENGINEERING IN EARLY WESTERN HISTORY

Technology has impacted the way that we live since we began to record history. Long before the first engineering college was founded, we had intelligent, hard-working men[1] working as engineers developing new products and processes. We can look at history and cite many developments that changed society, and for the most part, benefited society.

Prior to about 6000 BC, there was little evidence of technology. Men and women were hunters and gatherers. They scrounged for food and lived in caves if they were lucky and shacks or huts if they were not. The significant development in the period from 6000 to 3000 BC was in agriculture with the domestication of animals and the cultivation of grains. With time, it was possible to insure a relatively stable—if limited—food supply and people began to congregate together in villages and small cities.

The first technology involved the building of roads, bridges and boats for local transportation. People clustered in villages and small cities wanted to transport food, water and other materials over short distances. The geographic regions where technology was evident were extremely local. Early Western history involved only Mesopotamia, Egypt, Greece and then Rome.

---

[1] Historical accounts of engineering achievements indicate that the profession was male dominated. The author is sensitive to the gender issue in engineering education, but it must be recognized that the presence of women in the profession is very recent when considered relative to a historical span of about 8000 years.

In a few select cities, there were aqueducts to transport water and limited sewerage. In most regions, only the ruling class lived well. The great monuments of Egypt and Greece were built with slave labor. In fact, slaves (people who were on the loosing side in a war or those born into slavery) provided most of the power required for building, mining and agriculture for many thousand more years. The rise of religion, after the coming of Christ, is often credited with the reduction in the prevalence of slavery. However, the steam engine developed by Thomas Savery, Thomas Newcomen and James Watt in the early 18th century signaled the beginning of the end of the need for vast amounts of human power. As we enter the 21st century, the development of the computer and software programs portends another beginning of the end for the need for vast amounts of clerical efforts.

15.2.1 The Egyptians

The Egyptians built many huge monuments and were masters in moving large blocks of stone from distant quarries to the construction site and then elevating blocks to their respective positions in the structure. Some estimates indicate that 20 – 50,000 men were required to drag the 20 to 50 ton blocks of stone used in the construction of these monuments. In spite of the majestic grandeur and durability of the monuments, the new innovations introduced by the Egyptian engineers were not outstanding. They constructed temples in a simple fashion using only uprights (columns) and cross pieces (short, deep beams). While the arch was known during this period and constructed with mud bricks, it was never adapted to stone construction. Classical treatments of the history of technology tell us very little about the impact of these developments on society. However, it is apparent that the construction of a pyramid or a temple requiring a significant segment of the working population for many decades. This effort probably did little to improve the welfare of even the **free** segment of the population. It is also evident that the effort required of the slaves to move the countless large, heavy stones must have been brutal.

15.2.2 The Greeks

Although they had very capable engineers, the Greeks are better known for their contributions to science. Thales, Euclid and Aristotle are well known for their contributions to the development of the basics of geometry and physics. This new knowledge of geometric relations was applied almost immediately in architectural engineering. However, the scientific knowledge available during this period was so limited that nearly 2,500 years elapsed before this knowledge was expanded sufficiently to be useful to the engineering community.

Greek engineers built cities, roads, water (hydraulic) systems and some machinery. Archimedes and Hero were innovators who used the pulley, screw, lever, and hydraulic pressure to develop cranes, catapults, pumps, and several hydraulic devices. The Hellenistic period brought better buildings and local roads which improved living conditions for the upper class. However, the construction of the city infrastructure was still performed largely by slaves because the use of other forms of power was nonexistent. The Hellenistic world was divided into extremely independent city-states. Politics of the day did not foster cooperation between these governmental entities; as a consequence, few long roads were built. Transportation was largely by sea in boats equipped with oars and a galley of slaves. Some of the boats were equipped with a square sail, but these were effective only on down-wind tacks.

15.2.3 The Romans

The Roman engineers followed the Greeks, and while they were less inventive, they were masters of detail. As the Roman Empire spread from the Middle East to Scotland, the Roman engineers built long roads connecting the distant cities; roads so well constructed that they were in service for many centuries. The large cities constructed in the conquered lands had paved streets and heated homes with water supplies and sewer systems. The Roman engineers had little or no theoretical knowledge, but they had developed practical methods of construction that served their purposes well. During this period, the subjects of statics and mechanics of materials were not understood. Beam theory would not be developed for another 1,700 years. The Roman engineers based their construction on experience, and they used very large safety factors. These empirical methods, while crude by today's standards, sufficed to produce many bridges, roads, aqueducts and cities. A great bridge over the Tagus river in Spain supported a road nearly 200 m long with six spans 60 m above the river. Considering it was completed at about 100 AD, this bridge was a remarkable engineering feat. The durability of the Roman construction is well documented by their structures some of which are still in use today—2,000 years after they were placed in service.

The Roman engineers deviated from the Egyptian and the Greeks in their construction by using much smaller building stones and bricks. They were still able to produce massive structures with the smaller blocks by utilizing mortar and cement. The use of smaller stones and bricks and cement clearly impacted society. Buildings could be constructed with much less effort since thousands of slaves were not necessary to move a single block of material. Life for the slaves clearly improved. Also the upper classes benefited from the more rapid construction of buildings and houses for their cities.

While the Roman engineers were outstanding in the construction of civil infrastructure, they did very little to advance the use of power in driving the few machines in use during that period. Humans were still used in most cases to power pumps, mills and cranes. In rare cases, water wheels were used at power mills to grind grain, but these were of the undershoot design and much less effective than the overshoot design that was developed later. While references on the history of technology are quiet on the issue, the life of a worker powering a pump or a mill must have been very difficult.

Eventually, the vast Roman Empire crumbled, and we entered a period that the historians classify as medieval civilization from 325 to 1300 AD. The fact that barbarians stripped and destroyed the cities does not imply that progress stopped, but it sure did slow progress to a snails pace[2].

---

[2] When Rome collapsed, the Byzantine Empire followed, and significant engineering achievements were made in developing dome structures and curved dams. The author has omitted discussion of these developments to keep this historical treatment brief.

15.2.4 Early Agriculture and the Use of Animals

In spite of the slow development of engineering and the lack of the construction of monuments during the Middle Ages, we did experience an early agricultural revolution. White [4] has described the changes that occurred as agriculture moved from the dry sandy soil of Italy into the heavy alluvial and wet soil of northern Europe. The plows that were effective in the light sandy soils would not turn the sod in the heavy, moisture-laden soil of the north. A new, heavy-wheel-plow was invented that would cut through the grass sod and turn over the soil. However, this plow required eight yoked oxen to provide the large thrust necessary to pull it through the heavy sodded soil. Very few farmers of the day owned eight oxen. They had to combine their land, share their oxen and cooperate in plowing and planting. The combination of the farmers and their land holdings eventually led to a manorial society.

It may be difficult to imagine that the invention of a new plow made such a major change on the way people lived; today, we are able to produce all of the food that we need in the U.S. exporting excesses in significant quantities, with less than 2% of our population. However, before the advent of engines and motors (stream, gasoline and electrical), men and women struggled to grow enough food to live. In the Middle Ages, 90% of the population worked in agriculture, and they had much less to eat than we do today.

Oxen were used as a source of power for plowing and hauling on the farms. However, oxen are very slow and they consume large quantities of food. At that time, horses were of limited use because of inadequate harnesses and saddles and the frequent problem of splitting their hoofs. (Early harnesses fitted about the horse's neck tended to strangle the animal.) A newly designed horse collar which transferred the load to the shoulders of the horse was introduced—together with iron horse shoes—alleviated these problems. Horses gradually replaced the oxen because they were 50% faster, worked longer and ate less.

The horse was not very effective in frontal attacks in early warfare because stirrups did not exist. Consequently, horse mounted warriors were of limited value because they could not adequately thrust their lance and stay mounted on the horse. With the advent of the stirrup, the horsemen could stand-up, lean forward and thrust the lance through the defenders' shields without loosing their seat. The simple addition of stirrups to saddles had a profound influence on society. Feudalism evolved with an aristocracy based on warriors conquering and defending landholdings. A new battle strategy with armored knights mounted on large strong horses was dominant for many centuries[2].

In early history, small technological improvements produced major changes in the form of government, the type of society and the way that people of all classes lived. It is remarkable that for about 5,000 years, humans were the prime source of power for most activity. Progress in agriculture and engineering proceeded together to provide more food, better housing, roads and bridges and eventually horse- and oxen-powered plows and wagons.

15.3 ENGINEERING AND THE INDUSTRIAL REVOLUTION

Civilization has always been power limited; even today, our space travel is severely constrained by the inadequacies of rocket power. Until the last three centuries, animals, windmills and waterwheels provided the only power available for agriculture, construction, milling, mining, and transportation. Progress was often slow because power was not available when and where it was needed. Life, in the power starved 17th century, was rather bleak for the general population. They worked from dawn to dark as farmers or craftsmen. Only those endowed with landed estates and/or money lived what we would consider a comfortable life.

15.3.1 The Development of Power

The first of several breakthroughs in developing new sources of power occurred in the 18th century. In Great Britain, Savery, Newcomen and Watt developed stationary steam engines. While these early steam engines were woefully inefficient (only 0.5%), they were produced in relatively large numbers (500 existed in 1800) and were employed to pump water and power mills (textile, rolling and flour).

Since the steam engines were replacing horses as power sources, James Watt conducted experiments to measure the power that a "brewery horse" could provide. His measurements indicated this horse generated 32,400 (foot-pounds/minute) of power. The horse's output was rounded to 33,000 (foot-pounds/minute) and this value is still used today as the conversion factor for one horsepower (HP).

Most of the 18th century was devoted to innovations improving several different models of steam engines. Applications were largely limited to stationary power sources because engines, of that day, were large, heavy and inefficient. However, with improved metallurgy and machining methods, it was possible in the 19th century to reduce the size and the weight of the engines permitting them to effectively power river boats, ocean-crossing ships, rail locomotives and steam-powered automobiles.

The development of the steam engine and its adaptation to several modes of transportation formed the foundations of the industrial revolution. The life style of almost everyone was markedly changed with the emergence of steam-powered factories and much more effective transportation. Fewer people were employed in agriculture, but many more were working under appalling conditions in factories and mines. Twelve hour days, seven days a week, with only Christmas as a holiday during an entire year was the norm. Manufactured products, such as clothing and housewares, were available to a larger share of the population at reduced prices, but the margin between earnings and the money necessary for a factory worker, miner, or farmer to stay alive remained very narrow. Steam power had relieved many thousand of men and horses from brutally hard labor; they were released from one dull job to another. On a more positive note, the progress in agriculture and engineering enabled the population to triple during this period (1660 to 1820).

15.4 ENGINEERING—19TH AND 20TH CENTURIES

For the eight thousand years of recorded history prior to about 1800, the advance of technology was extremely slow. We lived in an agricultural society with most of the population growing food which was consumed locally. Factories were being developed and simple products needed on the farm, in the home, or by the military were produced. Transportation of goods was usually limited to sailing ships on the seas or horse drawn barges on rivers or canals.

During the 19th and 20th centuries technology, literally exploded. The many scientific discoveries of the preceding millennium sufficiently expanded knowledge to allow countless engineering applications. To illustrate this explosion, let's define modern technology to consist of the following four elements:[3]

1. Abundant available power.
2. Transportation by air, land and water with associated infrastructure to provide rapid, convenient and inexpensive access.
3. Communication between everyone, anywhere, anytime, and at a reasonable cost.
4. Information (knowledge) storage, retrieval, and processing in seconds at any location, anytime, and at reasonable costs.

While the advances in technology have been astonishing, particularly in the last 50 years, there are still significant deficiencies. Rocket power for launching space vehicles currently constrains space travel. Traffic jams in large cities extend the working day for many millions of people every workday. We still suffer nearly 45,000 deaths by accident on the highways every year. Communication in many parts of the world is still not available to the public. (Three years ago the author vacationed near a small fishing village in Mexico and learned that the radio-phone at the remote hotel there was the only phone in the entire area.) We are just beginning to explore the different ways we use the massive amount of information that has recently become available to people fortunate enough to own a personal computer and a modem.

Take some time during the summer or during the spring recess and read about the technological developments of the last two centuries. Several key developments with approximate dates are listed below. Think about these inventions, and how they have affected your life.

- Batteries 1800.
- Steam locomotive 1825.
- Electric generator 1831.
- Steamship 1835.
- Electric motor 1835.
- Telegraph 1837
- Baltimore to Wheeling railway 1850.

---

[3] We assume in this listing that housing, food and housewares are available to all from the technological-based industrial complex already existing in developed countries. A significant different listing would be necessary for third-world countries.

- Telephone 1876.
- Internal combustion engine 1876.
- Incandescent lamps 1880.
- Central electric power generation 1882.
- Steam turbines 1889.
- Alternators (ac generators) 1893.
- Radio 1896.
- Automobile 1900.
- Airplane 1903.
- Vacuum tubes 1906.
- Television 1935.
- Jet engines 1940.
- Atomic bomb 1945.
- Nuclear bomb 1950.
- Digital computer 1950.
- Transistor 1953.
- Integrated circuit (IC) 1959.
- Nuclear power 1960.
- Communication satellite 1970.
- Microprocessor 1971.
- Internet 1990.

Obviously, the list above is not complete as many more innovations are important and could have been included. The idea is to show you that the period from 1800 to 1990 was packed full of new technology. This technology markedly impacted the way that we live, work and play.

We can also debate the exact year of the invention. However, the exact year is rarely important since an invention usually evolves into a commercially successful product after several iterations with improvements or modifications by several designers. You should not worry about the exact dates. Instead, you should recognize the continuous progress and observe the shift in emphasis from one period to another. In power, we moved from animals, to steam, to internal combustion (gasoline) and to nuclear energy. Power, from gasoline fueled internal combustion engines, became mobile, and electricity was widely distributed. In transportation, we improved from the horse and buggy and sailing ships to automobiles, airliners and nuclear powered submarines. In communications, we progressed from the pony express to cellular telephones and the Internet. In information, we are evolving from a hard copy library system to a digital multi-media information bank available at any time, to everyone, at nearly zero cost.

Have these advances in technology changed society? I believe that they have had a profound beneficial effect. We produce an overabundance of food with less than 2% of our population working in agriculture. We manufacture most of our consumable products with about 20% of the work force. Our work week is shorter, and we have many more affordable products. Most families own at least two cars and home ownership is at an all time high. We do hear concerns regarding the improvements in our standard of living over the past 20 years, particularly for those folks at the lower end of the income spectrum. Technology has probably not helped this group as much as others. Global trade has permitted many of the semi-skilled tasks to be moved to third world countries where the cost of labor is about 10 to 15 % as much as in the developed countries. Many of the previously well-paid factory jobs in manufacturing in the U.S. have been moved off-shore where labor costs are much lower.

15.5 NEW UNDERSTANDINGS

In the discussion of the historical development of technology, we have been optimistic about the very positive effects of technology on society. In the past, the public accepted, without serious questions, all forms of technology and the risks that were involved. However, the situation has changed to remarkable extent in the past 40 years. We (engineers included) have lost the public trust. An excellent example of this loss of trust pertains to the technology employed in the generation of electricity using nuclear power. Since the Three Mile Island incident in 1979, the general public in the U.S. wants no part of nuclear power. There is a valid public perception that the technological risks associated with nuclear power were significantly understated. John O'Leary, a deputy director for licensing at the Atomic Energy Commission (AEC) that "the frequency of serious and potentially catastrophic nuclear incidents supports the conclusion that sooner or later a major disaster will occur at a major generating facility" [6]. This statement was made several years prior to the accident at Three Mile Island. In spite of this warning of the likelihood of a serious accident, federal, state or local governments have not planned for orderly emergency evacuation of nearby residents or for containment measures to limit the contamination of many of our reactors.

The Chernobyl accident in 1986 proved that O'Leary was correct. The No. 4 reactor at the Chernobyl facility exploded sending a radioactive plume in the air so high that it was detected by an alert nuclear plant operator in Sweden. Many died in this radioactive explosion and many more will die in the future from radiation poisoning. Countries as far away from Chernobyl as Italy were affected by high radiation levels.

The nuclear industry and the government agencies regulating it, world wide, have not been candid with the public. Nor has the public been very interested in intelligent debate resorting instead to large, ugly demonstrations to express their displeasure. We need an accurate and honest assessment of the risks of any nuclear power project by the industry, the regulators and experts representing the public. The risks must be weighed against the benefits to society. Sometime in the future fossil fuels will be come unavailable, scarce or costly. When the fuel crises occurs, we may find that nuclear power is the only viable alternative for stationary power generation.

We are not here to argue the case for or against nuclear power. We use it as an example of the fact that understating or poorly assessing risk brings a severe penalty—loss of public trust followed by a ban on technical development in the subject area.

As engineers, we must significantly broaden our perception regarding technological risk. Today, many engineers dismiss failure due to human, operator or pilot error because they are not machine errors. This attitude is absolutely wrong. The human operator and the machine constitute a system and affect public safety accordingly. The machine and its operator must be included in system design since they constitute an interacting set of weaknesses and capabilities [6].

Another new understanding pertains to geography. Prior to the Second World War (WW2), international trade was very limited. We produced what we consumed, mined our minerals, and pumped our own oil. Geography was very important because it constrained trade by law and by distance. After the war, laws were changed and trade agreements (GATT, NAFTA, etc.) were arranged. International trade was encouraged so much so that the U.S. now consistently imports more goods and materials than it exports. Our current deficit in the trade balance, in excess of one hundred billion dollars, is the equivalent of 2,500,000 high-income jobs at $40,000 a year.

Previously the cost of shipping limited trade between countries separated by large distances. However, following WWII many technological improvements were implemented in the shipping business: we developed the super tanker—greatly improving the efficiency of shipping huge quantities of oil. We also developed very large containerized ocean vessels reducing the cost and time of loading and unloading cargo. Large efficient diesel engines were installed on the ships permitting increased speed and significant reductions in shipping time.

With the time and costs of shipping greatly reduced and the legal barriers to trade removed, we evolved into a global marketplace. Engineers compete to develop world-class products or services. Products designed in the U.S. may be produced anywhere in the world. Or products designed in Europe or Asia may be produced in the U.S. Today, design and production activities are located to minimize costs and to maximize benefits to the customers and/or to the corporate entities.

The third new understanding is in the role of the governments (federal, state, county and city). Prior to WWII, governments were small and taxes were relatively low. The primary role of the federal government was to insure national security. The state governments worried mostly about transportation infrastructure, and the local governments concerned themselves with education. Today, the situation is markedly different. Governmental agencies, at all levels and in response to their interpretation of the law (either old or new), issue regulations which pertain to issues of concern to society including the protection of the environment, safety, health and energy conservation.

We can debate the wisdom of many of the regulations, but that is not the point. The regulations currently exist and new ones will continue to be issued with an alarmingly high frequency. As an engineer, you must be prepared to serve society-at-large as reflected by these governmental regulations. If you question the wisdom of some of the regulations, the best approach is to become involved in the political process so that you may have the opportunity to influence them. Engineers tend to resist constraints because they limit the freedom of their designs and add to the cost of the product. We tend to think of regulations as government imposed constraints, but they are really barriers imposed by society-at-large. Society controls the regulatory process because it has the voting power to change the decisions of the politicians drafting the laws and of the bureaucrats formulating the regulations.

15.6 BUSINESS, CONSUMERS AND SOCIETY

Engineers serve three constituencies—business, the customer and society-at-large. The business leaders (management) look to engineers to develop products, services, and processes that meet global competitive challenges. Consumers seek more convenient, reliable, enhanced, and value-laden products at reduced prices. Society-at-large, through elected politicians and public interest groups, demand action leading to solutions of problems involving safety, health, energy conservation and preserving or improving the environment. It is an overlapping set of constituencies, as shown in Fig. 15.1, because society encompasses business, governments, and consumers.

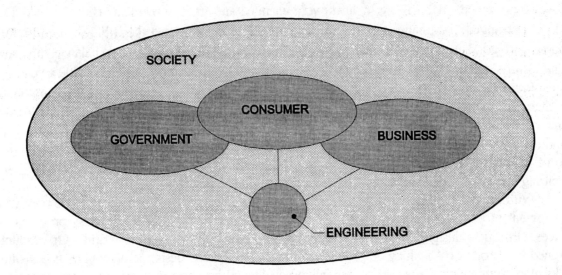

Fig. 15.1 Schematic illustration of the overlapping constituencies served by engineers.

To discuss the issues raised by serving three constituencies, we will use the U.S. automotive industry as an example. Marina Whitman has described many aspects regarding the three-way demands. Because we will draw from her excellent contribution, you may want to study the paper [8] in depth.

Business demands are simple although they are often difficult to meet. Any business must be profitable to stay alive. Losses can occur over the short term, but if the red ink is prolonged the company is forced by its creditors into bankruptcy proceedings. Making a well defined profit every quarter is the usual goal of management. For many years following WWII, it was relatively easy for U.S. businesses to make money. Bombing and shelling during the war had destroyed essentially all the manufacturing capability in every developed country in the world except in the U.S. It was easy to be a leader when the competition was without factories or infrastructure. For the automobile industry, the golden years began to erode in the late 1960s, and survival became a life or death battle in the 1980s. Their problems came from global competition (mostly Japan). The Japanese firms provided much more reliable automobiles at lower costs than those produced in the U.S.

To show the extent of the market penetration by imported automobiles, recall that the imports accounted for only 4% of the market in 1962, but they controlled 30% of the market by 1990. This devastating loss of market share was not limited to the auto industry; several other industries including optics, consumer electronics, ship building, steel, etc. suffered even more serious losses.

Engineers and management in the U.S. firms had lost the competitive edge after decades of modest competition from the Europeans. The Japanese automobile companies (Toyota, Honda, Nissan, etc.) provided more affordable, reliable, appealing, fuel efficient and value-laden cars than either the American or European companies. In recent years, the loss market share has been reversed by improved participative management and technical methods (concurrent design and systems engineering[4]), but regaining the market share of the 1960s will be a very long and difficult process.

The second constituency, the consumers, has been well recognized in the past decade. The importance of keeping the customer pleased is foremost in the minds of both management and engineering. Systematic methods for establishing the consumers needs have been developed and placed into practice [9]. A product development method that begins with the customer interviews and proceeds through all of the phases of product development is currently followed by many companies in the U.S. This procedure is briefly described in another chapter in this text. We do not want to underestimate the importance of the second constituency, but engineers have usually kept the customer in mind during the design of new products. The newer methods [9] tend to insure that the customers' requirements are more accurately defined by the product specification before beginning a development.

The third constituency, society-at-large, generates the most challenging requirements. Society-at-large is represented by government agencies and a variety of public action groups. Officials of the government agencies (bureaucrats) produce regulations that define requirements for automobiles that may affect safety, fuel conservation or the environment. The public action groups create pressure and influence politicians in setting public policy that eventually generates new requirements.

Engineers often have some difficulties with regulations and/or proposals from the public action groups. The laws are written by lawyers, and the regulations are written by public servants. As neither lawyers nor public servants are always technically literate, they sometimes seek technical support prior to drafting binding regulations. At other times, the regulations are placed into effect prior to adequate study by knowledgeable representatives of the public and engineering communities. The difficulty with this procedure is that the demands of the customer, including style, comfort, affordability, reliability, performance, convenience, etc., may be in conflict with societal imposed regulations on safety, fuel efficiency and emissions.

A very recent example illustrates what many believe to be an ill-advised regulation [10]. The National Highway Transportation Safety Administration (NHTSA) mandated requirement for dual airbags in every vehicle licensed in the U.S. Most people consider air bags a valuable safety feature, and many lives (2,500 is the current estimate) have already been saved in serious frontal accidents. However, since 1990 many children have been killed by rapidly inflating airbags. As of November 1997, 55 children under the age of nine have died by rapidly deploying air bags since the use of this safety feature was mandated on the passenger side. More recent estimates[11]—which include small women in addition to children indicate that a total of 99 deaths are attributable to air bags. In addition, NHTSA is investigation another 31 probable air-bag deaths. The number of deaths of women, children and senior citizens continues to increase with time.

The killing of women and children by a safety device is a very serious—**OOPS**. Moreover, the regulation prohibits anyone other than the car's owner to disable even the passenger side air

---

[4]Systems engineering means many different things to different groups. We define it as a cost and time efficient method to specify, design, develop and integrate all of the individual subsystems so that the total system for the product meets the requirements of customers, business and society-at-large.

bag[5]. If you are a concerned parent, you can disable it yourself, but few owners have the required skills. Also, if you ruin the mechanism and/or the bag, the cost of replacement is about $1000. Because of their high costs, air bags have replaced stereo systems as the theft of choice if your car is robbed.

The issue is not whether we should have air bags or not. They have proven to be a very effective safety device for most adults either driving or riding as passengers. The issue is with the rigidity of the federal mandates and the failure of the regulators to:

1. Update the regulations to accommodate different behavior of drivers.
2. Provide more design flexibility in the requirements.
3. Provide a means of permitting the driver to activate or deactivate the passenger side bag.

The air bag regulation was written in 1977 when only 14% of the population even bothered to use their seat belts. By 1994, when the mandate was fully effective, 49 of 50 states were enforcing seat belt laws and seat belt usage had risen to 68%. While seat belts do not eliminate the need for air bags, they do increase the time allowed for the deployment of the air bag. The NHTSA did not change the regulation to accommodate the increased usage of seat belts until early in 1997. Responding to the controversy about the risk of air bags, NHTSA issued a new regulation permitting automakers to install airbags 25% to 30% less powerful than the current bags. The new regulation gives the automakers two years to implement the change. In the meantime, air bags continued to be deployed at a speed of 200 MPH to protect an unbelted adult male. This is great for the big guys; however, for kids or smaller women, a 200 MPH kick in the face is extremely dangerous. When seat belts are used, there is no need for the 200 MPH deployment of the bag because the belt constrains the passenger providing more time for deployment of the bag.

Industry representatives recognized these problems and warned the NHTSA of the dangers to children. Rather than change the regulation to provide more flexibility in the design of the air bag systems, NHTSA drafted a warning label and industry added instructions in the owner's manuals explicitly cautioning against placing a front facing child safety seat in the car's front seat. The manuals indicate that it is safe to place a child in the passenger seat if the child is buckled-in with a seat belt, **BUT** the seat must be in the full rear position. The manual also states that children should never lean over with their faces near the air bag cover while the car is in motion. The warning labels and instructions in the owner's manual are a result of poor design that was forced by an inflexible regulation. We now have in place a safety device (air bag) that is dangerous to small children. Warning labels and instructions in manuals reduce the consequences of litigation (there is another law indicating that the user must be warned of possible dangers), but they are totally inadequate to protect children, too young to read, riding in the passenger seat.

---

[5] A recent ruling from the NHTSA allows auto dealers to install an on-off switch to control the passenger side air bag; however, NHTSA requires an extremely complex process for drivers to obtain a waiver which is necessary before the dealer will perform the necessary modifications to the deployment system. First, you have to obtain a NHTSA request form. Then you must swear—under threat of criminal penalty—that you are too short to be safe with an air bag deployment, or can not always put the children in the back seat, or have one of five NHTSA approved medical conditions. If you option for the medical conditions you must have a letter from a medical doctor confirming the condition. To date only 30,000 people have fought the battle with NHTSA and obtained waivers.

It would be relatively easy to modify the design of the control system to include an on/off switch with a warning light on the control panel of the automobile. With this feature the driver could deactivate the passenger side air bag whenever he or she wished to do so. Another design approach entails a modification of the pneumatic system to reduce the explosive force with which air bags deploy. This modification would probably have taken place several years ago when seat belt usage finally became widespread. However, the federal regulation did not permit modification then, and only recently has it allowed modification after the death of 82 (and counting) helpless women and children.

Today (May 1998), the public is concerned, if not angry. The NHTSA has received many letters from owners requesting the waivers required to have trained mechanics at the dealers to disconnect the dangerous air bags. Many mechanics refuse to disconnect the airbags for fear of future litigation. A recent survey of 700 repair shops indicated that two-thirds refused to install the on-off switches. There appears to be a significant amount of confusion about the remedy.

The big three automakers do not have a ready solution. GM proposes cut-off switches in existing vehicles. Ford is in favor of permitting only certain categories of people to disconnect the system. Chrysler is reconsidering its opposition to cut-off switches. The dealers and independent mechanics do not want to touch the problem because they fear they might be sued sometime in the future. With the new regulation, it appears that all of the automakers will be reducing the power of the deployment of the airbags which should mitigate the problem in time.

The current disarray is alarming. The interaction of the government agencies representing the public-at-large, business and the consumer is woefully inadequate. We need a solution, and we need it quickly. Yet, the regulatory system is too awkward to provide rapid solutions. The awkwardness comes from inflexibility on the part of the government and inadequate cooperation between the automakers. We can understand competition on the part of business to increase market share, but safety features should be excluded from the normal competitive secrecy. The government agencies must become much more flexible and recognize that they cannot decree away all danger.

The difficulties between the government and business are not new. An interesting summary of the regulatory relationship between government and business was recently published by Jasanoff [12]. We quote her directly:

> Studies of public health, safety and environmental regulation published in the 1980s reveal striking differences between American and European practices for managing technological risks. These studies show that U.S. regulators on the whole were quicker to respond to new risks, more aggressive in pursuing old ones, and more concerned with producing technical justifications for their actions than their European counterparts. Regulatory styles, too, diverged sharply somewhere over the Atlantic Ocean. The U.S. processes for making risk decisions impressed all observers as costly, confrontational, litigious, formal, and unusually open to participation. European decision making, despite important differences within and among countries, seemed by comparison almost uniformly cooperative and consensual; informal, cost conscious, and for the most part closed to the public.

The assessment by Sheila Jasanoff gives us clear direction. We must reduce the adversarial relationship between business and government. As engineers working for both government and business, we should be an important element in the future to produce a regulatory system with more flexibility, cooperation, cost consciousness, and consensus.

15.7 CONCLUSIONS

The standard of living in the U.S. will be determined by the interplay of three powerful influences:

- New and rapid technological advances.
- Business (management) response to global consumer demands.
- Social demands as evidenced by legislative and regulatory requirements.

Engineers have always provided the leadership in technological advances, and we expect to continue to provide business and the public-at-large with cutting-edge, world-class technology.

Business requires capital, technology and good management to remain competitive. In addition to providing the technological base for a company, engineers frequently serve in management. If this career path appeals to you, plan on extending your education in a business school. The combination of a Bachelor of Science degree in engineering and a Master of Science in business provides a very solid foundation for a career path in a technically-oriented business.

In the past, engineers, in general, have not been actively engaged in societal issues. We lack patience in dealing with the public and become irritated by their lack of technical literacy. We fume at the inefficiencies of government agencies and their lack of flexibility. We become outraged at the arrogance of bureaucrats and at their lack of concern for time and costs. We retreat into our caves and work on new developments and new products. In the meantime, the public-at-large grows wary of many important sociotechnical systems and the government resorts to litigious processes which impedes real progress.

It is unfortunate that engineers have not been a significant force in dealing with societal issues. As society, business and the customer create conflicting demands, engineers are essential in the crafting of well-balanced solutions. The conflicting demands provide new opportunities and complex challenges for engineers. To take advantage of them, the engineers of tomorrow will require enhanced communication skill, greater disciplinary flexibility, a better understanding of the mechanisms of regulatory agencies and a much wider perspective relative to societal demands.

REFERENCES

1 Singer, C. E. J. Holmyard, A. Hall, and T. Williams, Eds. A History of Technology, Oxford University Press, New York, five volumes, 1954-1958.

2 Hughes, T. P., "From Deterministic Dynamos to Seamless-Web Systems," Engineering as a Social Enterprise, ed. Sladovich, H. E., National Academy Press, Washington, D. C. 1991, pp. 7-25.

3 Sladovich, H. E., ed. Engineering as a Social Enterprise, National Academy Press, Washington, D. C. 1991, pp. 7-25.

4 White, L., Jr., Medieval Technology and Social Change, Oxford: Clarendon Press, 1962

5 Kirby, R. S., S. Withington, A. B. Darling, and F. G. Kilgour, Engineering in History, Dover Publications, New York, 1990.

6 Ford, D. F., Three Mile Island: Thirty Minutes to Meltdown, Viking, New York 1982

7 Adams, R. McC. "Cultural and Sociotechnical Values," Engineering as a Social Enterprise, ed. Sladovich, H. E., National Academy Press, Washington, D. C. 1991, pp. 26-38.

8 Whitman, M. v. N., "Business, Consumers, and Society-at-Large: New Demands and Expectations," ed. Sladovich, H. E., National Academy Press, Washington, D. C. 1991, pp. 41-57.

9 Clausing, D. Total Quality Development: World-Class Concurrent Engineering, ASME Press, New York, 1994.

10 Payne, H. "Misguided Mandate," Scripps Howard New Service, Knoxville News-Sentinel, January 12, 1997, p. F-1.

11 "Regulatory Arrogance Taking a Fatal Toll on Motorists", USA Today, Editorials, May 18, 1998.

12 Jasanoff, S. "American Exceptionalism and the Political Acknowledgment of Risk,' Daedalus, Vol. 119, No. 4, 1990, pp. 61-81.

EXERCISES

1. Recall a product that you or your family has purchased in the past month or so. Write a brief paper describing both the positive and negative impacts of that product on society.

2. Recently, the Federal Aviation Administration enacted a regulation affect the mailing of packages weighing more than 16 ounces. The regulation requires you to present the package to a postal clerk for mailing, and prohibits the mailing of the package from a post box. Write a paper covering the following issues:

    - Describe the regulation in more detail.
    - Why is the FAA writing regulations affecting the U.S. Post Office?
    - What effect (cost) does this regulation have on the individual and on small business.
    - Do you believe that the regulation will be effective for its intended purpose? Please give arguments supporting your viewpoint.
    - What actions will the post offices (40,000 of them) have to take to make the regulation effective?
    - What actions do the post offices actually take with regard to the regulation?
    - Will these actions be costly? Estimate the costs to both the public and the individual. Assume that a postal clerk is paid $10.00/hour and that an individual considers his or her free time worth $10.00/hour.

- Give your assessment of the cost to benefit ratio for this regulation?

3. Suppose that you were a slave in the time of Rameses II and were assigned to the task of constructing his statue. For those not up to date on Egyptian statues, this one weighs a 1000 tons and was 56 feet tall. Write a paper describing your daily tasks.

4. Horses were not used extensively to relieve man from brutal work or to provide a significant advantage in military efforts until nearly 1000 AD. Write a brief paper describing both the social and technical reasons for the very long time required to effectively employ the horse in either military or commercial enterprise.

5. Suppose that you were living on a manor in England in about 1300 AD. Describe your lifestyle and indicate how technology affects your work and play. Select in which of the two classes of society that you existed.

6. Describe where power is available in the U.S. today. Define the type of power in your response. Also include an example where you personally suffered because of a lack of power.

7. Write a paper comparing the lifestyles, as you imagine them, for a man living in 1998 and 1298. Did technology make a difference in the quality of life.

8. Repeat exercise 7 replacing the man with a woman.

9. Why is the public-at-large turned off on power generation with nuclear energy? Is the public-at-large correct in their collective assessment?

10. Why do we have a global marketplace today? Does global trade improve our standard of living? What does our current trade deficit have to do with wages for factory workers? Does technology help or hinder our balance of trade deficit?

11. If you were the director of the National Highway Transportation Safety Administration (NHTSA), what action would you take regarding the public concern about air bags.

- Explain the reasons for your actions.
- Explain how you would convince the big three automakers to agree with you.
- Describe the outline that you would follow in the press conference announcing this action?
- Describe how you would handle the public interest groups that disagree with your ideas.
- Describe how you would handle the public interest groups that agree with you.

12. What non technical courses can you take during your undergraduate program that will broaden your prospective and aid you in dealing with sociotechnical issues?

CHAPTER 16

SAFETY, RISK AND PERFORMANCE

16.1 LEVELS OF RISK

When engineers design a new product or modify an existing product, we often introduce an element of risk in society-at-large. Sometimes the risk is minimal, and the resulting damage from a malfunction or failure is small. However, in some instances, the risk may be large and the damage may be catastrophic.

Some simple examples are in order to illustrate the concept of risk with attendant benefits to society. Suppose we design a new model of an electric, pistol-grip drill that is powered from the standard 120 volt, 60 cycle single-phase power supply from the local utility company. The operator is always at risk for an electric shock. It is our responsibility as engineers to minimize this risk while maintaining the advantage in performance of an electric-powered drill.

16.1.1 Acceptable Risk

How do we protect the operator from electric shock? The operator is holding an electric motor with 120 volts across the armature coils and the field coils adjacent to his or her hands. The answer in this instance is to build the case for the drill from a tough, durable plastic that is structurally strong, but also an excellent electrical insulator. The case keeps the operator from touching any part of the electrical circuit powering the motor. In fact, Black & Decker® a major manufacturer of small power tools, uses double insulation to provide two independent insulation barriers to prevent the operator from making contact with any part of the electrical circuit. Operators may still manage to receive an occasional shock, but it will not be easy. Accidental contact of the operator's hands with a live electrical circuit is a rare event. In fact, the risk of electrical shock is so small that we routinely pick up a power tool and employ it to perform some task without even a passing thought about the possibility of getting shocked. Operating a power tool does involve risk, but it is acceptable because the operator is not consciously concerned about his or her safety while using the tool.

16.1.2 Voluntary Risk

Let's move up the risk ladder and consider flying in a commercial airliner from point A to point B, some 1,200 miles distant. Is it safe for us to make this trip? Most of us appreciate that it is reasonably safe to fly commercial airliners, but realize that there is a slight probability of a crash[1]. The statistics show the probability of a fatality in an airline accident to be about 1 in a billion passenger miles flown. So if you make the round trip from point A to B and then return to A, your probability of getting killed due to an airline accident is:

$$P_k = (2)(1200)/10^9 = 2.4 \times 10^{-6} = 0.0002.4\%$$

This is a very low probability for a fatal accident, and consequently most people do not worry much about the possibility of dying in a crash when they board an airliner. However, if you are a sales engineer flying 100,000 miles a year, every year for 20 years, your total mileage accumulates to 2×10^6 miles. Your probability of getting killed in a crash increases to:

$$P_k = 2 \times 10^6/10^9 = 2 \times 10^{-3} = 0.2\%$$

The probability has increased significantly if you accumulate the miles flown by a traveling professional over a 20 year period. Knowing there is one chance in 500 that you will die in an airline crash, would you still want to be a sales engineer flying weekly to meet with customers? Could you tolerate this level of risk? Would you be apprehensive?

Everyone has some tolerance for risk. We weigh the speed, convenience, cost and risk of flying against that of traveling by train or by car. In almost all instances, travelers select the plane for long distances and the car for short distances. We select the train only in those rare instances when it combines relatively good service (high speed with frequent trains) to convenient (downtown) locations. Most of us do not contemplate the risks involved in travel because, except for driving a car, fatal accidents are rare events.

Unfortunately, fatal accidents in automobiles are not rare events. Each year about 45,000 fatalities occur on U.S. highways, and in addition hundreds of thousands of serious injuries occur. It is clearly more dangerous to drive than to fly, when you compare the probability of fatalities and/or injuries on a per mile basis. How do we handle the higher risk associated with driving? Rationalization for one—I am an excellent driver, and it will not happen to me. Not necessarily a true statement, but the rationalization of one's superior driving skills alleviates the worry and concern about a fatality or injury producing accident. The real risk still remains.

---

[1] The author is probably more sensitive to the risk of commercial air travel than most people. I knew, rather well, two colleagues who were killed in separate crashes—one on the East Coast and the other on the West Coast.

When we drive, we are an active factor in determining the risk. Our driving habits (high speed, reckless steering, tailgating, etc.) and driving skills affect—to a large degree—the level of risk involved. When flying, we are only a participant albeit a passive one. We have voluntarily decided to fly and sometimes even have a choice of airlines. We can pick the airline with the best safety record. (Southwest Airlines is the best with no accidents in their entire 26 year history). However, sometimes, even knowing the facts, we are placed at risk, and we have little or no choice in the matter.

16.1.3 Involuntary Risk

Suppose you live 40 miles northeast of a nuclear power plant. Is 40 miles sufficiently far away to avoid significant fallout from a serious explosion involving the nuclear reactor at the power plant? Remember the winds are usually from the Southwest, so that the fall-out plume will be pointed in your direction in the event of an accidental release of radioactive gasses or particles.

The risk associated with radioactive fall-out, air pollution, toxic chemicals, polluted ground water, etc., is a serious concern to almost everyone; people do not like to be exposed involuntarily. They expect governmental regulators to control the environment and to reduce the level of the risks of accidental exposure. Unfortunately, governmental agencies are not effective in protecting the population from these exposures. Accidents have occurred in the control of reactors, and citations for safety violations by inspectors of the Nuclear Regulatory Commission are commonplace. From a public perception, the risk of a serious malfunction of a nuclear reactor became unacceptably high in the U.S. and many other European countries. After the accident at the Three Mile Island facility in 1979, the commercial nuclear industry died in this country. Some plants that were under construction were completed and others were converted to fossil fuels, but no new nuclear plants have been started since. With the more recent and very serious accident at Chernobyl in 1986, any new plans for power generation in the U.S., using nuclear energy, have been placed on hold for the foreseeable future.

There is a very large difference between the acceptable level of risk depending on whether the risk-producing activity is voluntary or involuntary. The degree to which we personally control the risk is also extremely important. Activities like skiing, scuba diving, horseback riding, hang gliding, mountain climbing, dirt biking, etc. carry extremely high levels of risk. Yet, intelligent people swarm to the ski resorts and pay big bucks to potentially break their bones. Why? They have voluntarily decided that the thrill, or other pleasures, derived from the activity is worth the risk. Also, they control the level of risk. They can chose the "bunny" slope and minimize the risk or the "black diamond" slope to maximize the thrill.

As engineers, we must recognize that some level of risk is involved in almost all the products we produce. It is imperative that we minimize this risk while maintaining an acceptable level of performance. Also, it is also essential that this risk be acceptable to our customers and to society-at-large. An appropriate balance between risk and performance must be achieved. It is important that we cooperate with business and governmental regulatory agencies to provide a realistic assessment of the risk level. Finally, the public should be made aware of the risks involved and should not be surprised by news releases describing the gory details of victims involved in these accidents.

16.2 MINIMIZING THE RISK

An engineer minimizes risk by preventing the failure of each and every component of the system. This is not an easy task as there are several different ways in which components can fail. Parts fail by breaking and aging, by corrosion or fatigue, by overload and burning, etc. In some instances, we can anticipate these failures and replace the parts before they malfunction. We call this controlled replacement of finite life parts—scheduled maintenance. In most cases, we try to design each component so that failure will not occur during the anticipated life of the product.

16.2.1 Safety in Design

A complete treatment of methods to avoid failure would require more than a single chapter or even an entire textbook. However, an analytical procedure for designing tension members will be introduced to provide you with an illustration of a conservative design which insures safety. These tension members will be sized (made sufficiently large) so that they exhibit a safety factor. In other words, they will accommodate a service load higher than anticipated in service over their entire life cycle.

Let's begin by introducing a tension member as shown in Fig. 16.1.

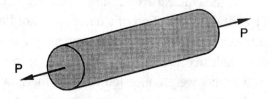

Fig. 16.1 A tension member is a long thin rod subjected to axial load P.

When the rod is subjected to an axial load P, an axial stress σ develops which is uniformly distributed over the cross section of the rod. This stress distributed over a section of the rod is illustrated in Fig. 16.2.

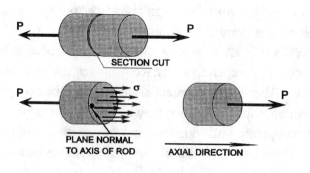

Fig. 16.2 The stress σ is distributed uniformly over the cross-sectional area of the tension rod.

We determine the magnitude of this stress from the formula:

$$\sigma = P/A \qquad\qquad (16.1)$$

where P is the load in lb or Newton
A is the cross sectional area of the rod in in^2 or m^2.

Two simple examples for computing stresses are given below:

If P = 10,000 lb, and the rod has a circular cross section with a radius of 0.25 in., determine the axial stress σ. The cross sectional area is given by:

$$A = \pi r^2 = \pi(.25)^2 = 0.1963 \ in^2$$

From Eq. (16.1), the axial stress σ is:

$$\sigma = P/A = 10,000/0.1963 = 50,930 \ psi$$

where the abbreviation psi stands for the units lb/in^2.

At this stage of the analysis, it is not possible to interpret the significance of our answer of 50,930 psi. We need more information pertaining to the strength of the material from which the tension member is fabricated. When the strength is known, we make a comparison of the magnitude of the stress with this strength and remark on its significance.

In this first example, we have used the U.S. Customary Units expressing the axial load in pounds, the cross sectional area in square inches, and the stress in psi or pounds per square inch. In the next example, we will use the International System of Units (SI) where the load is expressed in Newton, the cross sectional area in meters, and the stress in Pascal.

If the axial load P is 50,000 Newton (N), and the radius of the circular rod is 5 mm, find the axial stress σ. First, determine the cross sectional area A as:

$$A = \pi r^2 = \pi(5 \times 10^{-3})^2 = 78.54 \times 10^{-6} \ m^2$$

Then, from Eq. (16.1), we can determine the axial stress as:

$$\sigma = P/A = 50 \times 10^3 \ N / 78.54 \times 10^{-6} \ m^2 = 636.6 \times 10^6 \ Pa = 636.6 \ MPa$$

where the abbreviation Pa stands for Pascal—the unit used for stress in the SI system.

A Pascal is equal to a Newton per square meter (N/m^2). When calculating stresses in the SI system, we usually determine very large numbers for the stress when it is expressed in Pascal. For this reason, we usually use MPa (mega Pascal) where one MPa = 10^6 Pa = $N/(mm^2)$.

Okay! We can compute the stress on a tension member if we can ascertain its size and the load that it must resist. The stress we determine is a number with associated units—either psi or MPa. As a number, it is not very useful until we compare it to the strength of the material from which the tension member is fabricated.

Let's suppose that the tension rod used in this example is machined from a very strong alloy steel with a yield strength of 100,000 psi. Do you think that the tension rod is safe? Will it retain its shape under load? Will it fail by yielding (stretching under load and not recovering completely when the load is removed)? We answer these questions by a simple comparison of the applied stress σ and the strength of the material S. In this example, the axial tensile stress σ = 50,930 psi and the strength of the material from which the rod was machined is S = 100,000 psi. The comparison shows that the yield strength of the rod is greater than the applied stress; therefore, the rod will not fail by yielding or breaking. But how safe is the rod? We know it will not fail; however, some measure of the safety of the rod needs to be established.

16.2.2 Safety Factor

To respond to the safety issue for the rod, make a ratio of the strength divided by the stress, and define it as the safety factor (SF):

$$SF = S/\sigma \qquad (16.2)$$

In our example, the safety factor is given by Eq. (16.2) as:

$$SF = 100,000/50,930 = 1.96$$

The safety factor (SF = 1.96) indicates that the tension rod is almost twice as strong as it must be to resist yielding under the applied load. We have a comfortable safety factor and can be confident that the rod will perform safely in service. Our only concerns are the accuracy with which the load has been predicted and the quality control for the material used in manufacturing the tension rods. If we:

- have extensive experience with the application;
- certain of the magnitude of the load;
- confidence in our manufacturing division tracking their materials;
- verification of the strength of the materials received from suppliers.

then we usually can be satisfied that a safety factor of about two is adequate. However, if we are not certain about the load and/or the materials, a safety factor of two is not sufficient. We increase the safety factor to accommodate our ignorance of the applied load or our lack of control over the material employed. Safety factors ranging from two to four are commonly employed in design. Safety factors of less than two require considerable care and expense. The use of relatively low safety factors is justified only for very high performance applications. In these special situations, the engineering analysis, prototype testing, quality control inspections, maintenance inspections and documentation need to be extensive.

16.2.3 Margin of Safety

The margin of safety (MS) is sometimes used to describe the degree of safety incorporated into the design of a component. The margin of safety should not be confused with the safety factor as they are different quantities. The margin of safety is defined as:

$$MS = (S - \sigma)/\sigma = SF - 1 \tag{16.3}$$

In our example, the margin of safety MS = 1.96 – 1 = 0.96 or 96%. The tension rod has a margin of safety of 96% which implies that the strength of the material exceeds the applied tensile stress by 96%.

If the load on a tension member can be determined, it is easy to size (adjust the cross sectional area) the rod to provide any margin of safety that you deem necessary to satisfy management and to meet professional obligations to society-at-large.

16.3 FAILURE RATE

Sometimes components fail by burning out, aging or wearing out; not necessarily by breaking or yielding (excessive permanent deformation). In this type of a failure analysis, the safety factor is not relevant. We must cope with wear, aging, or burn out by using other methods. To begin, let's recognize that all components do not wear out at the same time. Some people drive their car 50,000 miles before the brakes wear out and others may drive 60,000 miles or more. Some people can use their personal computers for several years before a component fails; yet others have problems after only a few months in service. To analyze these differences in service before failure, we introduce what is known as a mortality curve for both mechanical and electronic components, as illustrated in Fig. 16.3.

To introduce the concept of failure rate FR, keep records on the performance with time in service of a large number of components that are placed in service. We will begin our record keeping with N_0 components that are placed in service at time t = 0. After some arbitrary time t, some number N_f will have failed and the remainder N_s will have survived. Clearly at any arbitrary time:

$$N_f + N_s = N_0 \tag{16.4}$$

With time, N_f increases and N_s decreases, but the sum remains constant and equal to N_0. The failure rate FR is defined as:

$$FR = N_f / (N_0 \, t) \qquad \text{when } t > 0 \tag{16.5}$$

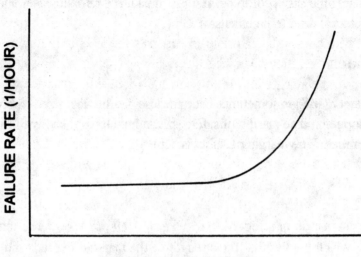

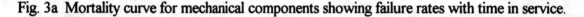

Fig. 3a Mortality curve for mechanical components showing failure rates with time in service.

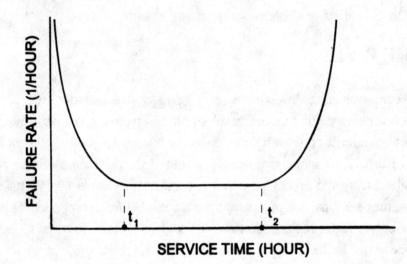
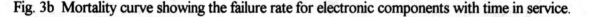

Fig. 3b Mortality curve showing the failure rate for electronic components with time in service.

When we plot the results of Eq. (16.5) with respect to time, the component mortality curves presented in Fig. 16.3 is obtained. For mechanical components (see Fig. 3a), where some degree of testing is performed on the assembly line, the failure rate is small when the component is new. However, the failure rate increases non-linearly with time in service because parts begin to wear, corrode, or they are abused.

The failure rate for electronic components can be characterized by the well known bathtub mortality curve presented in Fig. 16.3b. This curve exhibits three different regions or interest, each due to a different cause. The high failure rates in the first region, $t < t_1$, are due to components manufactured with minute flaws that were not discovered during inspection and testing at the factory. In the center region of the curve, for $t_1 < t < t_2$, the failure rate FR_0 is very low and nearly constant with time. Clearly, this is the best region to operate if we are to maximize the reliability of the product. Near the end of the service life, when t > t_2, the failure rate increases sharply with time as the components begin to fail due to the effects of aging.

For high-reliability systems, it is important to eliminate the inherently defective components associated with early times of operation. For mechanical systems, this is accomplished by inspecting and testing the performance to eliminate flawed components produced in manufacturing. Life is not so easy for the electronic component manufacturers because the feature sizes on the components are so small that inspection requires very high magnification—and even then not all of the flaws can be detected. A better procedure to eliminate the few inherently flawed components is to conduct what is know as burn-in testing prior to incorporating the components into an electronic product. With burn-in testing, all of the components that are produced are operated for a time t_1 (as defined in Fig. 16.3b). The flawed components burn out (fail) and are eliminated. The surviving components will then function with the much lower failure rate FR_0.

Clearly, we want to design components with low failure rates. If the failure rates are low, then the time between failures will be high. In fact, manufacturers sometimes cite what is known as the mean time between failure (MTBF) in their product specifications. The MTBF and the failure rate are related by:

$$MTBF = 1/FR \qquad (16.6)$$

Okay! We understand about failure rates of components and how they vary over the life of a component. How do we use this information about the failure rate to establish the reliability of our components to perform safely over the anticipated life of our product?

16.4 RELIABILITY

16.4.1 Component Reliability

To answer the question posed above, we must introduce the concept of probability or chance of occurrence. In considering probability, let's go back to the test conducted to determine the failure rate and make use of the data collected for N_f, N_s, and N_0. We can then determine the probability of survival P_s as:

$$P_s(t) = N_s(t)/N_0 \qquad (16.7)$$

Then the probability of failure P_f is:

$$P_f(t) = N_f(t)/N_0 \qquad (16.8)$$

Both the probability of survival and the probability of failure are functions of time. With increased time in service, the probability of survival decreases and the probability of failure increases.

From Eqs. (16.7) and (16.8), it is evident that:

$$P_s(t) + P_f(t) = 1 \qquad (16.9)$$

And that:

$$P_s(t) = 1 - [N_f(t)/N_0] \qquad (16.10)$$

If we differentiate Eq. (16.10) with respect to time t and rearrange the results, we obtain:

$$[N_0/N_s(t)][dP_s(t)/dt] = [-1/N_s(t)][dN_f(t)/dt] = -FR(t)dt \qquad (16.11)$$

Note, we consider the term $[1/N_s(t)][dN_f(t)/dt]$ to be the instantaneous failure rate FR(t) associated with a sample size N_s at the same instant.

When we integrate Eq. (16.11) and let FR(t) = FR_0, a constant, we obtain a relation for the probability of survival as a function of the failure rate.

$$P_s(t) = e^{-(FR_0 t)} \qquad (16.12)$$

where e = 2.71828 is the exponential number.

Let's consider another example using Eq. (16.12) to determine the reliability of a mechanical component. Suppose we have determined from our records of the number of failures of a given component over a long period of time. Analysis of the data indicates that our failure rate is essentially constant with 1 failure per 10,000 hours of service. Does that sound good to you? Will the reliability be adequate? Let's determine the probability of survival (the reliability) of our mechanical component with time. Substituting $FR_0 = 1/10,000h$ into Eq. (16.12), gives the results for the reliability shown in Table 16.1.

The reliability varies significantly depending on service life. For a short life, say a year, the reliability is reasonable with 81.9% chance of surviving. However, for long life, say 10 years, the reliability drops to only 13.5%. For very long life 50 to 100 years, failure is almost certain.

While the failure rate of 1 in 10,000 hours looks good in an initial assessment, the reliability which results from this failure rate is disappointing. You cannot be certain that you will have 10,000 hours of service prior to a failure. In fact, you have only a probability of 36.7 % to survive for 10,000 hours. If you want a 90 % reliability for a service life of 10,000 hours, the failure rate FR_0 must decrease to about 1 failure in 100,000 hours.

Table 16.1
Reliability $P_s(t)$ of a Mechanical Component
with Increasing Service Life (FR_0 =1/10,000 h)

Time (1000 h)	Time* (years)	$FR_0(t)$ (unitless)	Reliability $P_s(t)$
1	0.5	0.1	0.905
2	1	0.2	0.819
5	2.5	0.5	0.606
10	5	1.0	0.367
20	10	2.0	0.135
50	25	5.0	6.738×10^{-3}
100	50	10.0	4.540×10^{-5}
200	100	20.0	2.061×10^{-9}

*The conversion from hours to years of service life is based on 2,000 hours/year.

16.4.2 System Reliability

The failure of a component may or may not cause the failure of the system. A system may be comprised of several components and its reliability will depend on the probability of failure of the individual components and their arrangement.   There are two possible arrangements—series or parallel.   Consider your automobile to illustrate both the series and parallel arrangement of components. For lighting the highway during the night, an auto is equipped with two headlights. This is a parallel arrangement because we have two identical components (the headlights) which essentially perform the same function. If one headlight burns-out, you can still drive. Your visibility is impaired to some degree, but the system has not failed.

On the other hand, suppose you want to start your engine. Consider the components that are involved:

- Ignition switch
- Battery
- Solenoid relay
- Starter motor
- Starter motor clutch and gear
- Engine ignition components (spark plugs and points)

If any of these components should fail, the engine will not start when you turn the key. You need a complete series of successful components for the system to function. To start your engine, every component must properly function since the failure of a single component will result in a failure of the entire system.

16.4.2.1 Reliability of Series Connected Systems

Let's consider the reliability of a system involving three components which are in a series arrangement, as shown in Fig. 16.4.

Fig. 16.4 A system comprised of a series arrangement of three components.

Since all three of the components in this series arrangement must operate successfully for the system to perform correctly, the probability of survival of the system (P_s^s) is a product function given by:

$$P_s^s = P_{s1} \, P_{s2} \, P_{s3} \qquad\qquad (16.13)$$

where the superscript s refers to the entire system of components.

If we combine Eqs. (16.12) and (16.13), it is possible to express the system reliability in terms of the failure rate of the individual components as:

$$P_s^s(t) = e^{-(FR_1 + FR_2 + FR_3)t} \qquad\qquad (16.14)$$

As the number of components in a system is increased with a series arrangement, the system reliability is markedly decreased. Also the probability of survival continues to decrease with time in service.

Consider the case where we have n components in a series arrangement where n increases from 1 to 1000. Let's also assume that the reliability of each of the n components is the same ($P_{s1} = P_{s2} = \ldots\ldots = P_{sn}$) at some time during the operating life of the system. The results obtained from Eq. (16.13) are shown in Table 16.2.

Examination of the results presented in Table 16.2 clearly shows the detrimental effect of placing many components in a series arrangement. For components with a high reliability (0.999%), the system reliability drops to about 90% with 100 series arranged components. However, with components with a lower reliability (90%), the system degrades to a reliability of less than 1% when the number of components approaches 50. The lesson here is clear: if you connect a large number of components together in a series arrangement to develop a system, then the individual component reliability must remain extremely high over the entire service life of the system.

Table 16.2
Reliability of a System with n Series Arranged Components
as a Function of Component Reliability P_s

Number of Components, n	$P_s = 0.999$	$P_s = 0.990$	$P_s = 0.900$
1	0.999	0.990	0.900
2	0.998	0.980.	0.810
5	0.995	0.950	0.590
10	0.990	0.904	0.349
20	0.980	0.818	0.122
50	0.950	0.605	*
100	0.905	0.366	*
200	0.819	0.134	*
500	0.606	*	*
1000	0.368	*	*

* System reliability of less than 1%.

16.4.2.2 Reliability of Systems with Parallel Connected Components

Recall the description of a highway lighting system on an automobile which includes a pair of headlights. This system is an example of a parallel arrangement because both headlights must fail before the system fails. We can drive with one light although the State Troopers might warn us of the dangers of doing so. The human body has several parallel systems; we have two hands, two eyes, and two ears, which increase our reliability to function in a normal manner.

A system with a parallel arrangement of three components is represented with the model shown in Fig. 16.5.

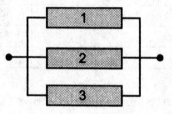

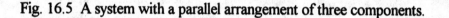

Fig. 16.5 A system with a parallel arrangement of three components.

These parallel systems are redundant because all of the components must fail before the system fails. Accordingly, we can write the relation for the probability of system failure as:

$$P_f^s = P_{f1} \, P_{f2} \, P_{f3} \qquad\qquad (16.14)$$

Substituting Eq. (16.9) into Eq. (16.14) gives:

$$1 - P_s^s = (1 - P_{s1})(1 - P_{s2})(1 - P_{s3}) \qquad\qquad (16.15)$$

Expanding Eq. (16.15) and reducing the resulting expression gives:

$$P_s^s = P_{s1} + P_{s2} + P_{s3} - P_{s1} \, P_{s2} - P_{s2} \, P_{s3} - P_{s1} \, P_{s3} + P_{s1} \, P_{s2} \, P_{s3} \qquad (16.16)$$

Let's examine Eq. (16.16) to determine how redundancy (the number of parallel components) improves reliability. Suppose we consider P_s for all the components to be the same, but we treat P_s as a variable increasing from 0.1 to 1.0. To show the effect of redundancy, we will consider a single component, two components and three components in parallel. The results obtained from Eq. (16.16) are shown in Table 16.3.

Table 16.3
Effect of the Degree of Redundancy on Reliability
$$P_{s1} = P_{s2} = P_{s3}$$

Component Reliability	Single Component	Two Components	Three Components
0.1	0.1	0.19	0.271
0.2	0.2	0.36	0.488
0.3	0.3	0.51	0.657
0.4	0.4	0.64	0.784
0.5	0.5	0.75	0.875
0.6	0.6	0.84	0.936
0.7	0.7	0.90	0.973
0.8	0.8	0.96	0.992
0.9	0.9	0.99	0.999
1.0	1.0	1.00	1.000

An examination of the results presented in Table 16.3 shows that the degree of redundancy improves the reliability of a system with a parallel arrangement of components. If you have a relatively poor component reliability, say 40%, and employ two of these components in a parallel arrangement, the system reliability improves to 64%. Add a third component to the parallel arrangement and the reliability increases further to 78.4%. Continue adding components, and you continue to improve the reliability, although it is a case of diminishing returns.

The improvement in reliability by employing parallel (redundant) components in building a system is a distinct advantage. However, we rarely, if ever, have a free lunch. The corresponding disadvantages are the increased costs and the added power, weight and size of the system. These are significant disadvantages, and redundant design is used only when component reliability is too low for satisfactory system performance or when a failure produces very serious consequences.

For a more complete discussion of component and system reliability, see reference [1].

16.5 EVALUATING THE RISK

When the space shuttle Challenger exploded during launch on January 28, 1986, a Presidential Commission was established to [2]:

1. Review the circumstances surrounding the accident to determine the probable cause or causes of the accident.
2. Develop recommendations for corrective or other actions, based on the Commission's findings and determinations.

Richard Feynman, a Nobel Prize winning physicist from the California Institute of Technology, was appointed to the Commission. Dr. Feynman, near the end of his career and his life, took the appointment very seriously and devoted his entire time and energies to the investigation. One of his many contributions to the review was to determine the probability of failure of the space shuttle system.

During the course of this investigation, Dr. Feynman asked three NASA engineers and their manager to assess the probability of failure P_f for a mission due to the failure of one component— the shuttle's main rocket engine. Secret ballots by the engineers indicated two estimates of $P_f = 1/200$ and one of $P_f = 1/300$. Under pressure, the manager finally estimated the risk at $P_f = 1/100,000$. There was such a large difference between the manager's and the engineers assessment that Dr. Feynman concluded that "NASA exaggerates the reliability of its products to the point of fantasy" [3].

The safety officer for the firing range at Kennedy Space Center, who had been under considerable pressure to remove the distruct charges from the shuttle, did not believe the reliability figures cited by NASA. He had collected data for all of the 2900 previous launches using solid rocket boosters. Of this total, 121 had failed. This data provide a very crude estimate of $P_f = (121)/(2900) = 0.042$ or about one chance for failure in every 24 launches. The safety officer considered this high risk of failure to be an upper bound because improvements made since the early launches improved the reliability of the solid booster motors. Also, the pre-launch inspections on the shuttle were very thorough. His estimate of risk accounting for taking these improvements was $P_f = 1/100$.

Dr. Feynman [3], after extensive interviews with engineers, reliability experts, managers from NASA and many subcontractors, concluded that the shuttle "flies in a relatively unsafe condition with a chance of failure of the order of 1%".

Risk assessment is a difficult task. Each component in a complete system must be evaluated to ascertain its failure rate. Then the components are placed in either a series or parallel arrangement, or some combination of the two, to model the system prior to determining its reliability. The system reliability may be lower or higher than the component reliability depending on the arrangement of the components.

Testing to determine component reliability is possible in some instances where the components are relatively inexpensive. However, establishing reliability estimates by testing requires a very large number of tests which often destroys the component. If you want to show a probability of failure less than $P_f = 1/1000$, you must test considerably more than 1000 components to failure. Obviously, it would not be possible to test more than 1000 booster rockets when the cost of a single test is of the order of several million dollars.

Often the probability of failure of a component must be estimated based on previous experience with similar applications. In some instances, we can compute probability of failure; however, these calculations require considerable knowledge of the spectrum of loading and the ability of the material to resist fracture.

When analytical methods are inadequate and engineering judgment is required to assess the probability of failure, the estimate should be made by a senior engineer with considerable experience and expertise. Even then, the estimate should be pessimistic rather than optimistic. A frank and honest estimate of P_f, based on all of the data and knowledge available, is much better than unrealistic appraisals that give an unwarranted feeling of safety.

16.6 HAZARDS

When we design a product, there are usually risks involved in either the production of the product or in its use in the marketplace. The public-at-large is exposed. What can you do to minimize the risk? We briefly discussed the concept of safety factor, component and system reliability, and evaluating the risk. There is one more topic to consider—developing an ability to recognize the hazards involved. You must clearly recognize the hazards before taking the necessary precautions in your designs to minimize the risks associated with them.

16.6.1 Listing of Hazards

Mowrer [4] has a complete list of hazards and an extended discussion of each. You are encouraged to read Mowrer's chapter in reference [4] and to use the extensive check lists incorporated in his coverage. A very brief excerpt from this material is provided to assist you in recognizing the many hazards that should be considered when in designing a product.

The list of hazards includes:

1. Dangerous chemicals and chemical reactions
2. Exposure to electrical circuits
3. Exposure to high forces or accelerations
4. Explosives and explosive mixtures
5. Fires and excessive temperature
6. Pressure
7. Mechanical hazards
8. Radiation
9. Noise

16.6.2 Dangerous Chemicals

Chemicals can be nasty and you must appreciate the extreme dangers of exposures to certain toxic substances. Do you remember the leak that developed in a storage tank in Bhopal, India, in 1984? The storage tank contained toxic the chemical methyl isocyanate used in the manufacture of pesticides. It was a big leak and about 80,000 pounds of the chemical was released to the environment. Three thousand nearby inhabitants were killed, 10,000 permanently disabled and another 100,000 injured. The problem was not in the design of the tank but in the training of the individuals responsible for the plant maintenance and operation. Nevertheless, a catastrophic accident occurred because several workers and managers, in positions of responsibility, did not understand the dangers of this very toxic chemical.

16.6.3 Exposure to Voltage and Current

What about exposure to electrical circuits? Most everyone has been shocked by the standard 120 volt, 60 cycle electrical power supplied by the local utility company. Why worry? You should worry because electrical shocks are **dangerous**. Yes—even the 120 volt supply can cause big problems. Your body acts like a resistor and limits the current flowing from the voltage source through your hands, arms, legs, etc. The problem is that the resistance of your body is variable. It depends on the moisture on your hands, or the type of soles on your shoes or even the moisture on the ground. If your hands are dry, your shoes have rubbers soles and you are standing on a dry floor when you touch one wire of the circuit with only one hand, then you will probably not feel much because you have arranged a very high resistance path to ground. The current flow through your body will be very small. However, if your hands are wet and you touched both of the wires (the white and the black) from the electrical supply, one with the left hand and the other with the right hand, you have placed 120 volts across your heart. You can be electrocuted quite well with 120 volts under these conditions where the moisture on your hands greatly reduces the effective resistance of the body.

Do not take chances with electricity. Insulate the operator from the circuits preferably with two independent layers of insulation. High voltages are even more serious than low voltages because the currents flowing through one's body increase dramatically. When the current flow through the human body increases to about 10-100 mA (milli-amperes), there are very serious consequences to the respiratory muscles. Higher currents of 75-300 mA produce problems in the operation of your heart. We can increase the electrical current (I) flowing through the body in two ways: first, by increasing the applied voltage (V), and second, by decreasing the resistance (R) offered by the body. The relation among the voltage, current and the resistance is given by Ohm's law as—$V = IR$. Using electrical insulation in the design of products with electrical power, markedly increases the resistance R and decreases the current I to negligible amounts.

Some people think that low voltages (5 or 10 volts) are safe because you barely feel a tingle when you touch a low voltage circuit. However, some low voltage circuits particularly on high performance computers carry substantial (100+ amps) currents. If you short a circuit with high current flows, you will strike an arc and generate significant amounts of heat. Also the flash of the arc can damage your eyes and the heat may produce serious burns. Insulate and shield even low voltage circuits in your designs.

16.6.4 The Effects of High Forces and Accelerations

High forces and high accelerations (or decelerations) go hand in hand. Newton's second law requires the connection ($F = ma$). If we suddenly apply the brakes on an automobile, the car decelerates and the passenger (without seat belt or air bags) is thrown into the windshield. You must always be concerned with acceleration or deceleration because of their effects on the human body. Military pilots are trained to withstand high accelerations (high Gs). A good pilot can pull 7 or maybe 8 Gs before they loss consciousness. However, a civilian will become irritated at less than 2 Gs. If you want to feel an acceleration thrill, go to an amusement park and ride the roller coaster. There is no need to incorporate high accelerations into the design of most new products. A reasonably hot sports car, that accelerates to 60 MPH in six seconds, requires an acceleration of only 0.46 Gs. You should be careful to keep both acceleration and decelerations low when you design products that move and change velocity.

16.6.5 Explosives and Explosive Mixtures

With the bombing of the Federal Building in Oklahoma City, we have all become aware of the disastrous effects of a large explosion. The gas pressures that are generated destroy very substantial buildings, break glass in a very large region and kill or injure many people. We also are aware of the common explosives like dynamite and ANFO (ammonia nitrate and fuel oil). In most designs, we do not encounter a need to accommodate these explosives. What we must be aware of are other less apparent agents that act like explosives under special circumstances. Fuels like natural gas propane, butane, etc. can leak and combine with air to produce an explosion when ignited. Boating accidents are common when gasoline leaks in an engine compartment. The resulting mixture of gasoline fumes and air explode when subjected to a small spark, destroying the boat and killing or injuring the passengers. Still another unusual source of fuel for an explosion is dust. When handling large quantities of a combustible solid, dust (fine particles) is generated. If these particles are suspended in air, the resulting mixture will certainly explode when ignited. A grain elevator exploded in Westwego, Louisiana, in 1977 killing 35 people when an explosive mixture of combustible particles (dust) from the grain and air was ignited. More recently, June 8, 1998 the DeBruce grain elevator in Haysville, Kansas exploded killing several workers.

16.6.6 Fires

Over a million fires occur in the U.S. every year. Some are vehicle fires (400,000) and others are structures (650,000) which begin to burn. Some people die in the fires (4700), and many more are injured (28,700). Clearly, fires are a serious problem. People are killed, injured or traumatized and property is lost (eight or nine billion dollars worth per year). Some of these fires are, simply put, stupid. For example, he or she who smokes in bed may some night fall asleep and set the house on fire. A surprisingly large number of fires (about 100,000 per year) are deliberately set apparently to collect the fire insurance, to take revenge or as a sick kind of diversion.

Engineers have a responsibility to decrease the number of fires which result from the products which we design. Did these fires start with an appliance or a motor that overheated for some reason? Determine the reason for the overheating and redesign the product so that excessive heating will not be generated. Why did the automobile burst into flames? Did a fitting on the gasoline line leak? Redesign to eliminate the fittings or specify a fitting that will not fail under the prevailing conditions. There are many solutions to the problem of fires in the U.S. We, as a nation, are far too casual about fires. We tolerate carelessness in personal practices, poor design in products intended for the home and fraud (arson) to collect fire insurance reimbursement for lost property.

16.6.7 Pressure and Pressure Vessels

We use pressurized fluids for many good reasons, and in most cases, our pressure vessels (the containers that hold the fluids) perform very well. It was not always the case. Back in the winter of 1919, in Boston, a huge tank about 90 feet in diameter and 50 feet high fractured. It contained two million gallons of molasses which flooded the local area. Twelve people died and another 40 were injured in the accident.

It is relatively easy to design a pressure vessel today. In fact, the American Society for Mechanical Engineers (ASME) has developed a code that engineers follow in their design to produce pressure vessels that can be certified as safe for service. However, we still occasionally encounter tanks that fail in service. The difficulty is usually with the steel plates that are welded together to manufacture the tank. Both the plates and the welds have to be tough at low temperatures and the welds must be free of large flaws. If you have the responsibility of designing a pressure vessel, follow the ASME code, make certain that the welding procedures followed in manufacturing produce crack-free welds, and employ a steel which is sufficient strong and tough. If you intend to become a mechanical engineer, you will have the opportunity to learn how to design pressure vessels, select suitable materials for their construction and specify welding procedures in courses presented later in the curriculum.

16.6.8 Mechanical Hazards

Mechanical hazards are features which exist on products which can cause injury to someone nearby. An example might be a fan used to cool the room. Is the fan blade adequately guarded, or can you stick your finger into the rotating blades? The cabinet that you have designed to hold a special tool has a sharp corner at hip level. Can someone walk into the cabinet and break skin on that corner? You have designed a center post crane to lift material and move it over a 25 foot diameter area. Will the center post be stable under all conditions of loading or will it collapse? You have designed a new pizza machine that rolls the dough into sheets exactly 0.120 inches thick. Have you provided protection for the entrance to the rolls that prevent the pizza person from inserting his or her fingers in the machine? You have designed a wonderful guard that prevents a person from exposing their hands and arms in operating a punch press. However, the guard is attached to the machine with two screws. Have you used locknuts and/or safety wires to insure that the screws will not loosen during the operation of the press?

There are many mechanical hazards that we encounter in designing equipment and products. Always examine each component, and look for sharp points, cutting edges, pinch points, rotating parts, etc. Do not count on peoples' good judgment. If they can stick a finger into the machinery, even if it is foolish to do so, you can be certain that sooner or later someone will do so.

16.6.9 Radiation Hazards

Radiation hazards are due to electromagnetic waves to which we are exposed. The damage produced depends on many of different factors such as:

- The strength of the source.
- The degree to which the emitting radiation is focused.
- The distance from the source.

- The shielding between the source and the object being radiated.
- The time of exposure to the radiation.

We have divided the radiation spectrum into several different regions --- very short wave length, visible light, infrared, and microwave radiation. The short wave length radiation is the most dangerous (x-rays, gamma rays, neutrons and etc.) with serious health risks for overexposures. The hazards due to UV, visible light and IR are usually due to excessive exposure where serious burns to the skin can occur. For very intense radiation, even short exposures are detrimental to the eyes. Microwave radiation is absorbed into the body and may result in the heating of one's internal organs. We are not certain how the organs dissipate this heat, nor of the effect of the localized increase in temperature because of this heating. It is prudent to avoid exposure to microwaves. In the U.S., OSHA (Offices of Safety and Health Administration) has issued a regulation limiting the power density of microwave energy to 10 mW/cm^2 for a period of more than 6 minutes.

16.6.10 Noise

Noise levels that occur in the environment may be damaging to our hearing, may interfere with our work or play and may degrade the quality of our life-style. Most of the noise we hear is man made although occasionally we encounter a storm and Mother Nature provides the sounds of thunder and wind. If you listen occasionally to a band playing rock and roll music, there is a temporary shift in the threshold of the hearing level. However, if you play in the rock and roll band almost every night for an extended period, the shift in the threshold hearing level becomes permanent.

Noise is a pressure disturbance that propagates through some medium such as air. The velocity of propagation through air is 344 m/s at room temperature. The pressure disturbance is oscillatory usually with many different frequencies present. The frequency content of the pressure waves depends on the source of the sound. A note from a violin will have much higher frequencies that a note from a tuba. We measure the intensity of the noise with a sound level meter which consists of a microphone, amplifier, and a display meter which provides a reading in decibels (dB).

As a general guideline, the threshold for audibility is less than 25 dB[2] before a person is considered handicapped. In addition, there is a threshold for feeling at about 120 dB and another threshold for pain between 135 and 140 dB. When we design a product the noise level is a serious consideration. Clearly, we must avoid the feeling and pain levels of noise intensity, but what levels are satisfactory. The U.S. Environmental Protection Agency (EPA) has established standards which provide guidance to engineers in the design of products. For example, the results presented in Fig. 16.6 show the relation among the sound pressure level, the communicating distance and the degree of speech intelligibility.

---

[2] What constitutes normal hearing differs from one authority to another. We are citing a reference by the American Academy of Ophthalmology and Otolaryngology.

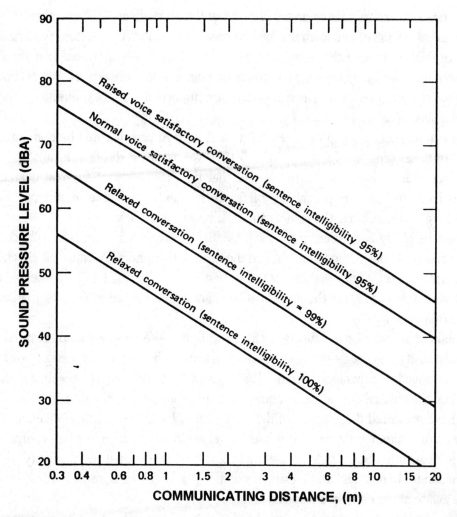

Fig. 16.6 Outdoor distances for intelligible conversation with a background of steady noise [6].

16.7 SUMMARY

When we introduce a product in the marketplace, there is usually some risk involved to the workers making the product, our customers or society-at-large. Hopefully, the risks will be small and acceptable to all concerned because of the significant benefits produced by using the product and/or the sociotechnical system. In determining if the risk is acceptable, the public needs an accurate and honest assessment of the consequences of a failure and the probability of the occurrence of an accident. It is the responsibility of the engineering profession to assist business and governmental agencies in these assessments.

While we must accept risk, we must also make every attempt to minimize it within reasonable constraints on cost and performance. There are many excellent engineering methods for ensuring safe design. We have briefly introduced the concepts of stress, strength, safety factor and margin of safety to illustrate one approach to minimizing risk. We have also recommended a range of commonly accepted safety factors and have provided the issues which should be considered in selecting the safety factor to employ when sizing components.

Sometimes we know that failures will occur and it is necessary to determine the probability of the failure event. To illustrate methods for determining probability of failure, we first introduced the concept of failure rate and the mean time between failures. We indicated that the failure rate could be determined simply by keeping records of the failures of components as a function of time after they were placed in service. These concepts are important, but they should not be confused with the probability of failure or survival.

We next introduced component reliability, and showed how it could be computed from the failure rate. In developing this relation [Eq. (16.12)], we assumed the failure rate was a constant over the service life. It is important to observe that the probability of survival decreases as the service life is increased and the probability of failure increases with time in service. For reliable service for a long period of time, the failure rate must be extremely low.

System reliability is different than component reliability. A system is usually composed of many components. If the components are arranged in series and they must all function for the system to operate correctly, the reliability of the system is given by Eq. 16.13. It is evident from the data presented in Table 16.2 that the number of components arranged in a series markedly lowers the reliability of a system.

Sometimes a system must function all of the time. We simply cannot consider a system failure because of its very negative consequences. In these instances, we design with redundant components arranged in a parallel system. Redundancy improves system reliability, as shown in Table 16.3, but at increased cost and the requirement for more power, weight and size.

We have presented the example of the explosion of the space shuttle Challenger to illustrate the importance of evaluating the risk. It is often a very difficult problem, and an analytical solution for the risk is frequently not possible. Nevertheless, it is a professional responsibility to prepare an intelligent, accurate, honest and frank estimate of the risk and to insure that all of the principals involved are aware of the consequences of a failure.

Finally, we have introduced a very brief listing of hazards which cause injury or death. Unfortunately the list is relatively long. It is important that you recognize the hazards and be vigilant in your designs to avoid them. Avoidance often does not require extensive calculation from elaborate formulae, but rather a detailed assessment of each component in the system and a good measure of common sense.

REFERENCES

1. Dally, J. W., Packaging of Electronic Systems: A Mechanical Engineering Approach, McGraw Hill, New York, 1990, pp. 288-296.
2. Lewis, R. S. Challenger: The Final Voyage, Columbia University Press, New York, 1988.
3. Feynman, R. "Personal Observations on the Reliability of the Shuttle," Appendix to the Presidential Commission's Report, Ayer Co., Salem, 1986.
4. Mowrer, F. W., Introduction to Engineering Design: ENES 100, McGraw Hill, New York, 1996, pp. 149-172.
5. Cunniff, P. F., Environmental Noise Pollution, Wiley, New York, 1977, pp. 101-115.
6. "Information on Levels of Environmental Noise Requisite to Protect health and Public Welfare with an Adequate Margin of Safety," EPA Report, March 1974.

EXERCISES

1. Have you been involved in one or more automobile accidents since you began to drive? Were you or anyone else injured or worse yet killed? Estimate the total mileage you have driven over this period and calculate your accident rate. What were the reasons for the accident or accidents? Comment on your driving behavior and its influence on the accident rate.

2. Are the benefits of driving worth the risks of injury or death? Determine your probability of being killed in a fatal accident while driving this year. Hint: Statistics on fatal accidents are always listed in the World Almanac. If you do not have a recent edition, use an older edition for your data. Make any assumptions necessary for your analysis, but justify each of them.

3. Your co-worker designs a tie rod (a tension member) from a ¾ inch steel bar that has a strength of 60,000 psi. If the team leader has indicated that he or she wants to maintain a safety factor of 2.8, determine the maximum load that can be applied to the tie rod.

4. If a component is designed with a safety factor of 2.6, what is its margin of safety?

5. If we are so smart and we have all of these gee whiz computer programs to determine stresses, why do we need to specify a margin of safety or a safety factor when we design a structural member?

6. Jane is an engineer in the transmission department of the Fink Motor Corp. Her job is to record data on the mileage prior to failure of the new light weight transmission that has been placed in 50,000 of the new 1998 model of the Clunker. She records the following data:

Mileage (1000 miles)	No. Failures, N_f
0 - 5	68
5 - 10	37
10 - 20	20
20 - 30	22
30 - 40	21
40 - 50	35
50 - 60	97
60 - 70	564
70 - 80	1,485
80 - 90	5,677
90 - 100	13,592

Determine the failure rate as a function of service life expressed in terms of mileage, and prepare a graph showing your results.

7. Using the data from the table in Exercise 6, determine the probability of survival and the probability of failure of the transmissions as a function of service life measured in terms of mileage. Prepare a graph of your results.

8. If the transmission in the Clunker was under warranty for 100,000 miles, what would be the consequences for the Fink Motor Corporation?

9. NASA's space shuttle has two solid-propellant, booster-rockets to provide the thrust required for launch. Is this a redundant system? State your reason for this conclusion.

10. If the probability of failure of a solid-propellant, booster-rocket on the space shuttle is 1/500, determine the reliability of the solid booster system.

11. You are designing a very large computer system for containing the data base for a huge reservation system. It is estimated that at any instant 600 operators will be accessing the data base, and another 3400 operators will soon be ready with their requests for computer availability. The main-frame computers that you plan to employ each have a MTBF of 6,000 hours. Present a design of the computer system that will insure that the 600 operators have a 99.9 % probability of being served. In your design show all the calculations, and carefully list your assumptions. Justify the costs involved if each mainframe computer employed in the system is valued at $400,000.

12. Prepare a paper discussing your reaction to the risk assessment of the space shuttle system by NASA. Give arguments for and against NASA's position and practices regarding safety of the shuttle and its crew both before and after the Challenger accident.

13. Describe an incident where you received an electrical shock. Something obviously went wrong to cause this incident. Please indicate the problem. Was there a design deficiency? What could be done to prevent the incident from reoccurring?

14. You are designing a cabinet-mounted, self-cleaning oven that requires very high power levels to heat its interior surfaces until they are free of all the burnt-on splattered grease and grime. What precautions can you take, as the lead engineer on this design team, to insure that the oven will not be the source of a fire during its anticipated 20 year life?

CHAPTER 17

ETHICS, CHARACTER AND ENGINEERING

17.1 INTRODUCTION

This chapter on ethics was written shortly after the 1996 Presidential election. The newspapers and TV news and talk shows covered many ethical issues emanating from both Congress and the White House. Ethical lapses were occurred on nearly a weekly basis keeping the news services busy reporting on questionable behavior of our elected officials. These officials are the people we have entrusted and empowered to write and approve the laws governing the country. The news media has pointed out numerous instances where our leaders have skirted the truth if not the law. Questionable fundraising activities by both parties lead to questionable practices prompted the Governmental Affairs Committee, chaired by Senator Fred Thompson, to request $6.5 million to cover the extensive cost of the investigations. (The Committee was finally authorized to spend $5 million on the investigation and sought additional funds to extend the investigation). How do you feel when learning that $50,000 to $100,000 is the price for an invitation to the White House and an overnight stay in the Lincoln Bedroom?

During the period from the initial writing of this chapter to its revision in April of 1998, the ethical situation has deteriorated. Ethical issues continue to be reported by the news media with alarming frequency. The Secretary of the Interior has been indited. A special prosecutor for the Secretary of the Department of Labor has been appointed. The White House in a state of siege with a small army of attorneys working continuously to mitigate several alleged indiscretions of the leaders of our government.

The lax ethical practices by many of our elected officials have led to a very serious skepticism about the merits of many of the government institutions and agencies. The fact that fewer than 50%

of the eligible voters took time to cast their ballots in the 1996 elections is testimony to the deep skepticism toward our government and our elected officials.

As engineers, we want to preserve, if not enhance, the confidence of the public. To gain public confidence, it is essential that we all behave in an ethical manner every day of every year. Whether we act as individuals or as professionals, consistent ethical behavior is mandatory. Our primary employer, business, must also act in an ethical manner. Since engineering developments are sponsored, financed and controlled by businesses, corporate and engineering behavior are inseparable. Our actions and words always must be above reproach.

17.2 CONFUSION ABOUT ETHICAL BEHAVIOR

Many people are confused about ethical behavior. The results of several polls indicate that Americans are less ethical today than in previous generations. Sixty-four percent of a sample of 5000 individuals admits to lying if it does not cause real damage. An even larger percentage of those sampled (74%) will steal providing the person or business that is being ripped off does not miss it [1]. Many students (75% high school and50% college) admit to cheating on important exams [2]. Most students have not developed a moral code to use as a guide for their behavior. Poor behavior when I was in high school, 1944-47, included making too much noise, chewing gum in class, talking, littering, running in the halls, etc. Today, these relatively minor transgressions have been replaced with more serious problems involving murder, drug and alcohol abuse, pregnancy, suicide, rape, robbery and assault. Extremely negative changes. Why?

There are probably many reasons for the degradation of ethical behavior over the past 50 years. Let's discuss two key reasons. First, institutions (family, church and schools) that once taught ethics are much weaker today than 50 years ago. Many families have been torn apart by an extremely high divorce rate and single parent homes are common. The concept of "love, honor and cherish, in sickness and in health" is often ignored.

The influence of our churches has deteriorated due to serious declines in membership. Unfortunately, church finances often limit their activities to current, not new, membership. Some social problems are with persons who are not well known to the church. The public has also become much more secular with no church affiliation. Our public school administrators are so concerned with possible litigation that they have largely removed the teaching of moral values from their curriculum. The consequence is that students graduating from high school have not been given an adequate opportunity to develop a personal code of ethics. The result of the failure of our institutions to teach ethics is that many of our young people are not aware of what is right and what is wrong. All actions are not gray. Many are plainly right, and many others are obviously wrong.

The second reason for the deterioration in ethics on the part of our younger population is the current state of ethical standards in society. We are exposed to cynical behavior and selfish attitudes on TV, magazines, and newspapers every day. The "bad guy" often walks away from severe crimes with minor penalties. Minor infractions are frequently ignored. Killers and child sex offenders are paroled after remarkably short periods of incarceration. Popular TV programs often show examples of lax ethics and self indulgences to generate a popular but sick kind of humor. The result is to create confusion in our youth as they try to develop some system for judging right from wrong. Our youth mirror the behavior of society leading to cynicism, selfish attitudes, and dishonesty and irresponsible conduct.

17.3 RIGHT, WRONG OR MAYBE

Christina Sommers [2] develops an interesting viewpoint regarding what is considered right, wrong and controversial by society-at-large. She argues that there are some ethical issues that are not clearly defined, and that society has not reached a consensus regarding the correctness of these issues. We can develop cogent arguments for and against a specified viewpoint for controversial issues such as abortion, affirmative action, capital punishment, etc. Sommers refers to these controversial issues as "dilemma" ethics, which should be distinguished from "basic" ethics. With "dilemma" ethics, society-at-large is not certain about what is right, wrong or gray. There are several attitudes prevalent in society, and sometimes the issue is so important to different segments of our population that demonstrations are organized to elevate one view or another.

We will make no attempt in this textbook to resolve any "dilemma" issues. These issues have been unresolved for decades, and they will require much more time and understanding before our society is willing to reach a consensus necessary for resolution. It is much more productive to treat "basic" ethics where right and wrong or white and black are much more clearly recognized by the majority of society.

17.3.1 Right versus Wrong

It is easy to define right and wrong. In fact, the understanding of right and wrong dates back to Aristotle [3] and fundamental doctrine has not changed much since then. Let's start with the four classic virtues [3]:

- Prudence
- Justice
- Fortitude
- Temperance

Prudence refers to careful forethought, good judgment and discretion. In other words, think before you act; exercise care and wisdom. If you act, will your action cause problems, now or later, for you or anyone else? Are you careful about your remarks about others? One of my rules is not to speak ill of others under any circumstances. Is the action that you are about to undertake in your best interest? Will that action be appreciated by your family, friends, manager and peers?

Justice involves fairness, honor, keeping your word, honesty, truthfulness, etc. It is clearly wrong to lie, cheat, steal and break promises. It is also wrong to tolerate those about you who do so[1]. My behavior with regard to justice is governed by the golden rule. Do unto others as you would have them do unto you. It is a very simple rule, and it works.

Fortitude is about courage and persistence. We all know, or imagine, about fortitude with regard to warfare and/or battle. Warriors stand their ground and exhibit a very special brand of courage when their lives are at extreme risk. However, there are other brands of courage which

---

[1] Many people are willing to tolerate about them those who cheat, steal and break promises. They do not believe in imposing ethical constraints on friends or associates. This no tolerance concept is difficult to establish in many honor codes and the author may be in the minority on this issue.

must be exhibited in more typical circumstances. Will you stay with an idea, even if it is not popular, if you know that it is the "right" approach? Showing determination and resolution is sometimes difficult when the risks of failure are high or when peers encourage you to abandon your idea.

Temperance, of course, refers to moderation in drinking alcoholic beverages. However, temperance has much wider implications with regard to ethics. Temperance can also mean control of human passions such as anger, lust, hostility, exasperation, and lechery. With regard to food and beverages, it implies self control and moderation in consumption. Clearly, much is to be gained by avoiding overindulgence in food, drink, sex, and in restraining your emotional extremes.

There are other virtues such as loyalty and obedience that are less commonly discussed these days. With the restructuring and massive downsizing that has taken place in the business world in the past two decades, the concept of company loyalty to the employee and vice versa has been destroyed. Hopefully, loyalty to the family is an invariant, at least in those situations where a family, in the true sense, even exists. The concept of obedience was seriously damaged—if not destroyed—in this country by the Vietnam conflict. The government ordering young people to fight in a war of questionable value was too much for many to endure. Many young men refused to obey the orders to report for duty and fled the country to avoid persecution.

Because of these events (many others can also be added), the virtues of loyalty and obedience have shifted from the "basic" to the "dilemma" category, and arguments can be advanced on both sides of these questions. What do you think of being loyal to an employer who will eliminate your position during a period of **increasing** profits? How do you feel about completing your tax return for the Internal Revenue Service? Talk about straining one's patience!!!

17.3.2 Theological Virtues

In addition to the four basic virtues, prudence, justice, fortitude and temperance, we have three religious (theological) virtues. These are:

- Faith
- Hope
- Charity

The theological virtues are well known, so we will not elaborate on their importance. We will again emphasize that their merits are definite and irrefutable. It is good to be faithful, hopeful and charitable. If you want to be even more complete, add other ethical constants such as humility and respect to the list.

While the emphasis in most college courses on ethics is with social dilemmas where arguments and counter arguments are to be developed, almost all of our individual behavior can be judged by very clear and well understood virtues that are part of "basic" ethics. There is no need to be confused, nor is there any reason to impair your "basic" code of ethics with issues raised in the study of moral relativism.

17.4 LAWS AND ETHICS

The governments (federal, state and local) play a role in ethical behavior, because they generate laws and set public policy. Laws (mostly at the state level) are written to aid people in their pursuit of a morally correct and safe life. The purpose of these laws is to prohibit a well defined set of vices. For example, it is illegal to sell drugs, rob banks, commit burglary, kill or abuse another human, engage in prostitution, or pursue deviant sexual practices (rape, among other offenses).

These laws, and many more like them, serve to create a moral ecology in which we exist [4]. The laws require visible or outward conformity to a publicly-accepted moral code of behavior, but do nothing to inhibit illegal behavior unless it is observed, and then reported and prosecuted. An unscrupulous person can sell drugs if he or she is not observed in the act. Even if they are observed, they can avoid penalty if they are not arrested and fully prosecuted for one reason or another. Even if they are observed, detained, prosecuted, and convicted they may receive a suspended sentence. The law does not insure that infractions **will not** occur; the law simply indicates that certain behavior **may not be tolerated**.

In some instances, laws are not a suitable approach for creating a moral ecology. An example of an unsuitable law in this country was prohibition. When it became evident that the law was causing more problems than it solved, it was repealed. The government wanted to make it illegal to consume alcoholic beverages, but society-at-large wanted to drink. Bootlegging, racketeers, and speakeasies followed bringing crime and violence to the neighborhoods. It soon became apparent that the law was not a suitable approach to solve a temperance problem. In such cases, the government uses public policy to discourage people from pursuing a vice. Policies are adapted that limit (but do not preclude) the consumption of alcoholic beverages. Hours of sale are restricted, licenses are required, taxes are imposed, the number of outlets are constrained, etc.

Public policies are conveyed by means of rules and regulations— not laws. The policies should be crafted to strengthen families, communities and churches by discouraging, but not prohibiting, irresponsible conduct.

17.5 ETHICAL BUSINESS PRACTICES

Corporations are entities that under law are treated as persons. A corporation, acting like a person, has the right to conduct business and the obligation to pay taxes. Just as individuals are judged by their ethical standards, corporations are judged by their ethical conduct [5]. The chief executive officer (CEO) for a corporation sets the moral and ethical tone for the organization. Any large business is organized with several layers of management. If the CEO operates the business with the highest ethical standards, the middle and lower level managers will follow the example set at the highest level in the corporation. However, if lax ethical standards are permitted, a pattern is set and sloppy ethics will become the corporate style and standard.

In an excellent paper, Baker [5] has developed a list of issues that occur daily in a typical corporation. The manner in which a corporation deals with these issues establishes its character. Baker's list is shown below:

1. How people, employees, applicants, customers, shareholders, and the families who live near our facilities are treated.
2. Are we free of systemic, or individual, practices of discrimination?
3. How do we spend our shareholders' money on our expense accounts as an institution and as individuals?
4. What is the level of quality that goes into our products? Do we meet our customers' expectations for quality? Do we meet our own standards for quality?
5. What is our concern for safety, not only for our employees, but also for our customers and our neighbors in communities in which we operate?
6. How "ethically" do we compete?
7. How well do we adhere to the laws of the locales, regions, and nations in which we do business?
8. What is acceptable gift-giving and gift-taking?
9. How honest are our communications to our employees and our public advertising?
10. What are our corporate and personal positions on public policy issues and how do we promote those positions?

This is a long list and I am certain that we could add more items to Baker's catalog of concerns. But adding more items is not as important as recognizing that corporate decisions are necessary on a daily basis by the managers and employees. These day-to-day decisions determine the corporate character. The ethical judgments that serve as the basis for most of the decisions will depend on the individuals' personal values (virtues) and experience. The quality and consistency of the ethics applied in these decisions will markedly affect the success of the corporation, its management and its employees.

17.6 HONOR CODES

We will not lie, steal or cheat, nor tolerate among us anyone who does [6]. This is a fourteen word honor code that cadets learn almost as soon as they step foot onto a military academy or college campus that has strong ties to the military (Virginia Military Institute and the Citadel). Our military academies (Navy, Army and Air Force) are educational institutions that serve to train career officers for the services. Honor and integrity is a very large and important element in their educational program.

Honor codes are much more widespread than just the military institutions. There is a growing trend to introduce honor codes on campuses where they are absent and to strengthen them on campuses where they already exist [7]. An honor code typically forbids lying, cheating and stealing. When an infraction of the code is noted by a peer, the code requires that student to bring the case forward. The case is heard by a student committee. Although faculty members may testify if called upon, they have no control over the proceedings. The penalty imposed on those found guilty of breaking the honor code is determined by the student committees, and university administrators implement their decisions.

The honor codes are very well respected by the student body and fully appreciated by the faculty. Students generally support the honor codes because they also are concerned with cheating by their peers. They want to play the game fairly and on a level playing field. When the honor system is established, the students handle the responsibility for academic integrity with careful consideration. Punishments are handed out only after complete investigations involving representation of both sides of the case. In some instances, lawyers are retained by the individual being judged by the student committee.

The faculty is also very much in favor of the honor system. They never have appreciated proctoring exams. With the increase in cheating over the past 25 years, proctoring has evolved into policing, which is very distasteful. The advantage of the honor code to the faculty is that the unproctored examination transfers the responsibility for policing to the students. With an effective honor system, students often are permitted to take an exam at a time and place of their choosing.

Of course honor codes are not perfect. Some students cheat regardless of the code. The military academies, where the code is the strictest and the most rigorously enforced, have suffered the most notorious lapses in standards. Scandals at the U.S. Naval Academy have been the most prevalent and widespread in recent years with as many as 133 midshipmen recently involved in cheating on an electrical engineering exam. To make matters worse, the administrators at the academy appear to have interfered with the system, and in doing so, lost the trust of the midshipmen.

The lesson from the academies with regard to the effectiveness of the honor codes is that rigid codes and rules reduce cheating. However, if the students see a way of beating this very rigid system, they will occasionally take advantage of the opportunity. Honor codes are most effective when they produce an ambience of trust between the students, faculty, and the administration. Cheating, lying, and stealing will generally be reduced. For honest students, the scoring on exams will be fairer. For faculty the distasteful task of policing is eliminated. The feeling of trust between the faculty and the students is enabling. With time and prolonged success, the honor code and the trust that it engenders becomes an essential part of the educational process.

17.7 CHARACTER

I had the privilege of teaching for a year (1995-96) at the U.S. Air Force Academy. It was a wonderful experience for me for many reasons—both personal and professional. One of the most important reasons was the opportunity to participate in a well-planned approach to incorporate character building into the educational process. Character was a key educational outcome in every class that we taught at the Academy.

It is not possible to transplant character into an individual, and it is not possible to accomplish much in efforts to explicitly "teach" character. Character comes implicitly as we as individuals develop a personal set of ethical responsibilities. In a book like this one, we can list the attributes that form the foundation for a person's character. Let's consider some of these attributes:

- **Be committed to excellence**. This attribute simply means to do the best that you can do in both your personal and professional endeavors. To make this commitment, you have to assess your capabilities. How good are your skills? How determined are you in achieving your goals? How much time can you commit without endangering your health? If you know yourself, you can measure your achievements on a realistic scale. Use the full measure of that scale.

- **Respect the dignity of everyone**. Respect is the foundation for all achievements. We work, live and play in a diverse world. We must appreciate everyone that we encounter in life. Race, gender, ethnicity and religion are not criteria for judgment. Recognize the potential of those with whom you work and study. Support and encourage them and carefully avoid demeaning criticism. Teamwork, so essential in today's workplace, requires that you accept the differences inevitable in our diverse population and that you fully embrace fellow team members.

- **Integrity**. A single word describes a vitally important attribute of character. Integrity means that you decide to do the right thing solely because it is the right thing to do. You often face decisions that test your integrity. Do you instinctively make the right (honorable) decision? For example, as you parked your car last night, your bumper scraped the fender of the adjacent car. No one observed your accident. Do you leave a note with your name, address, and phone number? Or do you split and drive to another nearby parking lot? How many parking lot scrapes can you find on your car? Did you find a note from the party inflicting the damage claiming responsibility? A person with integrity will consistently make the correct decision, not because it is right for their personal benefit, but because it is clearly the "right" course of action. They "walk the talk."

- **Be decisive**. When you encounter a problem in engineering or your personal life, there is an information gathering period followed by a decision. Some folks do not want to make that decision. They prolong the information gathering period; they procrastinate; they hum and haw; they pass the buck. When you establish the facts, evaluate them, and make a timely decision. It will be your decision if you made it in isolation. If you made it within the framework of a team, it should represent a consensus of the members of the team.

- **Take full responsibility**. Decisions are made and actions follow. Often your decisions produce a winner, but sometimes they result in a loser. If the outcome was a loser and you participated in the decision process, it is your responsibility. Step up, accept that responsibility and lead the effort to fix the problem. Never—under any circumstances—begin to participate in a finger-pointing exercise. Finger pointing does not resolve problems; it exacerbates them.

- **Be temperate**. Self-discipline is an essential part of character. Eat, drink and be merry in moderation. Overindulgence is disgusting. No one appreciates a drunk or a glutton. Self-discipline insures control of the human passions, sensual pleasures, anger, rage and frustration. With self-control, it is possible to attain consistently high levels of achievement.

- **Exhibit fortitude**. On many occasions, we encounter significant difficulties in completing the task at hand. It is easy to give up and divert our attention to more pleasurable undertakings. We can quickly forget the missing assignment, the incomplete solution, the late review, or the unfinished drawing. Who is to know? Stamina, mental toughness and discipline are attributes that keep us on task. Stay with the job, and not only complete the work, but do it well and with dispatch.

- **Understand the significance of spiritual values**. Many, but not all, of us have a faith. We endorse a set of theological beliefs. Your beliefs and your church may be different than mine. Nevertheless, it is essential that I respect your convictions and you should respect mine. We all need to be sensitive to the important role that religion occupies in the mental comfort of many people. We must accommodate a diversity of beliefs by supporting the right of an individual to choose his or her faith and to pursue the ceremonies offered by the church representing this faith.

17.8 ETHICS OF ENGINEERS

We have devoted most of this chapter to a discussion of issues pertaining to individual behavior, ethics, virtues, morals, and character. A personal code of ethics is the cornerstone to achieving a meaningful sense of moral values. Professional ethics are also important, but they must begin only after a person has developed a well-understood personal code of ethics. Hopefully, the preceding sections of this chapter will be helpful in any attempt that you make to establish a sense of right and wrong.

There are several codes of ethics for engineers. Most of the founding societies of engineering have developed and distributed codes, guidelines for professional conduct and faith statements. The codes of ethics for the different engineering societies are all similar. The fact that each society has their own code, is more to insure a complete distribution of the code than to pursue unique ethical issues.

Let's consider the Code of Ethics of Engineers, presented in Fig. 17.1, that is sponsored by the Accreditation Board for Engineering and Technology (ABET). The code is divided into two parts. The first part deals with four fundamental principles, and the second part contains seven fundamental canons. This code was first advanced by the Engineers' Council for Professional Development in 1977. The fact that the code remains without modification for more than 20 years indicates that professional ethical values are as constant as personal ethical values.

The fundamental principles seek to insure that the engineer will uphold and advance the integrity, honor and dignity of the profession. These goals are common to our personal goals that were discussed in a previous section; however, the approach to achieve the professional goals is different as indicated below:

1. We are selective in the use of our knowledge and skills so as to ensure that our work is of benefit to society.
2. We are honest, impartial and serve our constituents with fidelity.
3. We work hard to improve the profession.
4. We support the professional organizations in our engineering discipline.

The seven fundamental canons in the ABET Code of Ethics of Engineers, presented in Fig. 17.1, are explained in considerable detail in guidelines that are used to expand and clarify the relatively brief statements that represent the "regulations" that shape our professional behavior. With the permission of ABET, we have reproduced the complete guidelines in Appendix B.

Accreditation Board for Engineering and Technology*

CODE OF ETHICS OF ENGINEERS

THE FUNDAMENTAL PRINCIPLES

Engineers uphold and advance the integrity, honor and dignity of the engineering profession by:

I. using their knowledge and skill for the enhancement of human welfare;

II. being honest and impartial, and serving with fidelity the public, their employers and clients;

III. striving to increase the competence and prestige of the engineering profession; and

IV. supporting the professional and technical societies of their disciplines.

THE FUNDAMENTAL CANONS

1. Engineers shall hold paramount the safety, health and welfare of the public in the performance of their professional duties.

2. Engineers shall perform services only in the areas of their competence.

3. Engineers shall issue public statements only in an objective and truthful manner.

4. Engineers shall act in professional matters for each employer or client as faithful agents or trustees, and shall avoid conflicts of interest.

5. Engineers shall build their professional reputation on the merit of their services and shall not compete unfairly with others.

6. Engineers shall act in such a manner as to uphold and enhance the honor, integrity and dignity of the profession.

7. Engineers shall continue their professional development throughout their careers and shall provide opportunities for the professional development of those engineers under their supervision.

111 Market Place, Suite 1050, Baltimore, MD 21202-4012

*Formerly Engineers' Council for Professional Development. (Approved by the ECPD Board of Directors, October 5, 1977)

AB-54 2/85

Fig. 17.1 Code of ethics recommended by ABET.

17.9 ETHICS IN LARGE ENGINEERING SYSTEMS

Engineers design and build many different products each year that are included in large and complex sociotechnical systems. Usually these products and the systems are conservatively designed with adequate safety factors, carefully tested, and perform well in service for extended periods of time. They provide a much needed service without endangering either individual or public safety. However, from time to time mistakes are made in the initial design. These mistakes are usually detected in prototype testing and eliminated prior to releasing the product to the market place. In very rare circumstances, a mistake or several mistakes are overlooked, for a variety of reasons, and the system is released and placed in service with an unacceptably high probability for failure. However, the result may lead to a catastrophic accident. The space shuttle system clearly falls into this category. If you are a space supporter, you may argue that the risks are worth the benefits. But if this is the case, unprepared and untrained high school teachers should not be invited to ride along to enhance the image of the space program. It is one thing to order a career astronaut into peril, but totally a different proposition to invite an uninformed civilian to participate in a very dangerous project.

We will be discussing, in considerable detail, the probability of failure (or an accident) in another chapter. Here you need to understand that there is always some risk of failure when designing high performance systems. We must accept a trade-off between risk, safety and performance. In most complex sociotechnical systems (air transportation for example), there is a small but finite risk for failure with a subsequent loss of life and property. This risk must be known to the public. It is the responsibility of the engineering community, industry and the government to alert the potential customers to the dangers involved. Knowing the risk, you can decide whether or not you want to use the system. Some people worry more than others and place a very high value on safety. They require a probability of failure of nearly zero. Do you know someone who will not travel by flying? (John Madden, the popular football announcer for the Fox network, travels from game to game each week on a special bus). Other folks love the thrill of a risk, and they are willing to accept a much higher probability of an accident. They think hang-gliding is great.

17.9.1 The Challenger Accident—a Case Study

With this background on risk, safety and performance, let's begin our discussion of the Challenger accident [9]. The Challenger was one of the original four orbiters that were built by National Air and Space Administration (NASA) to serve the space shuttle system. The rocket fuel supply (liquid hydrogen) on the Challenger 51-L mission exploded 73 seconds after launch on Tuesday, January 28, 1986. The crew of six and a civilian passenger were killed, and the space shuttle was lost.

The Challenger accident has been selected as a case study because it illustrates several ethical issues in the engineering and management of large, complex and inherently dangerous systems: Some issues that will be raised are:

1. The design of the shuttle and the selection of the contractors involved a lot of political considerations [9].

2. The lack of communications between people and organizations was a significant factor in the accident.

3. The interface between the upper-level administrators (business managers) and the engineers was an important element in the decision to launch on that disastrous morning.

4. Public attention and opinion, not safety, affected the decision-making process.

5. The risk potential was not known by the public and not appreciated by key administrators directly involved in the launch decision.

6. Christa McAuliffe, a high school teacher and mother of two children, was killed in the accident. Why?

7. Dr. Judith Resnik, a career astronaut and an electrical engineer, was also killed. She is very special to us because she earned her Ph. D. from the University of Maryland at College Park[2].

17.9.2 Background Information

To set the stage for the accident, we have to go back to the early 1970s. NASA had been very successful with the Apollo missions (in spite of Apollo 13), and looked forward to larger and more aggressive space endeavors. They proposed an integrated space system that would include a space station, space shuttle, space tug and manned bases on both Mars and the moon. Sounds great until you look at the price tag. The public did a quick look and wanted no part of it. A poll indicated that the public believed that Apollo had been too costly. The politicians, always driven by the polls, took note and reduced NASA's budget. NASA recognized the need for a new, cost-effective project to follow Apollo that the public (and politicians) would buy. Responding to these political pressures, NASA proposed the space shuttle which would serve the military, the scientific community and the rapidly growing commercial business of placing satellites in orbit. The space shuttle system, illustrated in Fig. 17.2, was marketed as a relatively routine space transport system. Even the name "shuttle" connected with the airline shuttle services that routinely fly every hour on the hour from one large city to another.

To make the space shuttle system a commercial success, NASA proposed a fleet of four orbiters [10] which would eventually fly on a weekly basis (NASA initially set a goal of 160 hours for the turnaround time for an orbiter.) They planned on nearly 600 flights in the period from 1980 to 1991. The early estimate of the cost of a launch was $28 million with a payload delivery cost of $100 to $270 per pound. After some operational experience, the cost estimates proved to be much too optimistic. Launches actually cost on the order of $ 280 million, and the cost to place a pound of payload in orbit on the shuttle was in excess of $5,200 [11]. Early experience with the space shuttle system indicated that NASA could not hold their schedules and their costs were running more than a factor of 10 higher than the original estimates. Not a very good report card, and NASA was struggling to improve its image as 1985 ended.

Prior to the launch of the Challenger on January 28, 1986, twenty-four shuttle flights had been made. These flights were all successful in that they returned to earth with all crew members

---

[2] The author served on the faculty at the University of Maryland at College Park for more than 20 years. Dr. Judith Resnick was an outstanding role model for our program in engineering until her untimely death.

safe. They also showed the operational capabilities of the shuttle system in placing commercial satellites into orbit, repairing satellites in space and salvaging malfunctioning satellites. However, these flights also showed that the shuttle system had many serious problems. The three main liquid rocket engines were too fragile with many critical components. Some of the tiles in the heat shield needed repair and/or replacement after every flight. The computers and the inertial navigation systems experienced occasional failures. The brakes and landing gear were stressed to the limit when the orbiter (an 80 ton dead stick glider) landed at speeds ranging from 195 to 240 MPH. And finally, the seals in the solid fuel booster rockets showed distress (sometimes extensive) in 12 of the 24 previous launches.

The record of the shuttle during the 1981-1985 period showed a consecutive series of successful launches, but with many prolonged delays to repair failing components in a very large, highly stressed system. The maintenance records showed so many problems that NASA estimated it took three man-years of work preparing for a launch for every minute of mission flight time. Clearly, the word shuttle to describe such a transportation system is a misnomer.

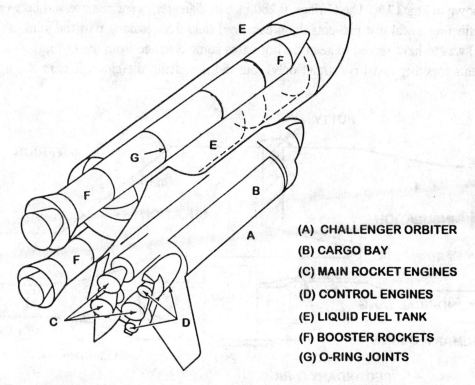

(A) CHALLENGER ORBITER
(B) CARGO BAY
(C) MAIN ROCKET ENGINES
(D) CONTROL ENGINES
(E) LIQUID FUEL TANK
(F) BOOSTER ROCKETS
(G) O-RING JOINTS

Fig. 17.2 Illustration showing the main components on the NASA space shuttle.

17.9.3 The Solid Propellant Booster Rockets

The explosion of the main fuel (liquid hydrogen) tank on the Challenger was due to the failure of the O-ring seals on the solid fuel boosters which were adjacent to the hydrogen fuel tank. To understand the seals and their purpose, it is essential that we appreciate the size and function of the two booster rockets used to provide much of the thrust necessary for the launch. They are enormous cylinders—12 feet in diameter and 149 feet tall. Each cylinder is filled with 500 tons of a solid propellant, which consists of a rubber mixture with aluminum powder and the oxidizer ammonium perchlorate. When ignited, the propellant burns to produce an internal pressure of about 450 psi (lbs/in^2) at a temperature of about 6000 °F. The expanding gasses exit the rocket motor case through a nozzle, to produce about 2.6 million pounds of thrust from each booster. These booster rockets are essentially very big Roman candles!

Big was the problem. The solid rocket boosters were made by Morton Thiokol in Utah, but launched from the Kennedy Space Center in Florida. They were much too long to ship across the country as a single cylinder. To circumvent this problem, the motor casing was fabricated in segments—each 27 feet long. The segments were filled with propellant and shipped by train and assembled in a special facility near the launch pad. This procedure solved the shipping problem. However, it created another. The rocket casing is a pressure vessel that must contain very hot gasses at a pressure of about 450 psi. The joints, where the cylindrical segments of the motor case were fitted together, needed to be sealed so that these hot gasses would not leak and cause damage to adjacent components of the launch vehicle.

The seal was made with a pair of rubber O-rings fitted over one finger of a clevis type of joint as shown in Fig. 17.3. The O-rings, 0.280 inch in diameter, were compressed between the two surfaces effecting a seal that prevents the pressurized fluid from leaking past the joint. A putty like compound was to have served to keep the hot gases some distance from the O-rings. The pins lock the segments together (axial constraint only), but did not clamp the clevis fingers about the center finger.

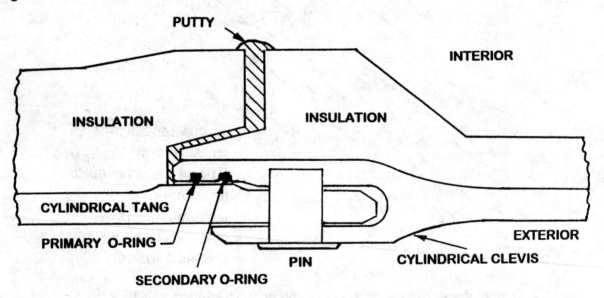

Fig. 17.3 Design of the O-ring seal used at the circumferential joints of the solid rocket motor cases.

The sealing of the segmented cylinders with the O-rings was not a new concept. A single O-ring had been employed previously on the Titan III rocket effectively sealing the circumferential joints on a smaller diameter motor case. Morton Thiokol, the contractor building the boosters, introduced the second O-ring to provide a margin of safety in the event the first O-ring failed. It all sounded good and the initial test firings of the booster in Utah were apparently successful.

Unfortunately, one test does not validate the safety of even a simple system. To insure high reliability of a component, many tests are required. Moreover, this test firing had little or no bearing on problems associated with the reuse of the solid rocket boosters (up to 20 times). After splash down and recovery from the Atlantic, the cylinders were shipped back and forth between Florida and Utah.

17.9.4 Failure of the O-ring Seal

Photographs taken at the final launch of the Challenger indicated that the O-ring seals failed almost immediately after ignition of the booster. The hot gasses cut through the O-rings and the joint of the booster. The hole spread and a flaming jet of white hot gasses escaping from the booster rocket cut through the wall of the adjacent tank containing the liquid hydrogen. The launch was effectively over. While many things happened in the last five seconds before the big explosion— all bad—the actual penetration of the adjacent tank which contained liquid oxygen and liquid hydrogen spelled the disastrous end of the mission.

It was a terrible day, and we as a nation were in shock as we mourned the loss of the crew and the schoolteacher/mother. It was sometime later, when the facts were brought to the public about the space shuttle program, that we learned the shuttle should not have been permitted to fly that day. The very high probability of failure of the O-ring seals was predictable. Moreover, key engineers at Morton Thiokol tried without success to prevent the launch.

The story containing all of the facts about the failure of the O-ring seal is too long to be covered here, but you are encouraged to read references 9 – 14, where very complete and well written accounts are given. We describe here the essential elements leading to the catastrophic failure.

1. The circumferential joint changed shape when the motor case was pressurized, and the gap which the O-rings filled increased markedly in size. A schematic illustration of the new gap geometry is presented in Fig. 17.4.

2. The new gap opening was so large that the back-up O-ring probably could not seal the joint. When the booster case was pressurized, the seal depended on a single O-ring—not two.

3. The hot exhaust gases had eroded the O-rings on 12 of the previous 24 launches indicating some leakage about half of the time, and on a few occasions, considerable erosion of the rings. There was also clear evidence that the putty failed to keep the hot gasses from attacking the rubber O-rings.

4. The joints moved during the early launch sequence as the orbiter engines and then the booster engines were ignited. This motion required the O-ring seals to be flexible and to reseat and reseal continuously during these movements.

5. The temperature the night before the launch dropped to 22°F and had only increased to about 28°F by the time of the launch.

6. The O-ring seals were not certified to operate below a temperature of 53°F by the contractor responsible for the solid rocket boosters.

7. Tests conducted by the contractor Morton Thiokol in 1985, six months before the accident, indicated that the O-ring seal was not effective at low temperatures. (At 50°F the O-rings would not expand and follow the movement in the joint during the period of operation of the booster.) In fact, at a normal temperature of 75°F, it required 2.4 seconds for the O-rings to reseat and effect the seal when the gap was opened.

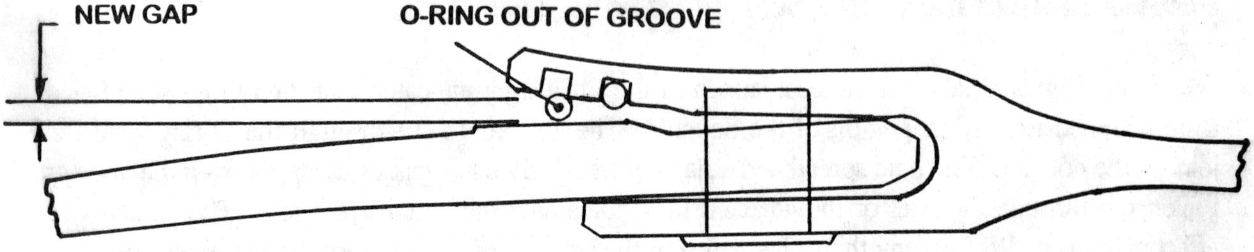

NEW GAP   **O-RING OUT OF GROOVE**

Fig. 17.4 Rotation of the joint due to cylinder pressurization opens the gap in the seal region.

The seven facts listed above show very clear evidence of two problems. First, the design of the seal in the circumferential joint was marginal and should have been fixed much earlier in the program. You do not look at a piece of burnt and eroded O-ring and walk away from the problem.

Second, the launch should have been postponed due to cold weather for several different reasons. The very low temperatures were well below the limits to which the system had been certified. The rubber in the O-rings becomes very stiff at these low temperatures and cannot respond quickly enough to accommodate joint movement. The O-rings were almost guaranteed to leak. Also, there was considerable ice on the rocket motors and the fuel tank. Pieces of this ice might damaged the rocket motors and the tiles on the heat shield of the orbiter when it fell off during the violent vibrations which always occur just after ignition but before lift-off.

17.9.5 Ignoring the Problem

Problems usually do not go away when we ignore them. They persist and sooner or later failure will occur. Although the seals had failed on 12 of the previous 24 flights, the failures were not catastrophic because the leaks were small and at locations which did not endanger the adjacent components. The boosters on the shuttle only operated for only a few minutes before they were cut loose to fall into the Atlantic for recovery at a later time. So the small leaks for short periods of time were tolerated by those concerned with the launch schedule. The trouble with the leak on the Challenger was that it was big, and the jet of hot gases issuing from the leak were pointed directly at the liquid hydrogen fuel tank. The jet of hot gases acted just like a cutting torch, and the huge fuel tank was penetrated within seconds.

The O-ring seals were inadequate and the joint needed to be redesigned and modified. In fact, during the development of the boosters in 1977 - 79, several memos were written by engineers at the Marshall Space Flight Center indicating their concern over the adequacy of the joint. They recommended that the joint be redesigned because "the adequacy of the clevis joint was completely unacceptable." The suppliers of the O-rings indicated "that the O-ring was being required to perform beyond its intended design and that a different type of seal should be considered."

Apparently, there was a major breakdown in communication, and none of these memos and reports from the Marshall Space Flight Center was forwarded to Morton Thiokol, the designers and builders of the boosters. A problem, properly detected by engineering at the NASA Center in charge of monitoring the technical aspects of the development of the boosters, was not pursued to its logical conclusion. Instead of insisting on a redesign of the joint, Marshall Space Flight Center approved the boosters in September of 1980. The pressures of cost overruns and repeated schedule slippage sometimes make managers (and engineers) accept flawed designs. It is very short sighted and not in the best interest of safe design to do so.

Okay, let's move on with the discussion. NASA accepted a booster rocket with inadequate seals, but the bird flew. On the first launch, with Columbia, there was no reported damage to the seals. The O-ring seals failed on the second flight. One of the O-rings was burnt with about 20% of the thickness of the ring vaporized. The ring failure did not affect the mission, but it was clear that the putty was not protecting the rings from the hot gasses. Marshall Space Flight Center reacted to the field experience by reclassifying the joint to a "Criticality 1 category." This was the official recognition on the part of NASA that a seal failure could result in "loss of mission, vehicle and crew due to metal erosion, burn through and probable case burst resulting in fire and deflagration." The failure of the O-rings by erosion continued with high frequency (50% of the missions), but NASA decided to accept the situation and did not recommend remedial action.

It appears that NASA decided to live with the seal problem at least until a new lighter booster motor could be designed with improved joints. These new boosters were to be ready about six months after the Challenger exploded. A very real lesson—too-little-too-late.

17.9.6 Recognizing the Influence of Temperature

From the initial development phase of the boosters, the O-ring seals had been recognized as a problem. A problem that NASA classified as very serious, but one that they must have erroneously believed did not elevate the risk to unacceptable limits. In January of 1985, a year before the accident, the Challenger was launched on a cold day at a temperature of 51°F. An examination of the joints after the recovery of the boosters showed that a number of O-rings were severely damaged and evidence of extensive blow by was observed. The O-ring seal problems were reviewed by the contractor, Morton Thiokol, and the monitor, Marshall Space Flight Center. During this review, Morton Thiokol noted that low temperatures were a contributing factor to the inability of the O-rings to seal properly. The condition was considered undesirable, but was acceptable. However, the O-ring problems persisted and NASA eventually placed a launch constraint on the space shuttle.

The launch constraint would prohibit launches until the problem had been resolved or reviewed in detail prior to each launch. Unfortunately, NASA routinely provided launch waivers for each subsequent flight to avoid delays in an unrealistic launching schedule. The problem of O-ring erosion was common, one or more joints on 50% of the launches exhibited significant erosion, and it had not caused any serious difficulty to date. It was expected and accepted.

17.9.7 A Management Decision

The very cold weather on the night before the fatal launch was no surprise. The weatherman (woman) was on the mark in predicting the temperature and the ice which formed on the structures that night. The engineers at Morton Thiokol, in Utah, were very concerned about the effects of the very low temperatures on the ability of the O-rings to function properly. Teleconferences took place between the Thiokol Moron engineers and the NASA program managers. The engineers argued to postpone the launch until the temperature increased to at least 53°F.

The NASA program managers were very unhappy about the engineer's concerns. They did not want the extended delay required to wait for warmer weather. Senior vice presidents from Morton Thiokol sensed the displeasure from the customer and took over the decision process. Senior management decided to keep the customer happy and reversed the decision of the Vice President of Engineering not to launch at these very low temperatures. Senior management at Morton Thiokol signed off on the launch ignoring the fact that the temperature at launch time was expected to be 25°F lower than the lowest temperature which engineering believed the seals would be effective (53°F).

The senior management at Morton Thiokol caved in to client pressure. The engineers were unanimous in their opposition to launch. Senior management asked the engineers to prove that the O-ring seals would fail. Of course, they could not state with 100% certainty that the rings would fail. Failure analysis is in terms of probabilities. The risk had elevated as the temperature decreased, but the engineers could not prove conclusively that the seals would fail in a manner that would detrimentally affect the mission. Senior managers at Morton Thiokol and program managers at NASA concluded erroneously that the risks were low enough to proceed with the launch.

17.9.8 Approvals at the Top

NASA has a four-level approval procedure to control the launch of the shuttle. This sounds good, but for a multilevel approval process to be of any value, information must flow freely from the bottom of the organization to the top of the chain of command. The technical discussions concerning the ability of the O-rings to function at the very low temperatures took place between principals at level 4 and Morton Thiokol, and also included level 3 program managers at Marshall Space Flight Center. While the discussion was extended, with clear polarization between engineering and management, not a word of these concerns was conveyed to the top two levels of management at NASA.

Approvals for the launch were given by the level 2 administrator for the National STS Program in Houston, TX, and the level 1 administrator at NASA's Headquarters in Washington D.C. These approvals were essentially automatic because the administrators in charge had been isolated. They were not privy to the management decision to fly with super cold O-ring seals. They were not informed of the engineering recommendation to postpone the launch and to wait until the temperature increased to at least 53°F.

The lesson here is clear—approvals by the very high level managers are worthwhile only if they are informed decisions. If complete information is not presented to the super bosses for their evaluation, then their approval is meaningless. These administrators are not knowledgeable and, as such, they add no value to a go—no-go decision.

17.10 SUMMARY

We have briefly described the degradation of ethical behavior during the past two or three generations and given some of the probable reasons for these changes. While there are many controversial issues which can be classified as "dilemma" ethics, there are several "basic" characteristics where the difference between right and wrong is clearly defined. The list of basic virtues include:

- Prudence
- Justice
- Fortitude
- Temperance
- Faith
- Hope
- Charity
- Humility
- Respect

We have laws written to prohibit, by punishment, a well-defined set of vices. These laws control behavior, but only the extremely repulsive actions on the part of individuals or corporations.

Individual behavior is important, but as we learned in the Challenger explosion, it is not sufficient. Corporations (business) must operate with the highest ethical standards. The chief executive officer (CEO) establishes the standards, and the lower level managers act to establish the corporate character.

Honor codes exist on some college campuses to guide ethical behavior (lying, cheating and stealing), and they are respected by the students and the faculty. Character development and honor codes go together in producing a meaningful educational experience. We list here some of the more important attributes in character development:

1. Excellence
2. Respect
3. Integrity
4. Decisiveness
5. Responsible
6. Temperate
7. Fortitude
8. Understanding

We treated professional ethics by introducing the Code of Ethics of Engineers that is sponsored by the Accreditation Board for Engineering and Technology (ABET). The code is short with four fundamental principles and seven basic canons. The canons are explained in an extensive set of guidelines which are reproduced verbatim in Appendix B.

Finally, we have discussed the fatal accident of the Challenger space shuttle which exploded during the launch in January 1986. Many ethical issues dealing with individual and corporate behavior are apparent as we review the events leading to this fatal accident. We hope that this example will aid you in developing both your individual and professional character.

REFERENCES

1. Patterson, J. and P. Kim, The Day America Told the Truth: What People Really Believe about Everything that Really Matters, Prentice Hall, New York, 1991.
2. Sommers, C. H., "Teaching the Virtues," *Public Interest*, No. 111, Spring 1993, pp. 3-13.
3. Woodward, K. L. "What is Virtue," *Newsweek*, June 13, 1994.
4. George, R. P., Making Men Moral: Civil Liberties and Public Morality, Oxford University Press, NY, 1994.
5. Baker, D. F. "Ethical Issues and Decision Making in Business," Vital Speeches of the Day, 1993
6. United States Air Force Character Development Manual, USAF Academy, C0, December 1994.
7. Abramson, R., "A Matter of Honor," *Los Angeles Times*, April 3, 1994.
8. Lickona, T. A. Educating for Character, Bantam Book, New York, 1991.
9. Jensen, C., No Down Link: A Dramatic Narrative about the Challenger Accident, Farrar, Straus and Gitoux, New York, 1996.
10. *Time*, February 10, 1986
11. Lewis, R. S., The Voyages of Columbia: The First True Space Ship, Columbia University Press, New York, 1984.
12. *The New York Times* April 23, 1986.
13. Feynman, R. P. What Do You Care What Other People Think? Bantam, New York, 1988.
14. Report to the President by the Presidential Commission on the Space Shuttle Challenger Accident, Ayer Co., Salem, 1986.

EXERCISES

1. Our political leaders are often attacked by the media for lax ethical behavior. Please write a short paper describing three recent lapses of ethical behavior on the part of the leadership in either the State or Federal government. Did these officials break the law? Is it important that they did or did not break the law?
2. List what you consider poor behavior in a college classroom.
3. Have you ever cheated on an important exam in high school? What about a college exam? Do you know of someone who cheated? What was your attitude when you observed this cheating?
4. Is there an honor code at the University where you are pursuing your studies? Have you read it? Do you abide by the rules?
5. Write a paragraph or two explaining your position regarding:

- Prudence
- Justice
- Fortitude
- Temperance

6. Write a paragraph or two describing your feelings about:

   - Faith
   - Hope
   - Charity

7. If you could write a law that would go on the books tomorrow and be strictly enforced, what behavior would it prohibit?

8. Why does the CEO of a corporation set the standard for ethical behavior? What are some of the issues that arise on a daily basis that develop corporate character? Is it possible for a corporation with tens of thousands of employees to develop a character like an individual?

9. Write a 400 word essay describing your character. Include a discussion of the weaknesses as well as the strengths of your character.

10. Examine the code of ethics for engineers given in Fig. 17.1, and describe you opinion of it. Can you live with the expectations if you become a professional engineer? Do you believe it is adequate, or should it be revised to reflect a more modern viewpoint of what is right and wrong?

11. Suppose that you were a lead engineer working for Morton Thiokol on the evening of January 27, 1986, involved in the discussion of the O-rings. What would you have done?

    - Early in the teleconference.
    - At the critical stage of the teleconference.
    - After management had taken over the decision process.

There is no right or wrong answer to these questions. We want to place you in a professional dilemma. Someday you may be placed in a similar situation where there is a trade-off between safety and corporate business interests. Think about your response in advance and be prepared to deal with such a dilemma and its consequences.

APPENDIX A

UNITED STATES POSTAL SERVICE ®

EXPRESS MAIL

RATEFOLD
Notice 123 Effective October 5, 1997

Weight Not Over (lbs.)	Same Day Airport	Custom Designed	Next Day & Second Day PO to PO	Next Day & Second Day PO to Addressee	Weight Not Over (lbs.)	Same Day Airport	Custom Designed	Next Day & Second Day PO to PO	Next Day & Second Day PO to Addressee
1/2	$9.00	$9.45	$10.25	$10.75					
1	10.50	14.00	12.05	15.00	36	$47.20	$57.55	$55.60	$58.65
2 [1]	10.50	14.00	12.05	15.00	37	48.10	58.65	56.70	59.70
3	11.95	16.15	14.20	17.25	38	49.10	59.80	57.85	60.85
4	13.05	18.30	16.35	19.40	39	50.00	60.85	58.90	61.95
5	14.15	20.45	18.50	21.55	40	50.90	62.00	60.05	63.05
6	15.30	24.30	22.35	25.40	41	51.90	63.05	61.15	64.15
7	16.40	25.40	23.45	26.45	42	52.80	64.20	62.25	65.30
8	17.55	26.50	24.55	27.60	43	53.75	65.35	63.40	66.40
9	18.70	27.60	25.65	28.65	44	54.70	66.40	64.50	67.50
10	19.75	28.75	26.80	29.80	45	55.60	67.55	65.60	68.65
11	20.90	29.80	27.85	30.90	46	56.55	68.65	66.70	69.70
12	22.05	30.95	29.00	32.00	47	57.50	69.75	67.80	70.85
13	23.15	32.00	30.10	33.10	48	58.45	70.85	68.90	71.95
14	24.30	33.15	31.20	34.25	49	59.35	72.00	70.05	73.05
15	25.40	34.25	32.30	35.30	50	60.30	73.05	71.10	74.15
16	26.50	35.35	33.45	36.45	51	61.25	74.20	72.25	75.30
17	27.65	36.50	34.55	37.60	52	62.15	75.30	73.35	76.35
18	28.80	37.60	35.65	38.65	53	63.15	76.40	74.45	77.50
19	29.90	38.70	36.75	39.80	54	64.05	77.55	75.60	78.60
20	31.00	39.80	37.85	40.90	55	65.00	78.60	76.70	79.70
21	32.15	40.95	39.00	42.00	56	65.95	79.75	77.80	80.85
22	33.25	42.00	40.05	43.10	57	66.85	80.85	78.90	81.90
23	34.40	43.15	41.20	44.25	58	67.80	81.95	80.05	83.05
24	35.55	44.25	42.30	45.30	59	68.75	83.15	81.20	84.25
25	36.60	45.35	43.40	46.45	60	69.65	84.45	82.50	85.55
26	37.75	46.45	44.50	47.50	61	70.65	85.85	83.90	86.95
27	38.75	47.55	45.65	48.65	62	71.55	87.15	85.20	88.25
28	39.70	48.65	46.70	49.75	63	72.45	88.45	86.50	89.55
29	40.65	49.80	47.85	50.85	64	73.45	89.85	87.90	90.95
30	41.60	50.90	49.00	52.00	65	74.35	91.15	89.20	92.25
31	42.50	52.00	50.05	53.10	66	75.30	92.55	90.60	93.65
32	43.45	53.15	51.20	54.20	67	76.25	93.85	91.90	94.95
33	44.40	54.20	52.25	55.30	68	77.15	95.25	93.30	96.35
34	45.30	55.35	53.40	56.45	69	78.10	96.55	94.60	97.65
35	46.30	56.45	54.50	57.50	70	79.05	97.85	95.90	98.95

[1] The 2-pound rate is charged for matter sent in a "flat rate" envelope provided by the USPS.

FIRST-CLASS MAIL

Matter Other Than Cards Weight Not Over (ozs.)	Nonautomation		Automation					
	Single-Piece	Presorted	Basic (Letter-Size)	3-Digit (Letter-Size)	5-Digit (Letter-Size)	Carrier Route (Letter-Size)	Basic (Flat-Size)	3/5 (Flat-Size)
1	$0.32 [1]	$0.295 [1]	$0.261	$0.254	$0.238	$0.230	$0.290 [1]	$0.270 [1]
2	0.55	0.525	0.491	0.484	0.468	0.460	0.520	0.500
3	0.78	0.709	0.675	0.668	0.652	0.644	0.704	0.684
4	1.01	0.939	0.905 [2]	0.898 [2]	0.882 [2]	0.874 [2]	0.934	0.914
5	1.24	1.169	—	—	—	—	1.164	1.144
6	1.47	1.399	—	—	—	—	1.394	1.374
7	1.70	1.629	—	—	—	—	1.624	1.604
8	1.93	1.859	—	—	—	—	1.854	1.834
9	2.16	2.089	—	—	—	—	2.084	2.064
10	2.39	2.319	—	—	—	—	2.314	2.294
11	2.62	2.549	—	—	—	—	2.544	2.524
Postcards[3] Stamped Cards								
Single	$0.20	$0.180	$0.166	$0.159	$0.143	$0.140	—	—
Double	0.40							

[1] Nonstandard surcharge may apply. [2] Weight not to exceed 3.3362 ozs; pieces over 3 ozs. subject to additional standards (DMM C810). [3] Rates shown apply to each single or double postcard when originally mailed; reply half of double postcard must bear postage at applicable rate when returned unless prepared as business reply mail.

United States Postal Service

Ratefold ■ Notice 123 ■ October 5, 1997

PRIORITY MAIL

| Single-Piece | | | | | | Weight Not Over (lbs.) | Presorted | | | | | |
| Zone | | | | | | | Zone | | | | | |
1, 2 & 3	4	5	6	7	8		1, 2 & 3	4	5	6	7	8
$3.00	$3.00	$3.00	$3.00	$3.00	$3.00	1	$2.89	$2.89	$2.89	$2.89	$2.89	$2.89
3.00	3.00	3.00	3.00	3.00	3.00	2[1]	2.89	2.89	2.89	2.89	2.89	2.89
4.00	4.00	4.00	4.00	4.00	4.00	3	3.89	3.89	3.89	3.89	3.89	3.89
5.00	5.00	5.00	5.00	5.00	5.00	4	4.89	4.89	4.89	4.89	4.89	4.89
6.00	6.00	6.00	6.00	6.00	6.00	5	5.89	5.89	5.89	5.89	5.89	5.89
6.35	6.90	7.10	7.20	7.80	8.00	6	6.24	6.79	6.99	7.09	7.69	7.89
6.65	7.50	8.10	8.40	9.20	9.80	7	6.54	7.39	7.99	8.29	9.09	9.69
6.95	8.00	9.00	9.50	10.40	11.60	8	6.84	7.89	8.89	9.39	10.29	11.49
7.40	8.60	9.80	10.60	11.30	13.00	9	7.29	8.49	9.69	10.49	11.19	12.89
7.80	9.30	10.55	11.40	12.15	14.05	10	7.69	9.19	10.44	11.29	12.04	13.94
8.25	9.90	11.35	12.20	13.00	15.10	11	8.14	9.79	11.24	12.09	12.89	14.99
8.70	10.55	12.10	13.00	13.90	16.15	12	8.59	10.44	11.99	12.89	13.79	16.04
9.10	11.20	12.80	13.80	14.75	17.20	13	8.99	11.09	12.69	13.69	14.64	17.09
9.55	11.85	13.60	14.55	15.60	18.25	14	9.44	11.74	13.49	14.44	15.49	18.14
10.00	12.45	14.35	15.35	16.50	19.30	15	9.89	12.34	14.24	15.24	16.39	19.19
10.40	13.15	15.05	16.15	17.35	20.35	16	10.29	13.04	14.94	16.04	17.24	20.24
10.85	13.75	15.80	16.95	18.20	21.40	17	10.74	13.64	15.69	16.84	18.09	21.29
11.30	14.35	16.50	17.75	19.05	22.45	18	11.19	14.24	16.39	17.64	18.94	22.34
11.70	15.05	17.25	18.55	19.95	23.50	19	11.59	14.94	17.14	18.44	19.84	23.39
12.15	15.65	17.95	19.30	20.80	24.55	20	12.04	15.54	17.84	19.19	20.69	24.44
12.60	16.35	18.70	20.10	21.65	25.60	21	12.49	16.24	18.59	19.99	21.54	25.49
13.00	16.95	19.40	20.90	22.55	26.65	22	12.89	16.84	19.29	20.79	22.44	26.54
13.45	17.55	20.15	21.70	23.40	27.70	23	13.34	17.44	20.04	21.59	23.29	27.59
13.85	18.25	20.85	22.50	24.25	28.75	24	13.74	18.14	20.74	22.39	24.14	28.64
14.30	18.85	21.60	23.25	25.15	29.85	25	14.19	18.74	21.49	23.14	25.04	29.74
14.75	19.50	22.30	24.05	26.00	30.90	26	14.64	19.39	22.19	23.94	25.89	30.79
15.15	20.15	23.00	24.85	26.85	31.95	27	15.04	20.04	22.89	24.74	26.74	31.84
15.60	20.80	23.75	25.65	27.70	33.00	28	15.49	20.69	23.64	25.54	27.59	32.89
16.05	21.40	24.45	26.45	28.60	34.05	29	15.94	21.29	24.34	26.34	28.49	33.94
16.45	22.10	25.20	27.20	29.45	35.10	30	16.34	21.99	25.09	27.09	29.34	34.99
16.90	22.70	25.90	28.00	30.30	36.15	31	16.79	22.59	25.79	27.89	30.19	36.04
17.35	23.40	26.65	28.80	31.20	37.20	32	17.24	23.29	26.54	28.69	31.09	37.09
17.75	24.00	27.35	29.60	32.05	38.25	33	17.64	23.89	27.24	29.49	31.94	38.14
18.20	24.60	28.10	30.40	32.90	39.30	34	18.09	24.49	27.99	30.29	32.79	39.19
18.60	25.30	28.80	31.20	33.75	40.35	35	18.49	25.19	28.69	31.09	33.64	40.24
19.05	25.90	29.55	31.95	34.65	41.40	36	18.94	25.79	29.44	31.84	34.54	41.29
19.50	26.55	30.25	32.75	35.50	42.45	37	19.39	26.44	30.14	32.64	35.39	42.34
19.90	27.20	31.00	33.55	36.35	43.50	38	19.79	27.09	30.89	33.44	36.24	43.39
20.35	27.80	31.70	34.35	37.25	44.55	39	20.24	27.69	31.59	34.24	37.14	44.44
20.80	28.45	32.40	35.15	38.10	45.60	40	20.69	28.34	32.29	35.04	37.99	45.49
21.20	29.10	33.15	35.90	38.95	46.65	41	21.09	28.99	33.04	35.79	38.84	46.54
21.65	29.75	33.85	36.70	39.85	47.70	42	21.54	29.64	33.74	36.59	39.74	47.59
22.10	30.35	34.60	37.50	40.70	48.80	43	21.99	30.24	34.49	37.39	40.59	48.69
22.50	31.05	35.30	38.30	41.55	49.85	44	22.39	30.94	35.19	38.19	41.44	49.74
22.95	31.65	36.05	39.10	42.40	50.90	45	22.84	31.54	35.94	38.99	42.29	50.79
23.35	32.35	36.75	39.85	43.30	51.95	46	23.24	32.24	36.64	39.74	43.19	51.84
23.80	32.95	37.50	40.65	44.15	53.00	47	23.69	32.84	37.39	40.54	44.04	52.89
24.25	33.55	38.20	41.45	45.00	54.05	48	24.14	33.44	38.09	41.34	44.89	53.94
24.65	34.25	38.95	42.25	45.90	55.10	49	24.54	34.14	38.84	42.14	45.79	54.99
25.10	34.85	39.65	43.05	46.75	56.15	50	24.99	34.74	39.54	42.94	46.64	56.04
25.55	35.50	40.35	43.85	47.60	57.20	51	25.44	35.39	40.24	43.74	47.49	57.09
25.95	36.15	41.10	44.60	48.50	58.25	52	25.84	36.04	40.99	44.49	48.39	58.14
26.40	36.80	41.80	45.40	49.35	59.30	53	26.29	36.69	41.69	45.29	49.24	59.19
26.85	37.40	42.55	46.20	50.20	60.35	54	26.74	37.29	42.44	46.09	50.09	60.24
27.25	38.05	43.25	47.00	51.05	61.40	55	27.14	37.94	43.14	46.89	50.94	61.29
27.70	38.70	44.00	47.80	51.95	62.45	56	27.59	38.59	43.89	47.69	51.84	62.34
28.10	39.35	44.70	48.55	52.80	63.50	57	27.99	39.24	44.59	48.44	52.69	63.39
28.55	40.00	45.45	49.35	53.65	64.55	58	28.44	39.89	45.34	49.24	53.54	64.44
29.00	40.60	46.15	50.15	54.55	65.60	59	28.89	40.49	46.04	50.04	54.44	65.49
29.40	41.30	46.90	50.95	55.40	66.65	60	29.29	41.19	46.79	50.84	55.29	66.54
29.85	41.90	47.60	51.75	56.25	67.75	61	29.74	41.79	47.49	51.64	56.14	67.64
30.30	42.50	48.35	52.50	57.10	68.80	62	30.19	42.39	48.24	52.39	56.99	68.69
30.70	43.20	49.05	53.30	58.00	69.85	63	30.59	43.09	48.94	53.19	57.89	69.74
31.15	43.80	49.75	54.10	58.85	70.90	64	31.04	43.69	49.64	53.99	58.74	70.79
31.60	44.45	50.50	54.90	59.70	71.95	65	31.49	44.34	50.39	54.79	59.59	71.84
32.00	45.10	51.20	55.70	60.60	73.00	66	31.89	44.99	51.09	55.59	60.49	72.89
32.45	45.75	51.95	56.50	61.45	74.05	67	32.34	45.64	51.84	56.39	61.34	73.94
32.90	46.35	52.65	57.25	62.30	75.10	68	32.79	46.24	52.54	57.14	62.19	74.99
33.30	47.05	53.40	58.05	63.20	76.15	69	33.19	46.94	53.29	57.94	63.09	76.04
33.75	47.65	54.10	58.85	64.05	77.20	70	33.64	47.54	53.99	58.74	63.94	77.09

[1] The 2-pound rate is charged for matter sent in a "flat rate" envelope provided by the USPS.
Parcels weighing less than 15 pounds but measuring more than 84 inches in length and girth combined are chargeable with a minimum rate equal to that for a 15-pound parcel for the zone to which addressed.

United States Postal Service

Ratefold ■ Notice 123 ■ October 5, 1997

PERIODICALS

Regular

Pound Rates (per pound or fraction)

Nonadvertising Portion — $0.161

Advertising Portion

Zone	Rate
Delivery Unit	$0.169
SCF	0.190
1 & 2	0.214
3	0.224
4	0.251
5	0.292
6	0.336
7	0.388
8	0.432

Piece Rates

Presort Level	Nonautomation	Automation[1] (Letter-Size)	(Flat-Size)
Basic	$0.240	$0.194	$0.209
3/5	0.202	0.173	0.175
Carrier Route	0.119	—	—
High Density	0.111	—	—
Saturation	0.095	—	—

Nonadvertising adjustment:
    For each 1% of nonadvertising content — $0.00057 per piece.

Entry discounts:
    Delivery unit zone piece discount of $0.021 applies to each addressed piece
        claimed in the pound rate portion at the delivery unit zone rate.
    SCF zone piece discount of $0.011 applies to each addressed piece claimed
        in the pound rate portion at the SCF zone rate.
[1]Lower maximum weight limits apply: letter-size at 3 ounces (or 3.3362 ounces
for heavy letters); flat-size at 16 ounces.

Preferred In-County

Pound Rates (per pound or fraction)

Zone	Rate
Delivery Unit	$0.112
All Others	0.122

Piece Rates

Presort Level	Nonautomation	Automation[1] (Letter-Size)	(Flat-Size)
Basic	$0.082	$0.082	$0.082
3/5	—	—	0.067
3-Digit	—	0.078	—
5-Digit	—	0.065	—
Carrier Route	0.044	—	—
High Density	0.039	—	—
Saturation	0.037	—	—

Entry discount:
    Delivery unit zone piece discount of $0.003 applies to each addressed piece
        claimed in the pound rate portion at the delivery unit zone rate.
[1]Lower maximum weight limits apply: letter-size at 3 ounces (or 3.3362 ounces for
heavy letters); flat-size at 16 ounces.

Preferred Nonprofit

Pound Rates (per pound or fraction)

Nonadvertising Portion — $0.142

Advertising Portion

Zone	Rate
Delivery Unit	$0.169
SCF	0.190
1 & 2	0.214
3	0.224
4	0.251
5	0.292
6	0.336
7	0.388
8	0.432

Piece Rates

Presort Level	Nonautomation	Automation[1] (Letter-Size)	(Flat-Size)
Basic	$0.217	$0.187	$0.193
3/5	0.172	—	0.148
3-Digit	—	0.149	—
5-Digit	—	0.149	—
Carrier Route	0.105	—	—
High Density	0.098	—	—
Saturation	0.084	—	—

Nonadvertising adjustment:
    For each 1% of nonadvertising content — $0.00042 per piece.

Entry discounts:
    Delivery unit zone piece discount of $0.012 applies to each addressed piece
        claimed in the pound rate portion at the delivery unit zone rate.
    SCF zone piece discount of $0.006 applies to each addressed piece claimed
        in the pound rate portion at the SCF zone rate.
[1]Lower maximum weight limits apply: letter-size at 3 ounces (or 3.3362 ounces
for heavy letters); flat-size at 16 ounces.

PERIODICALS

Preferred Classroom

Pound Rates (per pound or fraction)

Nonadvertising Portion — $0.143

Advertising Portion

Zone	Rate
Delivery Unit	$0.169
SCF	0.190
1 & 2	0.214
3	0.224
4	0.251
5	0.292
6	0.336
7	0.388
8	0.432

Piece Rates

Presort Level	Nonautomation	Automation[1] (Letter-Size)	Automation[1] (Flat-Size)
Basic	$0.211	$0.191	$0.184
3/5	0.160	—	0.142
3-Digit	—	0.148	—
5-Digit	—	0.140	—
Carrier Route	0.115	—	—
High Density	0.113	—	—
Saturation	0.107	—	—

Nonadvertising adjustment:
For each 1% of nonadvertising content — $0.00042 per piece.

Entry discounts:
Delivery unit zone piece discount of $0.006 applies to each addressed piece claimed in the pound rate portion at the delivery unit zone rate.
SCF zone piece discount of $0.004 applies to each addressed piece claimed in the pound rate portion at the SCF zone rate.
[1]Lower maximum weight limits apply: letter-size at 3 ounces (or 3.3362 ounces for heavy letters); flat-size at 16 ounces.

Preferred Science-of-Agriculture

Pound Rates (per pound or fraction)

Nonadvertising Portion — $0.161

Advertising Portion

Zone	Rate
Delivery Unit	$0.127
SCF	0.143
1 & 2	0.161
3	0.224
4	0.251
5	0.292
6	0.336
7	0.388
8	0.432

Piece Rates

Presort Level	Nonautomation	Automation[1] (Letter-Size)	Automation[1] (Flat-Size)
Basic	$0.240	$0.194	$0.209
3/5	0.202	0.173	0.175
Carrier Route	0.119	—	—
High Density	0.111	—	—
Saturation	0.095	—	—

Nonadvertising adjustment:
For each 1% of nonadvertising content — $0.00057 per piece.

Entry discounts:
Delivery unit zone piece discount of $0.021 applies to each addressed piece claimed in the pound rate portion at the delivery unit zone rate.
SCF zone piece discount of $0.011 applies to each addressed piece claimed in the pound rate portion at the SCF zone rate.
[1]Lower maximum weight limits apply: letter-size at 3 ounces (or 3.3407 ounces for heavy letters); flat-size at 16 ounces.

STANDARD MAIL (A)

Single-Piece		Keys and Identification Devices	
Weight (ozs.)	Rate	Weight (ozs.)	Rate
Not over 1[1]	$0.32	Not over 2	$0.99
Over 1 but not over 2	0.55	Over 2 but not over 4	1.54
Over 2 but not over 3	0.78	Over 4 but not over 6	2.09
Over 3 but not over 4	1.01	Over 6 but not over 8	2.64
Over 4 but not over 5	1.24	Over 8 but not over 10	3.19
Over 5 but not over 6	1.47	Over 10 but not over 12	3.74
Over 6 but not over 7	1.70	Over 12 but not over 14	4.29
Over 7 but not over 8	1.93	Over 14 but less than 16	4.84
Over 8 but not over 9	2.16		
Over 9 but not over 10	2.39		
Over 10 but not over 11	2.62		
Over 11 but not over 13	2.90		
Over 13 but less than 16	2.95		

[1]Nonstandard surcharge might apply.

United States Postal Service   Ratefold ■ Notice 123 ■ October 5, 1997

STANDARD MAIL (A)

REGULAR — Letter- and Nonletter-Size Pieces Weighing 0.2068 lb. (3.3087 ozs.) or Less
ENHANCED CARRIER ROUTE — Letter- and Nonletter-Size Pieces Weighing 0.2066 lb. (3.3062 ozs.) or Less

	Entry Discount [1]	Nonautomation		Automation [2, 4]			Automation [2]		High	
		Basic	3/5	Basic	3-Digit	5-Digit	Basic	Basic	Density	Saturation
Letter-Size Minimum Per Piece	None	$0.256	$0.209	$0.183	$0.175	$0.155	$0.146	$0.150	$0.142	$0.133
	DBMC (0.013)	0.243	0.196	0.170	0.162	0.142	0.133	0.137	0.129	0.120
	DSCF (0.018)	0.238	0.191	0.165	0.157	0.137	0.128	0.132	0.124	0.115
	DDU (0.023)	—	—	—	—	—	0.123	0.127	0.119	0.110
Nonletter-Size Minimum Per Piece	None	$0.306	$0.225	$0.277	3/5 $0.189		—	$0.155	$0.147	$0.137
	DBMC (0.013)	0.293	0.212	0.264	0.176		—	0.142	0.134	0.124
	DSCF (0.018)	0.288	0.207	0.259	0.171		—	0.137	0.129	0.119
	DDU (0.023)	—	—	—	—		—	0.132	0.124	0.114

Piece/Pound [3]

Piece/Pound Pieces Weighing More Than 0.2068 lb. (3.3087 ozs.) — Piece/Pound Pieces Weighing More Than 0.2066 lb. (3.3062 ozs.)

		Basic	3/5	Basic	3/5			Basic	High Density	Saturation
Per Piece Rates (for all entries)		$0.166	$0.085	$0.137	$0.049		—	$0.018	$0.010	$0.000
Per Pound Rates (by entry discount)		PLUS	PLUS	PLUS	PLUS		—	PLUS	PLUS	PLUS
	None	$0.677	$0.677	$0.677	$0.677		—	$0.663	$0.663	$0.663
	DBMC (0.064)	0.613	0.613	0.613	0.613		—	0.599	0.599	0.599
	DSCF (0.085)	0.592	0.592	0.592	0.592		—	0.578	0.578	0.578
	DDU (0.111)	—	—	—	—		—	0.552	0.552	0.552

NONPROFIT — Letter- and Nonletter-Size Pieces Weighing 0.2085 lb. (3.3362 ozs.) or Less
NONPROFIT ENHANCED CARRIER ROUTE — Letter- and Nonletter-Size Pieces Weighing 0.2084 lb. (3.3348 ozs.) or Less

	Entry Discount [1]	Nonautomation		Automation [2, 4]			Automation [2]		High	
		Basic	3/5	Basic	3-Digit	5-Digit	Basic	Basic	Density	Saturation
Letter-Size Minimum Per Piece	None	$0.135	$0.117	$0.102	$0.098	$0.085	$0.082	$0.093	$0.087	$0.081
	DBMC (0.013)	0.122	0.104	0.089	0.085	0.072	0.069	0.080	0.074	0.068
	DSCF (0.018)	0.117	0.099	0.084	0.080	0.067	0.064	0.075	0.069	0.063
	DDU (0.024)	—	—	—	—	—	0.058	0.069	0.063	0.057
Nonletter-Size Minimum Per Piece	None	$0.198	$0.149	$0.174	3/5 $0.125		—	$0.107	$0.100	$0.094
	DBMC (0.013)	0.185	0.136	0.161	0.112		—	0.094	0.087	0.081
	DSCF (0.018)	0.180	0.131	0.156	0.107		—	0.089	0.082	0.076
	DDU (0.024)	—	—	—	—		—	0.083	0.076	0.070

Piece/Pound [3]

Piece/Pound Pieces Weighing More Than 0.2085 lb. (3.3362 ozs.) — Piece/Pound Pieces Weighing More Than 0.2084 lb. (3.3348 ozs.)

		Basic	3/5	Basic	3/5			Basic	High Density	Saturation
Per Piece Rates (for all entries)		$0.100	$0.048	$0.076	$0.024		—	$0.013	$0.006	$0.000
Per Pound Rates (by entry discount)		PLUS	PLUS	PLUS	PLUS		—	PLUS	PLUS	PLUS
	None	$0.470	$0.484	$0.470	$0.484		—	$0.451	$0.451	$0.451
	DBMC (0.062)	0.408	0.422	0.408	0.422		—	0.389	0.389	0.389
	DSCF (0.088)	0.382	0.396	0.382	0.396		—	0.363	0.363	0.363
	DDU (0.114)	—	—	—	—		—	0.337	0.337	0.337

[1]Value of entry discount in parentheses included in rates shown. [2]Letter-size pieces over 3 ozs. subject to additional standards (DMM C810). [3]Each piece is subject to both a piece rate and a pound rate. [4]Automation rate for nonletters applies only to automation-compatible flats.

STANDARD MAIL (B)

Single-Piece Bound Printed Matter

Weight Not Over (lbs.)	Local	1 & 2	3	4	5	6	7	8
				Zone				
1.5	$1.11	$1.49	$1.52	$1.58	$1.66	$1.74	$1.84	$1.93
2.0	1.12	1.52	1.56	1.63	1.74	1.85	1.99	2.10
2.5	1.14	1.55	1.60	1.69	1.82	1.96	2.13	2.28
3.0	1.15	1.57	1.64	1.74	1.90	2.07	2.27	2.45
3.5	1.17	1.60	1.67	1.80	1.98	2.18	2.42	2.62
4.0	1.18	1.63	1.71	1.85	2.07	2.29	2.56	2.79
4.5	1.20	1.65	1.75	1.91	2.15	2.40	2.71	2.97
5.0	1.22	1.68	1.79	1.96	2.23	2.51	2.85	3.14
6.0	1.25	1.73	1.86	2.07	2.39	2.73	3.14	3.49
7.0	1.28	1.79	1.94	2.18	2.56	2.95	3.43	3.83
8.0	1.31	1.84	2.01	2.29	2.72	3.17	3.71	4.18
9.0	1.34	1.90	2.09	2.40	2.89	3.39	4.00	4.52
10.0	1.37	1.95	2.16	2.51	3.05	3.61	4.29	4.87

Basic Bulk Bound Printed Matter / Carrier Route Bulk Bound Printed Matter

Local	1 & 2	3	4	5	6	7	8	Weight Not Over (lbs.)	Local	1 & 2	3	4	5	6	7	8
$0.565	$0.765	$0.795	$0.849	$0.928	$1.014	$1.116	$1.203	1.5	$0.502	$0.702	$0.732	$0.786	$0.865	$0.951	$1.053	$1.140
0.576	0.786	0.826	0.898	1.004	1.118	1.254	1.370	2.0	0.513	0.723	0.763	0.835	0.941	1.055	1.191	1.307
0.588	0.808	0.858	0.948	1.080	1.223	1.393	1.538	2.5	0.525	0.745	0.795	0.885	1.017	1.160	1.330	1.475
0.599	0.829	0.889	0.997	1.156	1.327	1.531	1.705	3.0	0.536	0.766	0.826	0.934	1.093	1.264	1.468	1.642
0.611	0.851	0.921	1.047	1.232	1.432	1.670	1.873	3.5	0.548	0.788	0.858	0.984	1.169	1.369	1.607	1.810
0.622	0.872	0.952	1.096	1.308	1.536	1.808	2.040	4.0	0.559	0.809	0.889	1.033	1.245	1.473	1.745	1.977
0.634	0.894	0.984	1.146	1.384	1.641	1.947	2.208	4.5	0.571	0.831	0.921	1.083	1.321	1.578	1.884	2.145
0.645	0.915	1.015	1.195	1.460	1.745	2.085	2.375	5.0	0.582	0.852	0.952	1.132	1.397	1.682	2.022	2.312
0.668	0.958	1.078	1.294	1.612	1.954	2.362	2.710	6.0	0.605	0.895	1.015	1.231	1.549	1.891	2.299	2.647
0.691	1.001	1.141	1.393	1.764	2.163	2.639	3.045	7.0	0.628	0.938	1.078	1.330	1.701	2.100	2.576	2.982
0.714	1.044	1.204	1.492	1.916	2.372	2.916	3.380	8.0	0.651	0.981	1.141	1.429	1.853	2.309	2.853	3.317
0.737	1.087	1.267	1.591	2.068	2.581	3.193	3.715	9.0	0.674	1.024	1.204	1.528	2.005	2.518	3.130	3.652
0.760	1.130	1.330	1.690	2.220	2.790	3.470	4.050	10.0	0.697	1.067	1.267	1.627	2.157	2.727	3.407	3.987

Special Standard Mail

Weight Not Over (lbs.)	Single-Piece	5-Digit	BMC	Library Mail Single-Piece	Weight Not Over (lbs.)	Single-Piece	5-Digit	BMC	Library Mail Single-Piece
1	$1.24	$0.70	$1.04	$1.12	36	$13.23	$12.69	$13.03	$9.96
2	1.74	1.20	1.54	1.53	37	13.54	13.00	13.34	10.18
3	2.24	1.70	2.04	1.94	38	13.85	13.31	13.65	10.40
4	2.74	2.20	2.54	2.35	39	14.16	13.62	13.96	10.62
5	3.24	2.70	3.04	2.76	40	14.47	13.93	14.27	10.84
6	3.74	3.20	3.54	3.17	41	14.78	14.24	14.58	11.06
7	4.24	3.70	4.04	3.58	42	15.09	14.55	14.89	11.28
8	4.55	4.01	4.35	3.80	43	15.40	14.86	15.20	11.50
9	4.86	4.32	4.66	4.02	44	15.71	15.17	15.51	11.72
10	5.17	4.63	4.97	4.24	45	16.02	15.48	15.82	11.94
11	5.48	4.94	5.28	4.46	46	16.33	15.79	16.13	12.16
12	5.79	5.25	5.59	4.68	47	16.64	16.10	16.44	12.38
13	6.10	5.56	5.90	4.90	48	16.95	16.41	16.75	12.60
14	6.41	5.87	6.21	5.12	49	17.26	16.72	17.06	12.82
15	6.72	6.18	6.52	5.34	50	17.57	17.03	17.37	13.04
16	7.03	6.49	6.83	5.56	51	17.88	17.34	17.68	13.26
17	7.34	6.80	7.14	5.78	52	18.19	17.65	17.99	13.48
18	7.65	7.11	7.45	6.00	53	18.50	17.96	18.30	13.70
19	7.96	7.42	7.76	6.22	54	18.81	18.27	18.61	13.92
20	8.27	7.73	8.07	6.44	55	19.12	18.58	18.92	14.14
21	8.58	8.04	8.38	6.66	56	19.43	18.89	19.23	14.36
22	8.89	8.35	8.69	6.88	57	19.74	19.20	19.54	14.58
23	9.20	8.66	9.00	7.10	58	20.05	19.51	19.85	14.80
24	9.51	8.97	9.31	7.32	59	20.36	19.82	20.16	15.02
25	9.82	9.28	9.62	7.54	60	20.67	20.13	20.47	15.24
26	10.13	9.59	9.93	7.76	61	20.98	20.44	20.78	15.46
27	10.44	9.90	10.24	7.98	62	21.29	20.75	21.09	15.68
28	10.75	10.21	10.55	8.20	63	21.60	21.06	21.40	15.90
29	11.06	10.52	10.86	8.42	64	21.91	21.37	21.71	16.12
30	11.37	10.83	11.17	8.64	65	22.22	21.68	22.02	16.34
31	11.68	11.14	11.48	8.86	66	22.53	21.99	22.33	16.56
32	11.99	11.45	11.79	9.08	67	22.84	22.30	22.64	16.78
33	12.30	11.76	12.10	9.30	68	23.15	22.61	22.95	17.00
34	12.61	12.07	12.41	9.52	69	23.46	22.92	23.26	17.22
35	12.92	12.38	12.72	9.74	70	23.77	23.23	23.57	17.44

United States Postal Service

Ratefold ■ Notice 123 ■ October 5, 1997

STANDARD MAIL (B)

Intra-BMC/ASF Parcel Post All Parcels, Discounts Already Included					Weight Not Over (lbs.)	Destination BMC/ASF Parcel Post All Parcels, Discounts Already Included (BMC/ASF)			
Zone						Zone			
Local	1 & 2	3	4	5		1 & 2	3	4	5
$2.24	$2.31	$2.47	$2.55	$2.63	2	$2.10	$2.25	$2.30	$2.33
2.31	2.44	2.68	3.02	3.36	3	2.22	2.44	2.74	3.00
2.39	2.55	2.88	3.46	4.36	4	2.33	2.62	3.15	3.94
2.45	2.65	3.06	3.78	4.87	5	2.42	2.79	3.45	4.40
2.52	2.75	3.23	4.07	5.35	6	2.51	2.95	3.71	4.83
2.58	2.84	3.39	4.35	5.79	7	2.60	3.09	3.97	5.22
2.64	2.94	3.53	4.59	6.21	8	2.69	3.22	4.19	5.60
2.69	3.01	3.67	4.84	6.60	9	2.76	3.35	4.42	5.95
2.75	3.10	3.80	5.06	6.97	10	2.84	3.47	4.62	6.29
2.80	3.17	3.93	5.27	7.31	11	2.91	3.59	4.82	6.59
2.85	3.25	4.05	5.47	7.64	12	2.98	3.70	5.00	6.89
2.91	3.32	4.15	5.66	7.94	13	3.05	3.79	5.17	7.16
2.95	3.39	4.27	5.84	8.23	14	3.11	3.91	5.34	7.42
3.00	3.45	4.37	6.02	8.50	15	3.17	4.00	5.51	7.67
3.05	3.51	4.47	6.18	8.77	16	3.23	4.09	5.65	7.91
3.09	3.58	4.56	6.34	9.01	17	3.29	4.18	5.80	8.13
3.13	3.63	4.65	6.49	9.26	18	3.34	4.26	5.94	8.35
3.17	3.70	4.74	6.63	9.48	19	3.41	4.34	6.07	8.55
3.22	3.75	4.82	6.76	9.69	20	3.45	4.42	6.19	8.74
3.25	3.80	4.91	6.89	9.91	21	3.50	4.50	6.31	8.94
3.29	3.86	4.98	7.02	10.11	22	3.56	4.57	6.43	9.12
3.33	3.91	5.07	7.15	10.30	23	3.61	4.65	6.55	9.30
3.37	3.95	5.14	7.26	10.48	24	3.64	4.72	6.65	9.46
3.41	4.00	5.21	7.38	10.66	25	3.69	4.78	6.77	9.62
3.44	4.05	5.28	7.49	10.83	26	3.74	4.85	6.87	9.78
3.48	4.10	5.35	7.59	10.99	27	3.79	4.91	6.96	9.92
3.51	4.14	5.42	7.70	11.15	28	3.83	4.98	7.06	10.07
3.55	4.19	5.49	7.80	11.31	29	3.87	5.05	7.16	10.21
3.59	4.23	5.55	7.89	11.46	30	3.91	5.10	7.24	10.35
3.62	4.28	5.60	7.99	11.60	31	3.96	5.15	7.33	10.48
3.65	4.32	5.67	8.08	11.74	32	4.00	5.22	7.42	10.61
3.69	4.36	5.73	8.17	11.88	33	4.04	5.27	7.50	10.73
3.72	4.40	5.78	8.25	12.00	34	4.08	5.32	7.58	10.84
3.75	4.44	5.84	8.34	12.13	35	4.11	5.38	7.66	10.96
3.78	4.48	5.89	8.43	12.26	36	4.15	5.42	7.75	11.08
3.81	4.52	5.94	8.50	12.38	37	4.19	5.47	7.81	11.19
3.84	4.56	6.00	8.59	12.49	38	4.23	5.53	7.90	11.29
3.88	4.60	6.05	8.66	12.60	39	4.27	5.57	7.96	11.39
3.91	4.63	6.10	8.73	12.72	40	4.30	5.62	8.03	11.50
3.94	4.68	6.16	8.80	12.82	41	4.35	5.68	8.09	11.59
3.97	4.71	6.20	8.87	12.92	42	4.38	5.72	8.16	11.68
4.00	4.74	6.25	8.95	13.03	43	4.40	5.76	8.23	11.79
4.04	4.78	6.29	9.01	13.12	44	4.44	5.80	8.29	11.87
4.06	4.81	6.34	9.08	13.22	45	4.47	5.85	8.36	11.96
4.09	4.85	6.39	9.14	13.31	46	4.51	5.90	8.41	12.04
4.12	4.89	6.43	9.20	13.40	47	4.55	5.94	8.47	12.13
4.15	4.92	6.48	9.27	13.50	48	4.58	5.98	8.53	12.22
4.18	4.95	6.52	9.33	13.58	49	4.61	6.02	8.59	12.29
4.21	4.98	6.56	9.38	13.67	50	4.64	6.06	8.64	12.38
4.24	5.02	6.60	9.45	13.75	51	4.68	6.10	8.70	12.45
4.26	5.05	6.65	9.50	13.83	52	4.71	6.15	8.75	12.52
4.29	5.08	6.69	9.55	13.91	53	4.73	6.19	8.80	12.60
4.32	5.11	6.73	9.61	13.99	54	4.76	6.22	8.86	12.67
4.35	5.14	6.76	9.67	14.06	55	4.79	6.25	8.91	12.74
4.38	5.18	6.81	9.72	14.13	56	4.83	6.30	8.96	12.80
4.40	5.21	6.85	9.77	14.21	57	4.86	6.34	9.01	12.88
4.43	5.24	6.88	9.82	14.28	58	4.89	6.37	9.06	12.94
4.46	5.27	6.92	9.87	14.35	59	4.92	6.41	9.10	13.01
4.48	5.30	6.96	9.93	14.42	60	4.95	6.45	9.16	13.07
4.52	5.34	7.00	9.97	14.49	61	4.99	6.48	9.20	13.14
4.54	5.37	7.03	10.02	14.55	62	5.02	6.51	9.25	13.19
4.57	5.39	7.07	10.07	14.61	63	5.04	6.55	9.29	13.25
4.59	5.42	7.10	10.12	14.68	64	5.07	6.58	9.34	13.31
4.62	5.45	7.14	10.16	14.74	65	5.10	6.62	9.38	13.37
4.64	5.49	7.18	10.20	14.81	66	5.14	6.66	9.42	13.43
4.68	5.52	7.21	10.25	14.86	67	5.17	6.69	9.47	13.48
4.70	5.54	7.24	10.30	14.92	68	5.19	6.72	9.51	13.54
4.73	5.57	7.27	10.34	14.98	69	5.21	6.74	9.55	13.59
4.75	5.60	7.32	10.39	15.03	70	5.24	6.79	9.60	13.64

United States Postal Service   Ratefold ■ Notice 123 ■ October 5, 1997

STANDARD MAIL (B)

Inter-BMC/ASF Parcel Post — Machinable Parcels — No Discount No Surcharge

Weight Not Over (lbs.)	1 & 2	3	4	5	6	7	8
2	$2.63	$2.79	$2.87	$2.95	$2.95	$2.95	$2.95
3	2.76	3.00	3.34	3.68	3.95	3.95	3.95
4	2.87	3.20	3.78	4.68	4.95	4.95	4.95
5	2.97	3.38	4.10	5.19	5.56	5.95	5.95
6	3.07	3.55	4.39	5.67	6.90	7.75	7.95
7	3.16	3.71	4.67	6.11	7.51	9.15	9.75
8	3.26	3.85	4.91	6.53	8.08	9.94	11.55
9	3.33	3.99	5.16	6.92	8.62	10.65	12.95
10	3.42	4.12	5.38	7.29	9.12	11.31	14.00
11	3.49	4.25	5.59	7.63	9.59	11.93	15.05
12	3.57	4.37	5.79	7.96	10.03	12.52	16.10
13	3.64	4.47	5.98	8.26	10.45	13.07	17.15
14	3.71	4.59	6.16	8.55	10.84	13.59	18.20
15	3.77	4.69	6.34	8.82	11.22	14.08	19.25
16	3.83	4.79	6.50	9.09	11.58	14.55	20.30
17	3.90	4.88	6.66	9.33	11.92	15.00	21.35
18	3.95	4.97	6.81	9.58	12.24	15.42	22.40
19	4.02	5.06	6.95	9.80	12.55	15.83	23.25
20	4.07	5.14	7.08	10.01	12.84	16.21	23.84
21	4.12	5.23	7.21	10.23	13.12	16.59	24.41
22	4.18	5.30	7.34	10.43	13.39	16.94	24.96
23	4.23	5.39	7.47	10.62	13.66	17.28	25.47
24	4.27	5.46	7.58	10.80	13.90	17.60	25.97
25	4.32	5.53	7.70	10.98	14.14	17.91	26.45
26	4.37	5.60	7.81	11.15	14.37	18.21	26.91
27	4.42	5.67	7.91	11.31	14.59	18.50	27.34
28	4.46	5.74	8.02	11.47	14.81	18.78	27.77
29	4.51	5.81	8.12	11.63	15.01	19.05	28.17
30	4.55	5.87	8.21	11.78	15.20	19.30	28.57
31	4.60	5.92	8.31	11.92	15.39	19.55	28.94
32	4.64	5.99	8.40	12.06	15.58	19.79	29.30
33	4.68	6.05	8.49	12.20	15.76	20.02	29.66
34	4.72	6.10	8.57	12.32	15.94	20.24	30.00
35	4.76	6.16	8.66	12.45	16.11	20.46	30.33

FOR WEIGHTS OVER 35 POUNDS, USE NONMACHINABLE RATES.

Inter-BMC/ASF Parcel Post — Nonmachinable Parcels — Surcharge Included

Weight Not Over (lbs.)	1 & 2	3	4	5	6	7	8
2	$4.38	$4.54	$4.62	$4.70	$4.70	$4.70	$4.70
3	4.51	4.75	5.09	5.43	5.70	5.70	5.70
4	4.62	4.95	5.53	6.43	6.70	6.70	6.70
5	4.72	5.13	5.85	6.94	7.31	7.70	7.70
6	4.82	5.30	6.14	7.42	8.65	9.50	9.70
7	4.91	5.46	6.42	7.86	9.26	10.90	11.50
8	5.01	5.60	6.66	8.28	9.83	11.69	13.30
9	5.08	5.74	6.91	8.67	10.37	12.40	14.70
10	5.17	5.87	7.13	9.04	10.87	13.06	15.75
11	5.24	6.00	7.34	9.38	11.34	13.68	16.80
12	5.32	6.12	7.54	9.71	11.78	14.27	17.85
13	5.39	6.22	7.73	10.01	12.20	14.82	18.90
14	5.46	6.34	7.91	10.30	12.59	15.34	19.95
15	5.52	6.44	8.09	10.57	12.97	15.83	21.00
16	5.58	6.54	8.25	10.84	13.33	16.30	22.05
17	5.65	6.63	8.41	11.08	13.67	16.75	23.10
18	5.70	6.72	8.56	11.33	13.99	17.17	24.15
19	5.77	6.81	8.70	11.55	14.30	17.58	25.00
20	5.82	6.89	8.83	11.76	14.59	17.96	25.59
21	5.87	6.98	8.96	11.98	14.87	18.34	26.16
22	5.93	7.05	9.09	12.18	15.14	18.69	26.71
23	5.98	7.14	9.22	12.37	15.41	19.03	27.22
24	6.02	7.21	9.33	12.55	15.65	19.35	27.72
25	6.07	7.28	9.45	12.73	15.89	19.66	28.20
26	6.12	7.35	9.56	12.90	16.12	19.96	28.66
27	6.17	7.42	9.66	13.06	16.34	20.25	29.09
28	6.21	7.49	9.77	13.22	16.56	20.53	29.52
29	6.26	7.56	9.87	13.38	16.76	20.80	29.92
30	6.30	7.62	9.96	13.53	16.95	21.05	30.32
31	6.35	7.67	10.06	13.67	17.14	21.30	30.69
32	6.39	7.74	10.15	13.81	17.33	21.54	31.05
33	6.43	7.80	10.24	13.95	17.51	21.77	31.41
34	6.47	7.85	10.32	14.07	17.69	21.99	31.75
35	6.51	7.91	10.41	14.20	17.86	22.21	32.08
36	6.55	7.96	10.50	14.33	18.02	22.41	32.39
37	6.59	8.01	10.57	14.45	18.18	22.62	32.69
38	6.63	8.07	10.66	14.56	18.32	22.82	32.99
39	6.67	8.12	10.73	14.67	18.47	23.01	33.28
40	6.70	8.17	10.80	14.79	18.61	23.19	33.56
41	6.75	8.23	10.87	14.89	18.75	23.37	33.82
42	6.78	8.27	10.94	14.99	18.89	23.54	34.08
43	6.81	8.32	11.02	15.10	19.03	23.71	34.33
44	6.85	8.36	11.08	15.19	19.16	23.87	34.58
45	6.88	8.41	11.15	15.29	19.27	24.03	34.81
46	6.92	8.46	11.21	15.38	19.40	24.19	35.05
47	6.96	8.50	11.27	15.47	19.52	24.34	35.27
48	6.99	8.55	11.34	15.57	19.63	24.49	35.48
49	7.02	8.59	11.40	15.65	19.74	24.63	35.70
50	7.05	8.63	11.45	15.74	19.85	24.77	35.90
51	7.09	8.67	11.52	15.82	19.95	24.91	36.10
52	7.12	8.72	11.57	15.90	20.06	25.04	36.29
53	7.15	8.76	11.62	15.98	20.17	25.16	36.49
54	7.18	8.80	11.68	16.06	20.26	25.29	36.67
55	7.21	8.83	11.74	16.13	20.36	25.41	36.85
56	7.25	8.88	11.79	16.20	20.45	25.54	37.02
57	7.28	8.92	11.84	16.28	20.55	25.64	37.19
58	7.31	8.95	11.89	16.35	20.64	25.76	37.35
59	7.34	8.99	11.94	16.42	20.72	25.87	37.51
60	7.37	9.03	12.00	16.49	20.82	25.97	37.67
61	7.41	9.07	12.04	16.56	20.89	26.08	37.82
62	7.44	9.10	12.09	16.62	20.98	26.19	37.97
63	7.46	9.14	12.14	16.68	21.06	26.28	38.12
64	7.49	9.17	12.19	16.75	21.14	26.39	38.25
65	7.52	9.21	12.23	16.81	21.21	26.48	38.39
66	7.56	9.25	12.27	16.88	21.30	26.57	38.52
67	7.59	9.28	12.32	16.93	21.37	26.67	38.66
68	7.61	9.31	12.37	16.99	21.43	26.75	38.79
69	7.64	9.34	12.41	17.05	21.51	26.85	38.90
70	7.67	9.39	12.46	17.10	21.58	26.93	39.03

United States Postal Service   Ratefold ■ Notice 123 ■ October 5, 1997

FEES

REGISTERED MAIL

Declared Value	Fee (in addition to postage)[1]	Declared Value	Fee (in addition to postage)[1]	Declared Value	Fee (in addition to postage)[1]
$0.00 to $100[2]	$4.85	$7,000.01 to $8,000	9.00	$17,000.01 to $18,000	13.50
$0.00 to $100	4.95	$8,000.01 to $9,000	9.45	$18,000.01 to $19,000	13.95
$100.01 to $500	5.40	$9,000.01 to $10,000	9.90	$19,000.01 to $20,000	14.40
$500.01 to $1,000	5.85	$10,000.01 to $11,000	10.35	$20,000.01 to $21,000	14.85
$1,000.01 to $2,000	6.30	$11,000.01 to $12,000	10.80	$21,000.01 to $22,000	15.30
$2,000.01 to $3,000	6.75	$12,000.01 to $13,000	11.25	$22,000.01 to $23,000	15.75
$3,000.01 to $4,000	7.20	$13,000.01 to $14,000	11.70	$23,000.01 to $24,000	16.20
$4,000.01 to $5,000	7.65	$14,000.01 to $15,000	12.15	$24,000.01 to $25,000	16.65
$5,000.01 to $6,000	8.10	$15,000.01 to $16,000	12.60		
$6,000.01 to $7,000	8.55	$16,000.01 to $17,000	13.05		

[1]Fees for articles with declared value over $100 include insurance.
[2]Without insurance.

Declared Value	Fees (in addition to postage)[1]
$25,000.01 to $1,000,000	$16.65 plus handling charge of $0.45 per $1,000 or fraction over first $25,000.
$1,000,000.01 to $15,000,000	$455.40 plus handling charge of $0.45 per $1,000 or fraction over first $1,000,000.
Over $15,000,000	$6,755.40 plus additional charges may be made based on consideration of weight, space, and value.

Fees for articles valued over $25,000 are for handling only. $25,000 is the maximum amount of insurance coverage available.

Additional Services

COD COLLECTION CHARGE — $3.50
(Maximum amount collectible is $600.00)
RESTRICTED DELIVERY — $2.75

RETURN RECEIPTS:
Requested at time of mailing showing to whom, signature, date, and addressee's address (if different) — $1.10
Requested after mailing showing only to whom and date delivered — $6.60

INSURED MAIL

Insurance Coverage Desired	Fee
$0.01 to $50	$0.75
50.01 to 100	1.60
100.01 to 200	2.50
200.01 to 300	3.40
300.01 to 400	4.30
400.01 to 500	5.20
500.01 to 600	6.10
600.01 to 5,000 (Maximum liability is $5,000.00)	$6.10 plus $0.90 for each $100 or fraction over $600 in declared value

EXPRESS MAIL INSURANCE

Insurance Coverage Desired	Fee
$0.01 to $500	none
Merchandise only	
$500.01 to $600	$0.90
600.01 to 700	1.80
700.01 to 800	2.70
800.01 to 900	3.60
900.01 to 1,000	4.50
1,000.01 to 5,000 (Maximum liability is $5,000.00)	$4.50 plus $0.90 for each $100 or fraction over $1,000 in declared value

CERTIFIED MAIL Fee—$1.35

COLLECT ON DELIVERY (COD)

Amount to be collected or insurance coverage desired	Fee[1]
$0.01 to $50	$3.50
50.01 to 100	4.50
100.01 to 200	5.50
200.01 to 300	6.50
300.01 to 400	7.50
400.01 to 500	8.50
500.01 to 600	9.50
Notice of nondelivery	2.80
Alteration of COD charges or designation of new addressee	2.80

[1]Fee for registered COD is $3.50, regardless of insurance value.

BUSINESS REPLY MAIL (BRM)

	Fee
BRM permit fee, per 12-month period	$ 85.00
BRM accounting fee, per 12-month period	205.00
Per piece, in addition to single-piece rate First-Class postage:	
With BRM advance deposit account:	
Regular	0.10
BRMAS	0.02
Without BRM advance deposit account	0.44

PARCEL AIRLIFT SERVICE

Weight	Fee
Not more than 2 lbs.	$0.40
Over 2 but not more than 3 lbs.	0.75
Over 3 but not more than 4 lbs.	1.15
Over 4 lbs.	1.55

SPECIAL HANDLING

Weight	Fee
Not more than 10 lbs.	$5.40
More than 10 lbs.	7.50

FEES

MERCHANDISE RETURN Fee, per 12-month period — $85.00. Per item returned — $0.30.

RETURN RECEIPT (Form 3811 in conjunction with another service)

Service	Fee
Requested at time of mailing showing to whom, signature, date, and addressee's address (if different)	$1.10
Requested after mailing showing only to whom and date delivered	6.60

RETURN RECEIPT FOR MERCHANDISE (Form 3804)

Service	Fee
Requested at time of mailing showing to whom, signature, date, and addressee's address (if different)	$1.20
Delivery record	6.60

CERTIFICATE OF MAILING

Description	Fee
Individual article listing, per article	$0.55
Firm mailing books (Form 3877), per article listed	0.20
Duplicate copies of Form 3817 or mailing bill, per page	0.55

Bulk quantities

Up to 1,000 pieces (1 certificate (Form 3606) for the total number)	$2.75
Each additional 1,000 pieces or fraction thereof	0.35
Duplicate copy	0.55

MONEY ORDERS

Service	Fee
Postal military money order (issued by military facilities authorized by the Department of Defense)	$0.30
Domestic money order (issued at other post offices, including those with branches or stations on military installations)	0.85
Inquiry fee (includes the issuance of a copy of a paid money order)	2.75

POST OFFICE BOXES

Fee Group	Box Size and Fee Per Semiannual Period				
	1	2	3	4	5
Group A	$24.00	$37.00	$64.00	$121.00	$209.00
Group B	22.00	33.00	56.00	109.00	186.00
Group C	20.00	29.00	52.00	86.00	144.00
Group D	6.00	10.00	18.00	26.50	41.50
Group E	0.00	0.00	0.00	0.00	0.00

Deposit per key issued — $1.00

CALLER SERVICE

Fee Group	Fee Per Semiannual Period
A	$250.00
B	240.00
C	225.00
D	225.00

Reserved number fee, per postal calendar year, each — $30.00

RESTRICTED DELIVERY

Fee, per item, in addition to postage and other fees — $2.75

METER SERVICE

Fees for **on-site** meter setting or examination	First meter	Each additional meter	Surcharge for each meter checked in or out of service in addition to on-site fee
Scheduled basis	$27.50	$3.25	$7.50
Unscheduled basis	31.00	3.25	7.50

MISCELLANEOUS FEES AND CHARGES

PERIODICALS APPLICATION FEES
Original entry — $305.00
News agent registry — $50.00
Additional entry — $85.00
Reentry — $50.00

ADDRESS SEQUENCING SERVICE
Per card included by the mailer and that was removed by the USPS for an incorrect or undeliverable address — $0.17

ADDRESS CORRECTION SERVICE FEES
Per notice issued:
Manual address correction service — $0.50
Automated address correction service — $0.20

ANNUAL MAILING FEES (per 12-month period)
First-Class Presort fee, per office of mailing — $85.00
Standard Mail (A) — $85.00
Parcel post destination BMC rate — $85.00
Presorted Special Standard Mail — $85.00

MAILING LIST SERVICES
For correction of name and address on occupant lists, per name on list — $0.17
Minimum per list — $5.50
For sortation of mailing lists on cards by 5-digit ZIP Code, per 1,000 addresses or fraction — $60.00
For address changes furnished to election boards and voter registration commissions, per Form 3575 — $0.17

NONSTANDARD SURCHARGES
First-Class Mail:
Single-piece — $0.11
Presort and automation rates — $0.05
Standard Mail (A)
Single-piece — $0.11

PERMIT IMPRINT
Application fee — $85.00

PICKUP SERVICE FEE
Per occurrence — $4.95

VERIFIED DELIVERY RECEIPT
Per receipt — $6.60

APPENDIX B

ACCREDITATION BOARD FOR ENGINEERING AND TECHNOLOGY

SUGGESTED GUIDELINES FOR USE WITH THE FUNDAMENTAL CANONS OF ETHICS

1. Engineers shall hold paramount the safety, health and welfare of the public in the performance of their duties.

   a. Engineers shall recognize that the lives, safety, health and welfare of the general public are dependent upon engineering judgments, decisions and practices incorporated in structures, machines, products, processes and devices.

   b. Engineers shall not approve nor seal plans and/or specifications that are not of a design safe to the public health and welfare and in conformity with accepted engineering standards.

   c. Should the Engineers' professional judgment be overruled under circumstances where the safety, health, and welfare of the public are endangered, the Engineers shall inform their clients or employers of the possible consequences and notify other proper authority of the situation, as may be appropriate.

      (1) Engineers shall do whatever possible to provide published standards, test codes, quality control procedures that will enable the public to understand the degree of safety or life expectancy associated with the use of the design, products, and systems for which they are responsible.

      (2) Engineers will conduct reviews of safety and reliability of the design, products or systems for which they are responsible before giving their approval to the plans for the design.

(3) Should Engineers observe conditions which they believe will endanger public safety or health, they shall inform the proper authority of the situation.

d. Should Engineers have knowledge or reason to believe that another person or firm may be in violation of any of the provision of these Guidelines, they shall present such information to the proper authority in writing and they shall cooperate with the proper authority in furnishing such further information as may be required.

(1) They shall advise proper authority if an adequate review of the safety and reliability of the products or systems has not been made or when the design imposes hazards to the public through it uses.

(2) They shall withhold approval of products or systems when changes or modifications are made which would affect adversely its performance insofar as safety and reliability are concerned.

e. Engineers should seek opportunities to be of constructive service in civic affairs and work for the advancement of the safety, health and well-being of their communities.

f. Engineers should be committed to improving the environment to enhance the quality of life.

2. Engineers shall perform services only in areas of their competence.

a. Engineers shall undertake to perform engineering assignments only when qualified by education or experience in the specific technical field of engineering involved.

b. Engineers may accept an assignment requiring education or experience outside their own fields of competence, but only to the extent that their services are restricted to those phases of the project in which they are qualified. All other phases of such a project shall be performed by qualified associates, consultants, or employees.

c. Engineers shall not affix their signatures and/or seals to any engineering plan or document dealing with subject matter in which they lack competence by virtue of education or experience, nor to any plan or document not prepared under their direct supervisory control.

3. Engineers shall issue public statements only in an objective and truthful manner.

a. Engineers shall endeavor to extend public knowledge, and to prevent misunderstandings of the achievements of engineering.

b. Engineers shall be completely objective and truthful in all professional reports, statements or testimony. They shall include all relevant and pertinent information in such reports, statements or testimony.

c. Engineers when serving as expert or technical witnesses before any court, commission, or other tribunal, shall express an engineering opinion only when it is founded upon adequate knowledge of the facts in issue, upon a background of technical competence in the subject matter, and upon honest conviction of the accuracy and propriety of the testimony.

d. Engineers shall issue no statements, criticism, nor arguments on engineering matters which are inspired or paid for by an interested party, or parties, unless they have prefaced their comments by explicitly identifying themselves, by disclosing the identities of the party or parties on whose behalf they are speaking, and by revealing the existence of any pecuniary interest they may have in the matters.

e. Engineers shall be dignified and modest in explaining their work and merit, and will avoid any act tending to promote their own interests at the expense of integrity, honor and dignity of the profession.

4. Engineers shall act in professional matters for each employer or client as faithful agents or trustees, and shall avoid conflicts of interests.

a. Engineers shall avoid all known conflicts of interest with their employers or clients and shall promptly inform their employers or clients of any business association, interest, or circumstances which could influence their judgment or the quality of their services.

b. Engineers shall not knowingly undertake any assignments which would knowingly create a potential conflict of interest between themselves and their clients or their employers.

c. Engineers shall not accept compensation, financial or otherwise, from more than one party for services on the same project, nor for services pertaining to the same project, unless the circumstances are fully disclosed to, and agreed to, by all interested parties.

d. Engineers shall not solicit nor accept financial or other valuable considerations, including free engineering designs, from material or equipment suppliers for specifying their products.

e. Engineers shall not solicit or accept gratuities, directly or indirectly, from contractors, their agents, or other parties dealing with their clients or employers in connection with the work for which they are responsible.

f. When in public service as members, advisers, or employees of a governmental body or department, Engineers shall not participate in considerations or actions with respect to services provided by them or in their organization in private or product engineering practice.

g. Engineers shall not solicit or accept an engineering contract from a governmental body on which a principal, officer or employee of their organization serves as a member.

h. When, as a result of their studies, Engineers believe a project will not be successful, they shall so advise their employer or client.

i. Engineers shall treat information coming to them in the course of their assignments as confidential, and shall not use such information as a means of making personal profit if such action is adverse to the interests of their clients, their employers or the public.

   (1) They will not disclose confidential information concerning the business affairs or technical processes of any present or former employer or client or bidder under evaluation without his consent.

   (2) They shall not reveal confidential information or findings of any commission or board of which they are members.

   (3) When they use designs supplied to them by clients, those designs shall not be duplicated by the Engineers for others without express permission.

   (4) While in the employ of others, Engineers will not enter promotional efforts or negotiations for work or make arrangements for other employment as principals or to practice in connection with specific projects for which they have gained particular and specialized knowledge without the consent of all interested parties.

j. The Engineer shall act with fairness and justice to all parties when administering a construction (or other) contract.

k. Before undertaking work for others in which the Engineer may make improvements, plans, designs, inventions, or other records which may justify copyrights, or patents, they shall enter into a positive agreement regarding ownership.

l. Engineers shall admit and accept their own errors when proven wrong and refrain from distorting or altering the facts to justify their decisions.

m. Engineers shall not accept professional employment outside of their regular work or interest without the knowledge of their employers.

n. Engineers shall not attempt to attract an employee from another employer by false and misleading representations.

o. Engineers shall not review the work of other Engineers except with the knowledge of such Engineers, or unless the assignments/or contractual agreements for the work have been terminated.

   (1) Engineers in governmental, industrial or educational employment are entitled to review and evaluate the work of other engineers when so required by their duties.

(2) Engineers in sales or industrial employment are entitled to make engineering comparison or their products with the products of other suppliers.

(3) Engineers in sales employment shall not offer nor give engineering consultation or designs or advice other than specifically applying to equipment, materials or systems being sold or offered for sale by them.

5. Engineers shall build their professional reputation on the merit of their services and shall not compete unfairly with others.

   a. Engineers shall not pay nor offer to pay, either directly or indirectly, any commission, political contribution, or a gift, or other consideration in order to secure work, exclusive of securing salaried positions through employment agencies.

   b. Engineers should negotiate contracts for professional services fairly and only on the basis of demonstrated competence and qualifications for the type of professional services required.

   c. Engineers should negotiate a method and rate of compensation commensurate with the agreed upon scope of services. A meeting of the minds of the parties to the contract is essential to the mutual confidence. The public interest requires that he cost of engineering services be fair and reasonable, but not the controlling consideration in selection of individuals or firms to provide these services.

      (1) These principles shall be applied by Engineers in obtaining the services of other professionals.

   d. Engineers shall not attempt to supplant other Engineers in a particular employment after becoming aware that definite steps have been taken toward the others' employment or after they have been employed.

      (1) They shall not solicit employment from clients who already have Engineers under contract for the same work.

      (2) They shall not accept employment from clients who already have Engineers for the same work not yet completed or not yet paid for unless the performance or payment requirements in the contract are being litigated or the contracted Engineers' services have been terminated in writing by either party.

      (3) In the case of termination of litigation, the prospective Engineers before accepting the assignment shall advise the Engineers being terminated or involved in litigation.

   e. Engineers shall not request, propose nor accept professional commissions on a contingent basis under circumstances under which their professional judgments may be compromised, or when a contingency provision is used as a device for promoting or securing a professional commission.

f. Engineers shall not falsify nor permit misrepresentation of their, or their associates', academic or professional qualifications. They shall not misrepresent nor exaggerate their degree of responsibility in or for the subject matter of prior assignments. Brochures or other presentations incident to the solicitation of employment shall not misrepresent pertinent facts concerning employers, employees, associates, joint ventures, or their past accomplishments with the intent and purpose of enhancing their qualifications and work.

g. Engineers may advertise professional services only as a means of identification and limited to the following:

> (1) Professional cards and listings in recognized and dignified publications, provided they are consistent in size and are in a section of the publication regularly devoted to such professional cards and listings. The information displayed must be restricted to the firm name, address, telephone number, appropriate symbol, and names of the principal participants and the fields of practice in which the firm is qualified.
>
> (2) Signs on equipment, offices and at the site of the projects for which they render services, limited to the name of the firm, address, telephone number and type of service, as appropriate.
>
> (3) Brochures, business cards, letterheads and other factual representations of experience, facilities, personnel and capacity to render service, providing the same are not misleading relative to the extent of participation in the project cited and are not indiscriminately distributed.
>
> (4) Listings in the classified section of telephone directories, limited to the name, address, telephone number, and specialties in which the firm is qualified without resorting to special or bold type.

h. Engineers may use display advertising in recognized dignified business and professional publications, providing it is factual and relates only to engineering, is free of ostentation, contains no laudatory expressions or implications, is not misleading with respect to the Engineers' extent of participation in the services or projects described.

i. Engineers may prepare articles for the lay or technical press which are factual, dignified and free from ostentation or laudatory implications. Such articles shall not imply other than their direct participation in the work described unless credit is given to others for their share of the work.

j. Engineers may extend permission for their name to be used in commercial advertisements, such as may be published by manufactures, contractors, material suppliers, etc., only by means of a modest dignified notation acknowledging their participation and the scope thereof in the project or product described. Such permission shall not include public endorsement of proprietary products.

k. Engineers may advertise for recruitment of personnel in appropriate publications or by special distribution. The information presented must be displayed in a dignified manner, restricted to the firm name, address, telephone number, appropriate symbol, names of principal participants, the fields of practice in which the firm is qualified and factual descriptions of the positions available, qualifications required and benefits available.

l. Engineers shall not enter competitions for designs for the purpose of obtaining commissions for specific projects, unless provision is made for reasonable compensation for all designs submitted.

m. Engineers shall not maliciously or falsely, directly or indirectly, injure the professional reputation, prospects, practice or employment of another engineer, nor shall they indiscriminately criticize another's work.

n. Engineers shall not undertake nor agree to perform any engineering service on a free basis, except for professional services which are advisory in nature for civic, charitable, religious or non-profit organizations. When serving as members of such organizations, engineers are entitled to utilize their personal engineering knowledge in service of these organizations.

o. Engineers shall not use equipment, supplies, laboratory nor office facilities of their employers to carry on outside private practice without consent.

p. In case of tax-free or tax-aided facilities, engineers should not use student services at less than rates of other employees of comparable competence, including fringe benefits.

6. Engineers shall act in such a manner as to uphold and enhance the honor, integrity and dignity of the profession.

a. Engineers shall not knowingly associate with nor permit the use of their names nor firm names in business ventures by any person or firm which they know, or have reason to believe, are engaging in business or professional practices of a fraudulent or dishonest nature.

b. Engineers shall not use association with non-engineers, corporations, nor partnerships as 'cloaks' for unethical acts.

7. Engineers shall continue their professional development throughout their careers, and shall provide opportunities for the professional development of those engineers under their supervision.

a. Engineers shall encourage their engineering employees to further their education.

b. Engineers shall encourage their engineering employees to become registered at the earliest possible date.

c. Engineers shall encourage their engineering employees to attend and present papers at professional and technical society meetings.

d. Engineers should support the professional and technical societies of their disciplines.

e. Engineers shall give proper credit for engineering work to those to whom credit is due, and recognize the proprietary interests of others. Whenever possible, they shall name the person or persons who may be responsible for designs, inventions, writings, or other accomplishments.

f. Engineers shall endeavor to extend the public knowledge of engineering, and shall not participate in the dissemination of untrue, unfair or exaggerated statements regarding engineering.

g. Engineers shall uphold the principle of appropriate and adequate compensation for those engaged in engineering work.

h. Engineers should assign professional engineering employees with complete information on working conditions and their proposed status of employment, and after employment shall keep them informed of any changes.

Accreditation Board for Engineering and Technology, Inc.
111 Market Place, Suite 1050
Baltimore, MD 212302-4012